T0295495

Biological Nitrogen Fixation and Sustainable Agriculture

Biological Nitrogen Fixation and Sustainable Agriculture

Edited by Nash Medlin

New York

Published by Syrawood Publishing House,
750 Third Avenue, 9th Floor,
New York, NY 10017, USA
www.syrawoodpublishinghouse.com

Biological Nitrogen Fixation and Sustainable Agriculture
Edited by Nash Medlin

International Standard Book Number: 978-1-64740-350-8 (Hardback)

Cataloging-in-publication Data

Biological nitrogen fixation and sustainable agriculture / edited by Nash Medlin.
p. cm.
Includes bibliographical references and index.
ISBN 978-1-64740-350-8
1. Nitrogen--Fixation. 2. Nitrogen-fixing microorganisms. 3. Nitrogen-fixing plants.
4. Biological models. 5. Microbial ecology. I. Medlin, Nash.
QR89.7 .N58 2023
589.950 413 3--dc23

TABLE OF CONTENTS

PREFACE

itrogen fixation refers to a chemical process in which molecular nitrogen, having a strong triple covalent ond, is converted into ammonia in the air or related nitrogenous compounds found in aquatic systems soil. Atmospheric nitrogen is molecular dinitrogen that is generally a nonreactive molecule. Biological trogen fixation is a microbially mediated process, which utilizes the nitrogenase protein complex for ansforming dinitrogen gas into ammonia. Sustainable agriculture refers to a type of farming which ms to fulfill the society's current requirements without compromising the capability of the future nerations in meeting their needs. Nitrogen fixation is integral for sustainable food production and ricultural practices. This book provides comprehensive insights on biological nitrogen fixation and stainable agriculture. Its aim is to present researches that have transformed the study of this subject d aided its advancement. The book is a resource guide for experts as well as students.

ne researches compiled throughout the book are authentic and of high quality, combining several sciplines and from very diverse regions from around the world. Drawing on the contributions of any researchers from diverse countries, the book's objective is to provide the readers with the latest hievements in the area of research. This book will surely be a source of knowledge to all interested d researching the field.

the end, I would like to express my deep sense of gratitude to all the authors for meeting the set adlines in completing and submitting their research chapters. I would also like to thank the publisher r the support offered to us throughout the course of the book. Finally, I extend my sincere thanks to y family for being a constant source of inspiration and encouragement.

Editor

1

Effects of Pesticides, Temperature, Light and Chemical Constituents of Soil on Nitrogen Fixation

Shweta Nandanwar, Yogesh Yele, Anil Dixit, Dennis Goss-Souza, Ritesh Singh, Arti Shanware and Lalit Kharbikar

Abstract

Nitrogen is a vital component of atmosphere and plays important roles in the biochemistry of all life forms on the earth. Various mechanisms of biological nitrogen fixation and recycling in the environment have been evolved in all known ecosystems. For example, symbiotic nitrogen fixation is the major N2-fixing mechanism in the agroecosystems. Symbiotic nitrogen fixation is dependent on the biotic factors, such as host plant genotypes and the microbial strains. However, the interaction of these biotic factors is influenced by abiotic factors, such as climate and environmental conditions. The effects of various environmental variables, such as pesticides, temperature, and light as well as acidity, alkalinity, salinity, phosphorus, and water content status of the soils on the nitrogen fixation have been discussed briefly in this chapter.

Keywords: nitrogen fixation, salinity, ammonification, nitrite, denitrification

1. Introduction

The nutrient nitrogen is an essential element required by all life forms including the plants due to the presence in essential molecules such as proteins, amino acids, and enzymes. Although, this nutrient is in a large amount in the atmosphere, a small group of microorganisms is able to fix it becoming them available to plants.

The biological and physical processes in an ecosystem convert the nitrogen into multiple chemical forms and the phenomenon is called as nitrogen cycle. Biological nitrogen fixation accounts for about 70% of the atmospheric nitrogen. The four steps involved in a nitrogen cycle, namely nitrogen fixation, ammonification, nitrifications, and denitrification are described in **Figure 1**.

This process of nitrogen fixation is influenced by several aspects of biotic and abiotic variables. In this chapter, the effects of various factors especially the pesticide applications, temperature, and light as well as acidity, alkalinity, salinity, phosphorus, and water content status of the soils on the nitrogen fixation have been discussed briefly.

This chapter has addressed each environmental variable and discussed the effect of each one on the nitrogen fixation process and brings information related to it, for

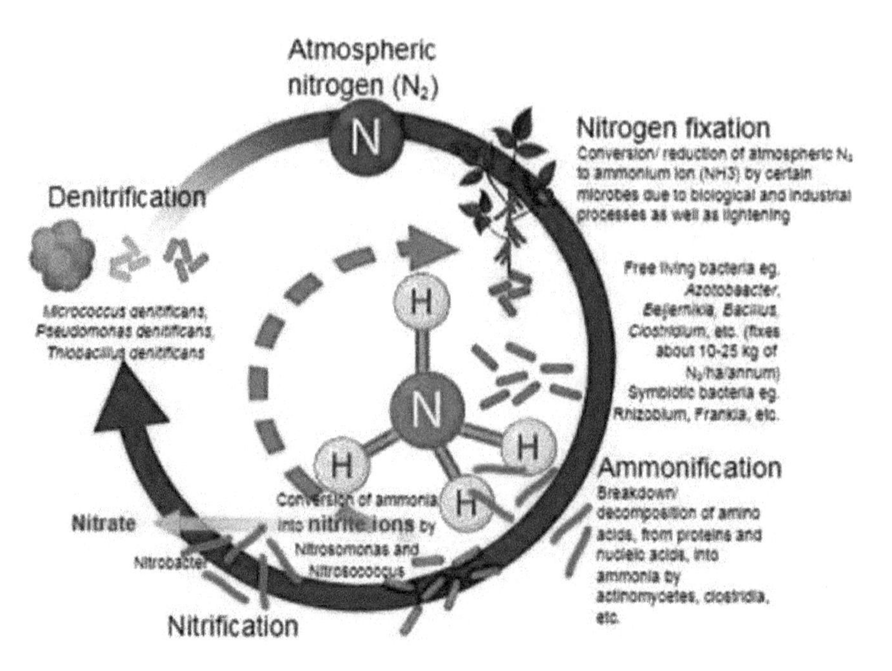

Figure 1.
The nitrogen cycle.

example, regarding the use of pesticides in an indiscriminate way. Usually, when the pesticides are used in the recommended dose, it does not harm the soil microbiota, but when the pesticides are overused, it may affect the physiological and growth characteristics of both rhizobia and legume plants. They also inhibit more or less the symbiotic nitrogen fixation by blocking biochemical communication and dialogues between the legumes and rhizobial symbionts.

2. Effects of pesticides on nitrogen fixation

In the current scenario of modern agriculture, agrochemicals play a key role in improving plant protection and production of food crops. Chemical pesticides are extensively used against a range of pests infesting agricultural crops worldwide. Due to the indiscriminate use of pesticides, the issue of the impact of these chemicals on the environment and soil has gained attention. Only about 0.1% of the total applied pesticides reach the target site/organism, while the remaining bulk contaminates the soil environment [1]. Consequently, sizable amounts of pesticides reach soil directly and indirectly, which affects the composition of soil microflora [2].

One of the most important and potentially limiting factors to biological nitrogen fixation is the use of chemical pesticides. Recommended doses of pesticides, in general, do not have any harmful effect on nitrogen fixation.

At higher doses, various chemical pesticides affect the physiological and growth characteristics of both rhizobia and legume plants. They also inhibit more or less the symbiotic nitrogen fixation by blocking biochemical communication and dialogues between the legumes and rhizobial symbionts [3].

Herbicides applied to leguminous crops constitute a potential hazard to the estab-lishment and performance of the N2-fixing root nodules. Soil and foliar application of herbicide at the recommended rates altered the morphology of root hairs and reduced nodule numbers and nitrogenase activity [4]. Some rhizobium strains also shown resistance towards herbicides when isolated and grown in laboratory media.

Legume seeds are inoculated with specific rhizobium to increase yield and at the same time they are also treated with fungicides to disinfect and guard it against seed and soil-borne pathogens. It is imperative to know the effect of seed disinfectants on the efficacy of rhizobia. Earlier studies show fungicidal seed treatment is safe as far as nitrogen fixation by rhizobium in symbiosis with legumes is concerned. Contradictory, Osman et al. reported a very strong negative effect of the dose of the thiram on *Rhizobium meliloti* [5].

3. Effects of temperature and light on nitrogen fixation

Temperature has a significant influence on survival and persistence of N-fixing microbes in soils [6]. Elevated temperatures may delay nodule initiation and development, and interfere with nodule structure and functioning in temperate legumes. Whereas in tropical legumes, competitive ability and N fixation efficiency of legume symbionts are mainly affected. Similarly, low temperatures reduce nodule formation and nitrogen fixation in temperate legumes.

In the extreme environment (−40 to −68°C) of the high arctic, native legumes can nodulate and fix nitrogen at rates comparable to those observed with legumes in temperate climates [7]. This indicates that both the plants and their rhizobia have successfully adapted to arctic conditions.

At constant temperature (20°C) under long photoperiod (14 hours), plants may not show a significant decrease in N fixation in terms of fixation per plant or per unit nodule mass [8]. However, under a short photoperiod (6 hours) at the same temperature, plants show a significant decrease in N fixation. Under short photoperiod, plants compensate for reduced photosynthesis by maintaining only half the root nodule mass and N fixation activity as that of long photoperiod. However, similar rates of N_2 fixation per unit mass of nodule may be observed in some plants under both the long and short photoperiods. This is because the shoot reserves for sustaining nitrogenase activity may compensate for reduced N fixation ability of plants.

4. Effects of acidity, alkalinity, and salinity on nitrogen fixation

A major problem associated with many acidic soils is metal toxicity. Highly acidic soils ($pH < 4.0$) frequently have low levels of phosphorus, calcium, and molybdenum and high levels of aluminium and manganese which are often toxic for both plants and symbiotic N-fixing bacteria [7]. Nodulation by this N-fixing bacteria is more affected than the normal growth of the host-plants in such conditions. It is a well-known fact that most leguminous plants grow less favourably in acidic conditions than in neutral or slightly alkaline conditions [9]. This is mainly due to a reduced nitrogen fixation as may be concluded from the improved growth at low pH upon the addition of combined nitrogen.

Highly alkaline soils ($pH > 8.0$) tend to be high in sodium chloride, bicarbonate, and borate, and are often associated with high salinity which reduces symbiotic nitrogen fixation [4]. However, the symbiotic N-fixing rhizobia, showing significantly higher salt tolerance, have been isolated from alkaline soils [10]. This suggests that the possession of high salt tolerance trait might be of some evolutionary significance for the survival of rhizobia in alkaline soils. Phosphate solubilising, alkaline tolerant rhizobia have also been isolated from wild legumes grown in dune systems of the southwest coast of India [11]. The stress tolerance traits of these rhizobia are of potential value in strain improvement of symbiotic bacteria for efficient N-fixing ability.

Soil salinity, which extremely increases in the protected cultivation, may occur when there are irregular irrigation schedules, inadequate drainage systems, wrong fertiliser applications, etc. [12]. The specific sensitivity of leguminous crops, where the N fixation is predominated by symbiotic bacteria, to soil salinity is well documented for initiation, development, and function of nodules [13]. An increase in salinity decreases the nodule permeability in these crops. This decrease is associated with a contraction of nodule inner-cortex cells and an increase in abscisic acid content of the nodule [14]. This phenomenon may lead to less survival of the symbiotic N-fixing bacteria associated with these crops. Even if the bacteria are survived, they may be less viable with decreased N-fixing ability. However, survival of the viable *Rhizobium* sp. under extreme conditions of salinity has been reported recently in root nodules of *Sesbania aculeata* [10].

5. Effects of phosphorus (P) status of soil on nitrogen fixation

Phosphorus deficiency in soils affects nodule functioning; however, it does not inhibit the other aspects of plant growth and metabolism. There exists an interaction between the plant's growth and N fixation in response to increasing P supply [15]. This interaction may be positive, zero, or negative. A negative interaction suggests a greater need for P by N2-fixing plants; a zero interaction indicates that the plants have the same P requirement for growth and N fixation, while a positive interaction indicates that higher P supply may be inhibitory to N2 fixation (**Figure 2**).

Legume tissues do not appear to have higher P content than those of other plants. Therefore a positive interaction exists between legume's growth and N fixation in response to P supply (**Figure 2**). For example, in both soybean and lucerne (alfalfa), nodules from plants grown under P deficiency may have a higher concentration of P than those grown with sufficient P and may fix more N2 per unit of P [16].

Negative interaction: > P = ↑ N_2 fixation

Zero interaction: P = N_2 fixation

Positive interaction: > P = ↓ N_2 fixation (In legumes: < P = ↑ N_2 fixation)

Figure 2.
Effects of phosphorus (P) status of soil on nitrogen fixation.

6. Effects of soil water availability on nitrogen fixation

Water influences the growth of soil micro-organisms through processes of diffusion, mass flow, and nutrient concentration [17]. Soil water retention is related to soil pore space, and soils containing larger pores and pore spaces retain less water. Thus, soil aggregates having smaller internal pore spaces offers more favourable environments for the growth of N-fixing microbes.

There is a high degree of correlation between soil water content and N-fixing activity [16]. Maximum N fixation occurs at about field capacity of soil and above this, the N-fixing activity may be reduced due to water logging. Slow natural drying of soil over a 6 week period may result in a progressive reduction of N-fixing activity, which can be restored by irrigation. This indicates that soil water availability is the major environmental factor affecting nitrogen fixation. It is plausible that water stress directly affects nodule activity. This effect may further be aggravated by reduced supplies of photosynthate due to wilted plant leaves [18].

7. Conclusion

Environmental variables such as pesticides, temperature, and light as well as acidity, alkalinity, salinity, phosphorus, and water content status of the soils have a significant impact on nitrogen fixation. Despite many decades of progress and the acquisition of ample information, the physiological and molecular bases of the effects of these environmental variables on the symbiotic and non-symbiotic N_2 fixation systems remains largely unknown and empirical in nature. Although understanding these processes was originally thought to be straightforward and tractable, we now have learned that the flavonoid nod-gene inducers are specific for a particular N_2 fixation system [19]. Needless to say that, the production of these inducers is influenced by environmental variables. Therefore, more work needs to be done to understand the underlying molecular bases for tolerance to environmental variables in both N_2 fixation systems. Further, the genomics and proteomics tools need to be combined with traditional plant breeding and microbial selection studies in order to rapidly define and utilise their genetic loci involved in tolerance to environmental stresses.

Acknowledgements

We would like to thank all colleagues who have done work on nitrogen fixation and related fields. We apologise to colleagues whose work in this rapidly changing field was not directly cited in this review due to space limitations and timing. We would also like to thank anonymous reviewers for helpful comments.

Author details

Shweta Nandanwar[1], Yogesh Yele[2], Anil Dixit[2], Dennis Goss-Souza[3], Ritesh Singh[4], Arti Shanware[5] and Lalit Kharbikar[2*]

1 Princess Margaret Science Laboratories, Harper Adams University, Newport, Shropshire, United Kingdom

2 ICAR—National Institute of Biotic Stress Management, Raipur, India

3 Santa Catarina State University—UDESC, Florianópolis, Brazil

4 Thakur Chhedilal Barrister College of Agriculture and Research Station, Bilaspur, India

5 Rajiv Gandhi Biotechnology Centre, RTM Nagpur University, Nagpur, India

*Address all correspondence to: lalitkharbikar@gmail.com

References

[1] Carriger JF, Rand GM, Gardinali PR, Perry WB, Tompkins MS, Fernandez AM. Pesticides of potential ecological concern in sediment from South Florida Canals: An ecological risk prioritization for aquatic arthropods. Soil and Sediment Contamination. 2006;**15**:21-45

[2] Andrea MM, Peres TB, Luchini LC, Pettinelli A. Impact of long-term pesticide application on some soil biological parameters. Journal of Environmental Science and Health, Part B. 2000;**35**:297-307

[3] Ahemad M, Khan MS. Pesticides as antagonists of rhizobia and the Legume- *Rhizobium* Symbiosis: A paradigmatic and mechanistic outlook. Biochemistry and Molecular Biology. 2013;**1**(4):63-75

[4] Ljunggren H, Martensson A. Herbicide effect on leguminous symbiosis. In: Proceedings of the *21st* Swedish Weed Conference. Uppsala: Sve-riges Lantbruksuniversitet; 1980. pp. 99-106

[5] Osman AG, Sherif AM, Elhussein AA, Mohamed AT. Sensitivity of some nitrogen fixers and the target pest *Fusarium oxysporum* to fungicide thiram. Interdisciplinary Toxicology. 2012;**5**:25-29

[6] Lie TA. Symbiotic nitrogen fixation under stress conditions. Plant and Soil. 1971;**35**(1):117-127

[7] Bordeleau LM, Prevost D. Nodulation and nitrogen fixation in extreme environments. Plant and Soil. 1994;**161**:115-125

[8] Murphy PM. Effect of light and atmospheric carbon dioxide concentration on nitrogen fixation by herbage legumes. Plant and Soil. 1986;**95**(3):399-409

[9] Hartwig UA. The regulation of symbiotic N2 fixation: A conceptual model of N feedback from the ecosystem to the gene expression level. Perspectives in Plant Ecology, Evolution and Systematics. 1998;**1**:92-120

[10] Kulkarni S. Crossing the limits of *Rhizobium* existence in extreme conditions. Current Microbiology. 2000;**41**(6):402-409

[11] Arun B, Sridhar KR. Growth tolerance of rhizobia isolated from sand dune legumes of the southwest coast of India. Engineering in Life Sciences. 2005;**5**(2):134-138

[12] Adil AA, Kant C, Turan M. Humic acid application alleviate salinity stress of bean (*Phaseolus vulgaris* L.) plants decreasing membrane leakage. African Journal of Agricultural Research. 2012;**7**(7):1073-1086

[13] Saadallah K, Drevon JJ, Abdelly C. Nodulation et croissance nodulaire chez le Haricot (*Phaseolus vulgaris* L.) sous contrainte saline. Agronomie. 2001;**21**:27-34

[14] Irekti H, Drevon JJ. Acide abcissique et conductance á la diffusion de l'oxygène dans les nodositeś de haricot soumises à un choc salin. In: Drevon JJ, Sifi B, editors. Fixation Symbiotique de l'Azote et Développement Durable dans le Bassin Méditerranéen. Vol. 100. INRA, Versailles Cedex, France; 2003. les colloques

[15] Robson AD. Mineral nutrition. In: Broughton WJ, editor. Nitrogen Fixation. Oxford: Clarendon Press; 1983. pp. 36-55

[16] Drevon JJ, Hartwig UA. Phosphorus deficiency increases the argon-induced decline in nodule nitrogenase activity in soybean and alfalfa. Planta. 1997;**201**:463-469

[17] Turco RF, Sadowsky MJ. Understanding the microflora of bioremediation. In: Skipper HD, Turco RF, editors. Bioremediation: Science and Applications. Soil Science (Special Publication). Vol. 43. Madison, WI: Soil Science Society of America Journal; 1995. pp. 87-103

[18] Sprent IJ. The effects of water stress on nitrogen-fixing root nodules. New Phytologist. 1972;**71**(4):603-611

[19] Schmidt PE, Broughton WJ, Werner D. Nod factors of Bradyrhizobium japonicum and Rhizobium sp. NGR 234 induce flavonoid accumulation in soybean root exudates. Molecular Plant Microbe Interaction. 1994;**7**:384-390

2

N-Fertilization Adjustment in Sugarcane Crop Cultivated in Intensive Mechanization

Sérgio G. Quassi de Castro
and Henrique C. Junqueira Franco

Abstract

Currently sugarcane is cultivated in Brazil in intensive mechanization during all cultural practices, since planting, harvesting until fertilizer applications. In this sense, to increase the sugarcane yield which it was stagnant along the last 5 years, it is necessary looking for alternatives to management the crop and maximize the yield gain. One alternative is trough the adjustment the N-fertilization in green cane crop according to the IPNI guidelines—"4R Nutrient Stewardship System" (International Plant Nutrition Institute—IPNI) program which seeks to apply fertilizer in the right location, at the right time, in the right amount and the right source. Therefore, the main goal of this chapter is shows these alternatives and the yield gains that it was obtained in researches during 2013–2017.

Keywords: nitrogen, *Saccharum* spp., green cane, fertilizer

1. Introduction

According to statistical surveys conducted in the last decade by the International Plant Nutrition Institute [1], the largest consumer market for NPK fertilizers was Asia (57%), followed by the Americas (25%), Europe (13%), Africa (3%), and Oceania (2%). In 2015, approximately 183.2 million tons of NPK fertilizers were consumed, being the most consumed nitrogen fertilizers with 60% of total (110.4 million tons), followed by phosphorus (22% (40.7 million tons)) and potassium (18% (32.1 million tons)). In the Americas, the largest consumers were the United States and Brazil, with Brazil showing a significant increasement in fertilizer consumption during the last decade, representing a consumption of approximately 14 million tons of NPK fertilizers in 2015. This value was 50% higher than obtained in the previous decade [1]. This large demand resulted in a financial movement of approximately R$ 19.5 billion per year, originating from the internal sale of NPK fertilizers [2].

In Brazil, the main crops (soybean, maize, and sugarcane) use approximately 6.2 million tons of NPK fertilizers during 2015. The sugarcane crop represents 22.6% of this amount, with consumption of 1.4 million tons of NPK fertilizers, which generated investment higher than R$ 2 billion per year. Within this billionaire market of fertilizers for sugarcane, potassic fertilizers were the most used (609 thousand tons), followed by nitrogen fertilizers (573 thousand tons) and phosphates fertilizers with 195 thousand tons used [2].

Most of the NPK fertilizer market in the Brazilian sugarcane sector is based on four sources, such as potassium chloride (source of K), urea and ammonium nitrate (N sources), and simple superphosphate (P source). Although the industry is a large and efficient waste recycler, such as filter cake that is rich in phosphorus and potassium-rich vinasse, the high demand for the importation of these raw materials shows the risk in our food and energy security.

Brazil is the world's largest producer of sugarcane with approximately 640 million tons [3], grown in an area of 9 million hectares, followed by India and China, respectively, with 352 and 126 million tons [4]. Despite this, the increase in the sugarcane area, which was 5.8 million hectares during the last 10 years, not took place increase in stalk productivity, which remains stagnant (72 Mg ha^{-1}—crop season 2016/2017) compared to the Brazilian historical series, which has already reached an average of 80 Mg ha^{-1} [3].

The IPNI has established a Best Practices Fertilizer Management (BPFM) program. This program shows the practical actions necessary to provide a better economic, social, and environmental performance of crops, in order to adapt the supply of nutrients to the needs of the crop minimizing the losses of its nutrients in the soil-plant-atmosphere continuum [5]. In the focus of plant nutrition and fertilizer use, BPFM encompasses the 4R principle: (1) applying the right nutrient source, (2) taking the right amount, (3) at the right place, and (4) in the correct time (right time). In this context, the present text aims to address aspects related to the nutrition of the sugarcane crop, focusing on nitrogen fertilization, showing the challenges and bottlenecks for increasing the efficiency of use of these nutrients by sugarcane.

2. Nitrogen fertilization: definition of the correct source, applied time, application mode, and right dose

The N is the nutrient that has the highest interaction in the environment due to the numerous reactions, mediated by microorganisms, that occur in the soil, being affected by temperature and humidity [6]. In addition, there are several routes of N losses (leaching, volatilization, and immobilization), promoting only about 26% of the N applied by fertilizer that is used by the plant [7]. For other crops this value usually is higher than 50% [8]. The large N stock in the soil, representing more than 95% of the total N, comes from organic matter. However, organic N is not directly utilized by plants, requiring their mineralization to produce ammonium, which can be absorbed or transformed into nitrate (nitrification process) that will be absorbed by plants [6].

In general, it is recommended to apply 30–60 kg ha^{-1} of N in plant cane, applied in the planting furrow. In ratoon the dose can vary from 80 kg ha^{-1} N, for yields between 60 and 80 tons of stalks per hectare (TSH), up to 140 kg ha^{-1} N, in areas where it is expected to produce above 140 TSH [9]. There is no doubt about the importance of nitrogen fertilization for productivity gains in sugarcane cultivation, especially in ratoon areas. In a very good literature review made by Otto et al. [7], the authors found that in 75% of the total number of papers reviewed, there was an increase in TSH due to nitrogen fertilization. In 30% of papers, the gain in stalk yield was higher than 25% in relation to the control.

The most common nitrogen fertilizers used in Brazilian sugarcane fields are ammonium nitrate (33% N), urea (45% N), and ammonium sulfate (21% N). It is interesting to note that each option has positive and negative aspects, such as (i) ammonium nitrate has N in two forms: nitrate (NO_3^-) and ammonium (NH_4^+), in which ammonium has been reported to be a mineral preferential form uptake by sugarcane [10]. In addition, there is not losses due the volatilization of ammonia by this N source (ammonium nitrate) in tropical soils. However, it can lead to nitrate leaching losses [11]. Also since it is a raw material for explosive devices, it must be restricted in its commercialization in the coming years around the world; (ii) urea is the fertilizer with the highest N content in its composition, making it possible to reduce the relative cost of its acquisition (R$ per kg of N). Nevertheless, with the advent of mechanized harvesting of sugarcane, which generated straw covers in sugarcane fields and currently 85% of all sugarcane cultivated area in Brazil is harvested this mode [12], the application of urea over the straw implies in higher losses of NH_3 by volatilization; (iii) ammonium sulfate, although it is the option with lower concentration of N, presents in its composition approximately 24% of sulfur (S). This fact makes this fertilizer an excellent option, especially in areas where there is no application of vinasse, as it is able to supply nitrogen and sulfur at the same time to the plant. However, it has a high salinity index. When it is applied locally, in high amounts and in periods of low soil moisture, it can cause problems in the plant growth.

One of the major limitations to increase the productivity of sugarcane in Brazil is the availability of adequate amounts of nutrients in soils, especially N [13]. This is due to the many factors that affect nitrogen utilization efficiency (EUN), such as soil characteristics (pH, cation exchange capacity (CEC), organic matter, texture, clay, aeration, and compaction), climatic conditions (temperature and rainfall), and agronomic practices (cultivation, soil preparation, and crop rotation) [14]. Due to the difficulty to define the dose of N to be applied in sugarcane due to the lack of a laboratory methodology that allows quantification of N available for plants into the soil, the dose of N is often defined according to the expected productivity [15]. This fact causes sub or super estimates of the N doses to be applied to the crop. In this scenario, new works [7, 16] seek the development of strategies for the management of nitrogen fertilization in sugarcane plantations, aiming to increase nitrogen fertilization efficiency.

In addition, the presence of straw over the soil surface and the harvesting season (wet or dry season) should be considered for the proper management of sugarcane. This can be justified by analyzing the average productivity obtained over four harvests—2015, 2016, 2017, and 2018 (**Figure 1**), where the same variety of sugarcane (IACSP95-5000) was harvested at different crop season (beginning, middle, and end of the harvest corresponding to the months of April, August, and October, respectively). Comparing the fertilizations (same dose) made in wet season and dry season, the results show that the fertilizations carried out in the humid season always promoted the highest yields (**Figure 1**). In general, the area harvested at the beginning of the harvest had the highest average yield of stalks (107 TSH), followed by the middle area (95 TSH) and the final harvest (77 TSH). Therefore, the best time to apply the N in the sugarcane ratoon should be when there is moisture in the soil, regardless of the period in which the harvest was performed. This contrasts with the traditional management of sugarcane, in which fertilization was done when the plant had a significant canopy (number of green leaves) and tillering [15].

Other possibilities for increasing the efficiency of nitrogen fertilization in sugarcane, together with increases in yield of stalks, are related to the application forms. Regarding the forms of nitrogen fertilizer application, this can be done in several ways (**Figure 2**): (1) application incorporated to the soil by cultivation in the interrow ("triple operation"), (2) superficial application in band, (3) application under the straw in band, (4) application made in both sides of the ratoon, (5) surface application in total area, and (6) surface application with liquid fertil-izer (a) and foliar application (b).

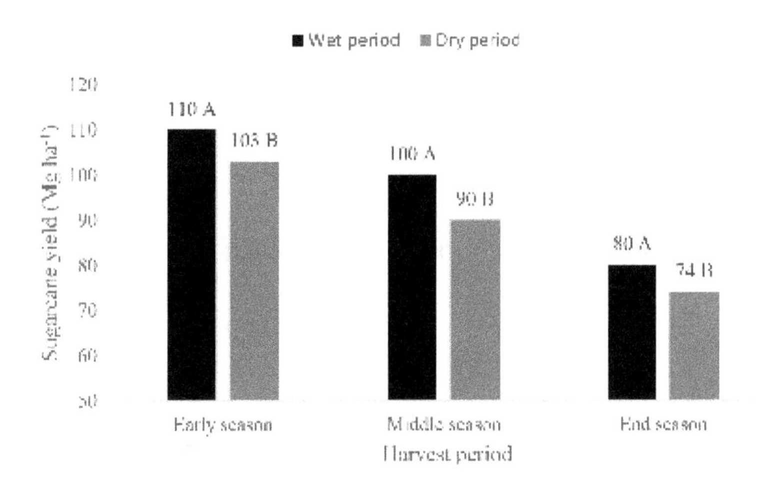

Figure 1.
Effect of the nitrogen fertilizer application time according to the sugarcane harvest season. Source: Castro [17]. Note: Capital letters differ from each other within each harvest season.

Figure 2.
Alternatives of N-fertilizer application methods in sugarcane ratoon. Source: Castro et al. [18].

The N-fertilizer applications made through the interrow are the most traditional in Brazil sugarcane fields. This is because the cultivation had been widely used in the burned cane—in the past. Regardless, in the currently years, with the adoption of the green cane cultivation (without fire in the harvest), the surface application has greater operational in sugarcane fields, where the straw layer resulting from the harvest can disrupt the scarification of the interweaving when opted for cultivation. Castro et al. [19] showed that no effect was obtained on crop productivity by the adoption of interlaced cultivation (**Table 1**).

The absence of alterations in TSH in sugarcane yield shows that the cultivation operation may not be necessary, mainly because sugarcane does not exploit the interleaving region [20, 21], even in areas where the traffic of machines in this region did not occur [22]. The application performed by the triple operation (number 1 in **Figure 2**), despite being incorporated in the soil, presents lower yields of TSH (**Figure 3**) due precisely to the distance at which the fertilizer is positioned in relation to the stump of the plant, and the area covered the root system, which in the middle of the interline is very small [20].

On the other hand, some studies show increases in TSH when incorporating the nitrogen fertilizer at 0.1 m depth [23], on both sides of the sugarcane ratoon: on average 0.2–0.3 m of the cane row (**Figure 2** (no. 4)). The deposition of N-fertilizer near the root

Treatments	2008	2009	2010	2011	Mean
With tillage	125[b*]	112[a]	85[a]	81[a]	101[a]
Without tillage	131[a]	112[a]	88[a]	81[a]	103[a]

** Means with the same letter in column did not differ according to the "Tukey" test ($p>0.05$)*

Table 1.
Sugarcane yield (TSH) associated to the performance or not of the mechanical cultivation of the interrow of crop (adapted Castro et al. [18]).

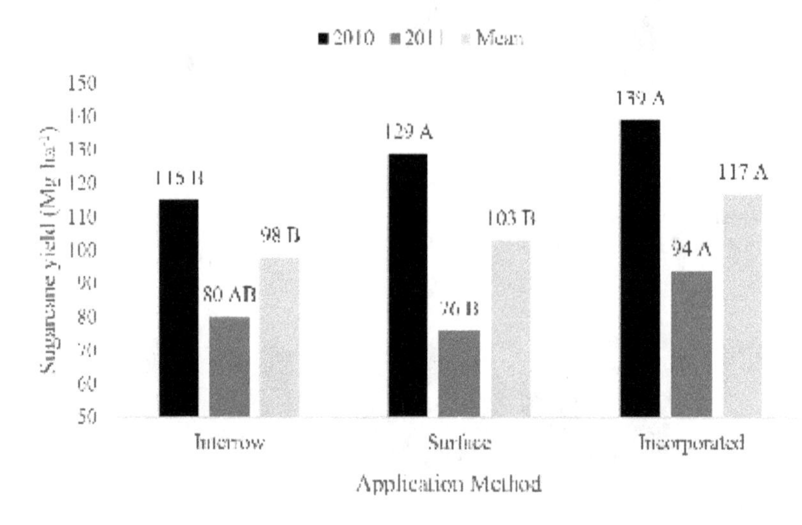

Figure 3.
Effect of the N-fertilizer application method in sugarcane yield. Source: Castro et al. [23]. Note: Capital letters differ from each other within each treatment evaluated in each year.

system of the plant can facilitate its absorption, improving the TSH of the crop. This aspect also explains the difference between the incorporated application and the surface application; because when applied on the soil surface, the fertilizer needs to transpose the layer of straw which often occurs with the rains, and because it is not constant, a temporal difference in N uptake and plant development when compared to the incorporated application. In general, the incorporation of N-fertilizer presented an increase in sugarcane yield of 14% (28 TSH) and 19% (38 TSH) when compared, respectively, to the superficial application and the interrow, which did not differ between them.

Due to the large extension of sugarcane cultivated areas and the short interval of time for fertilization (adequate soil moisture conditions), alternatives that allow higher operational yield (hectares h^{-1}) are necessary. In this sense, one of the options is the application of the nitrogen fertilizer by means of machines such as the Uniport of the Jacto manufacturer, which deposits the fertilizer below the layer of straw (**Figure 3** (n°3)), due to the fertilizer granules were applied with a pressure which these granules are able to transpose the straw layer. In a research conducted by Brazilian Bioethanol Science and Technology Laboratory (CTBE) [24], three forms of application of nitrogen fertilizer in sugarcane ratoon were compared, being application under straw and in strip and surface application in strip and application incorporated to the soil. The results showed that the yield of the crop was higher when the fertilizer was applied in an incorporated form or under the straw compared to the surface application (**Figure 4**). Considering the average productivity obtained in the two agricultural years (**Figure 4**), the application incorporated to the soil and under the straw provided increases in the TSH of 16% (26 TSH) and 13% (21 TSH), respectively, when compared to surface application.

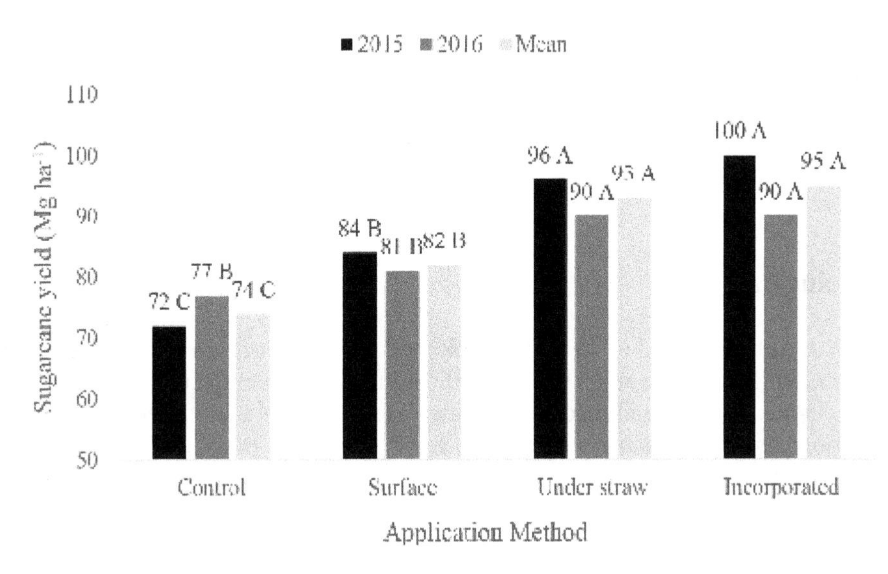

Figure 4.
Effect of N-fertilizer application forms in sugarcane yield. Source: Franco et al. [24]. Note: Letters compare the forms of application in each of the years.

Therefore, in nitrogen fertilization, one must consider the choice of a source that allows lower losses or minimized if such losses exist, for example, by adjusting the mode of application. It is also convenient to adapt the time of application of N-fertilizer, in which, if there are operational conditions, nitrogen fertilizer should be applied when there is moisture in the soil, recommending the application incorporated in both sides of the cane row or application under the straw. Considering the management of the time and method of application of the nitrogen fertilizer in sugarcane, associated to the fact that in the south central region of Brazil, crop harvest occurs from March to December, thus an extended period for the accomplishment of cultural

Sugarcane Price (U$$ Mg^{-1})	N fertilizer price (U$$ Mg^{-1})					
	200	225	250	275	300	325
	N economical rate (base line 120 kg N ha^{-1})					
	Early	Middle	**End**	Early	Middle	**End**
10						
13	-25%	-13%	**-25%**	-30%	-25%	**-30%**
15						
18						
20	+10%	0%	**+15%**	-15%	-5%	**-15%**
23						

Ps: Price of sugarcane according to Consecana Brasil; Nitrogen fertilizer price according to Anda [2].

Table 2.
N-economic rate to be applied in the sugarcane crop according to the harvesting season associated with the price of the raw material and the N-fertilizer.

treatments; it is possible to adjust the N dose to be applied (variation of 30%) without loss of productivity (**Table 2**). In this sense, the investment made in the acquisition of the nitrogen fertilizer must be removed by increasing productivity, then that high doses can obtain a gain which is not enough to pay off the investment [17].

3. Conclusion

The adoption of the best technologies (IPNI—Best Practices Fertilizer Management (BPFM)) to apply the N-fertilizer in sugarcane crop cultivated in intensive mechanization is possible to have an increase in the yield and a reduction in the application cost. It is possible with the choice of the right N rate, associated with the right application method and right time to perform this operation. In this context, the average yield gain is near 30%, as well as the production cost reduced near 15%.

Considering the sugarcane (green cane) cultivated in central south region of Brazil, where the harvest season occur between March and November, the best time to apply the N-fertilizer is in wet period. The best application method is incorporated at 0.1 m depth in both sides of the sugarcane row. With this adoption is possible have a reduction in the N rate applied in the sugarcane ratoon, as well as, there is an environmental sustainability in the nitrogen fertilization in sugarcane crop, due there is not adopt the high N rate in the fields.

The absence of alterations in TCH in sugarcane yield shows that the cultivation operation may not be necessary, mainly because sugarcane does not exploit the interleaving region [20, 21], even in areas where the traffic of machines in this region did not occur [22]. The application performed by the triple operation (number 1 in **Figure 2**), despite being incorporated in the soil, presents lower yields of TCH (**Figure 3**) due precisely to the distance at which the fertilizer is positioned in relation to the stump of the plant and the area covered by the root system, which in the middle of the interline is very small [20].

Author details

Sérgio G. Quassi de Castro[1*] and Henrique C. Junqueira Franco[2,3]

1 Laboratório Nacional de Ciência e Tecnologia do Bioetanol (CTBE), Centro Nacional de Pesquisa em Energia e Materiais (CNPEM), Polo II de Alta Tecnologia, Campinas, SP, Brazil

2 UNICAMP, Campinas, SP, Brazil

3 CROPMAN – Inovação Agrícola, Campinas, SP, Brazil

*Address all correspondence to: sergio.castro@ctbe.cnpem.br

References

[1] International Plant Nutrition Institute (IPNI). Relatório sobre consume de fertilizantes. 2016. Available from: http://brasil.ipni.net/article/BRS- 3132#aumentorelativo [Accessed: 10 January 2018]

[2] Associação Nacional para Difusão de Adubos (ANDA). Anuário Estatístico do Setor de Fertilizantes. São Paulo; 2016

[3] Companhia Nacional De Abastecimento—CONAB 2017. Acompanhamento da safra brasileira: Cana-de-açúcar. [WWW Document]. V.4—Safra 2017/2018 no1. Prim. Levant. Abril 2017. http://www.conab.gov.br/OlalaCMS/uploads/arquivos/17_04_20_14_04_31_boletim_cana_portugues_-_1o_lev_-_17-18.pdf [Accessed: 22 November 2017]

[4] Food and Agriculture Organization of the United Nations—FAO 2014. WWW Document. FAOSTAT. URL http://www.fao.org/faostat [Accessed: 22 November 2017]

[5] Bruulsema T, Lemunyon J, Herz B. Know your fertilizer rights. Crop and Soils. 2009;**42**(2):13-18

[6] Cantarella H, Montezano ZF. Nitrogênio e Enxofre. In: Prochnow LI, Casarin V, Stipp SR, editors. Boas práticas para o uso eficiente de fertilizantes - volume 2 - nutrientes. International Plant Nutrition Institute (IPNI): Piracicaba, SP; 2010. pp. 5-72

[7] Otto R, Castro SAQ , Mariano E, Castro SGQ , Franco HCJ, Trivelin PCO. Nitrogen use efficiency for sugarcane-biofuel production: what is the next? Bioenergy Research. 2016. DOI: 10.1007/s12155-016-9763-x

[8] Trivelin PCO, Franco HCJ. Adubação nitrogenada e a sustentabilidade de agrossistemas. Tópicos em Ciência do Solo. 2011;**7**:193-219

[9] Vitti GC, Otto R, Ferreira LRP. Nutrição e adubação da cana-de-açúcar: manejo nutricional da cultura da cana-de-açúcar. In: Belardo GC, Cassia MT, Silva RP, editors. Processos Agrícolas e Mecanização da Cana-de-Açúcar. Sociedade Brasileira de Engenharia Agrícola - SBEA; 2015. 608 p

[10] Nastaro Boschiero B. Adubação nitrogenada em soqueiras de cana-de-açúcar: influência do uso em longo prazo de fontes e/ou doses de nitrogênio. Tese (Doutorado Solos e Nutrição de Plantas). Piracicaba: Escola Superior de Agricultura "Luiz de Queiroz" - ESALQ - USP; 2017. 232 p

[11] Ghiberto PJ, Libardi PL, Trivelin PCO. Nutrient leaching in an Ultisol cultivated with sugarcane. Agricultural Water Management. 2015;**31, 149**:141

[12] Belardo GC, Cassia MC, Silva RP. Processos Agrícolas e Mecanização da Cana-de-Açúcar. Jaboticabal: Sociedade Brasileira de Engenharia Agrícola SBEA; 2015. 608 p

[13] Trivelin PCO. Utilização do nitrogênio pela cana-de-açúcar: três casos estudados com o uso do traçador ^{15}N. Tese (Livre-Docência). Piracicaba: Centro de Energia Nuclear na Agricultura, Universidade de São Paulo; 2000. 143 p

[14] Subbarao GV, Ito O, Sahrawat KL, Bery WL, Nakahara K, Ishikawa T, et al. Scope and strategies for regulation of nitrification in agricultural systems. Challenges and opportunities. Plant Sciences. 2006;**25**:303-335

[15] Penatti CP. Adubação da cana-de-açúcar: 30 anos de experiência. 1a ed. Piracicaba, SP: Editora Ottoni; 2013. 347 p

[16] Thorburn PJ, Biggs JS, Palmer J, Meier EA, Verburg K, Skocaj DM.

Prioritizing crop management to increase nitrogen use efficiency in Australian sugarcane crops. Frontiers in Plant Science. 2017;(8):1-16

[17] Castro SGQ. Manejo da Adubação nitrogenada em cana-de-açúcar e diagnose por meio de sensores de dossel. Tese (Doutorado em Engenharia Agrícola). Faculdade de Engenharia Agricola - FEAGRI/ UNICAMP; 2016. 129 p

[18] Castro SGQ , Magalhães PSG, Franco HCJ, Mutton MA. Harvesting systems, soil cultivation and nitrogen rate associated with sugarcane yield. Bioenergy Research. 2018;**11**:583-591

[19] Castro SGQ , Franco HCJ, Sanches GM. Nutrição e adubação da cana-de-açúcar – Macronutrientes. In: Campos CNS, Prado RM, editors. VI Simpósio Brasileiro sobre Nutrição de Plantas Aplicada em Sistemas de Alta Produtividade, FCAV-UNESP e CPCS-UFMS, Chapadão do Sul, MS. 2018

[20] Otto R, Trivelin PCO, Franco HCJ, Faroni CE, Vitti AC. Root system distribution of sugar cane as related to nitrogen fertilization, evaluated by two methods: Monolith and probes. Revista Brasileira de Ciência do Solo. 2009;**33**:601-611

[21] Otto R, Franco HCJ, Faroni CE, Vitti AC, Trivelin PCO. Sugarcane root and shoot phytomass related to nitrogen fertilization at planting. Pesquisa Agropecuária Brasileira. 2009;**44**:398-405

[22] Rossi Neto J, Souza ZM, Kölln OT, Carvalho JLN, Ferreira DA, Castioni GAF, et al. The arrangement and spacing of sugarcane planting influence root distribution and crop yield. Bioenergy Research. 2018. DOI: 10.1007/s12155-018-9896-1

[23] Castro SGQ , Decaro Júnior ST, Franco HCJ, Magalhães PSG, Garside AL, Mutton MA. Best practices of nitrogen fertilization management for sugarcane under green cane trash blanket in Brazil. Sugar Tech. 2016. DOI: 10.1007/ s12355-016-0443-0

[24] Franco HCJ, Castro SGQ , Borges CD, Kolln OT, Sanches GM. Formas de aplicação de N-fertilizante em soqueira de cana-de-açúcar. Anais. In: 10° Congresso Nacional da STAB, Ribeirão Preto, SP. 2016

3

Role of Biofertilizers in Plant Growth and Soil Health

Murugaragavan Ramasamy, T. Geetha and M. Yuvaraj

Abstract

Biofertilizers nowadays have been realised for shifting fortunes in agriculture. It has been proven successful technology in many developed countries while in developing countries exploitation of bioinoculants is hampered by several factors. Scientific knowledge on bioinoculants and its usage will pave way for its effective usage. At the same time overlooking the significance of ensuring and maintaining a high quality standard of the product will have negative impact. Hence a proper knowledge of bioinoculants and its functioning will pave way to tape the resources in a better way. Thus the chapter provide overview knowledge about different bacterial, fungal and algal biofertilizers, its associations with plants and transformations of nutrients in soil. Adopting a rational approach to the use and management of microbial fertilizers in sustainable agriculture thrive vast potential for the future.

Keywords: biofertilizers, microorganisms, diazotrophics, bioinoculants, biological nitrogen fixation

1. Introduction

One of the present day challenges in agriculture is eco-friendly practices. Though the benefits of Green revolution has been reaped by us in terms of production, the other side of it, i.e., over usage of chemical fertilizers and its subsequent deterioration of soil health has been realised these days [1]. Hence awareness of practicing organic agriculture has been taken to various spheres and products of organic agriculture are fetching up huge market. One of the organic agriculture practices includes usage of biofertilizers in farming. Biofertilizers are likely called as bioinoculants as they are the preparations containing living or latent cells of microorganisms that facilitate crop plants uptake of nutrients by their interactions within the rhizosphere once applied through seed or soil. It accelerate bound microorganism processes within the soil that augment the extent of convenience of nutrients in a very type simply assimilated by plants [2]. Use of biofertilizers has several other advantages as well like they are cost effective, eco-friendly and renewable source of plant nutrients hence forms one of the important components of integrated nutrient management. As of now we could not claim bio-inoculants as a right alternative to chemical fertilizers but in near future the scientific understanding of the same will pave way for its right use and reap full benefits [3]. In addition to this in global scale, recent published works on biofertilizers states about the varied role of bioinoculants viz., other than nutrient transformations in different crops. To mention few, increase in root growth has been observed in wheat due to inoculation of bioinoculant consortia. Likewise *Rhizobium* inoculation increases deam inase activity in pulses crops. Hence this chapter focuses on different bioinoculants and its uses in farming.

2. Importance of soil microbes in nutrient transformations

It is well established fact that soil microbes have versatile enzyme systems hence perform various nutrient transformations in soil which is very important for maintaining soil equilibrium and its health [4]. Among the nutrient transformations nitrogen and phosphors transformations forms significant importance, since they are the major plant nutrients derived from the soil.

3. Nitrogen transformations

Nitrogen cycle involves of transformations of nitrogen by particular group of soil microbes into organic, inorganic and volatile forms. In addition, a small part of the large reservoir of N_2 in the atmosphere is converted to organic compounds by certain free living microorganism or by plant microbe association that makes the element available to plant growth [5]. The atmospheric nitrogen constitutes about 78% in gaseous form which cannot be utilised by plant and other living organisms which is referred to as biological nitrogen fixation [6]. The details of nitrogen transformations occurring in soil with the role of microbes involved has been depicted below:

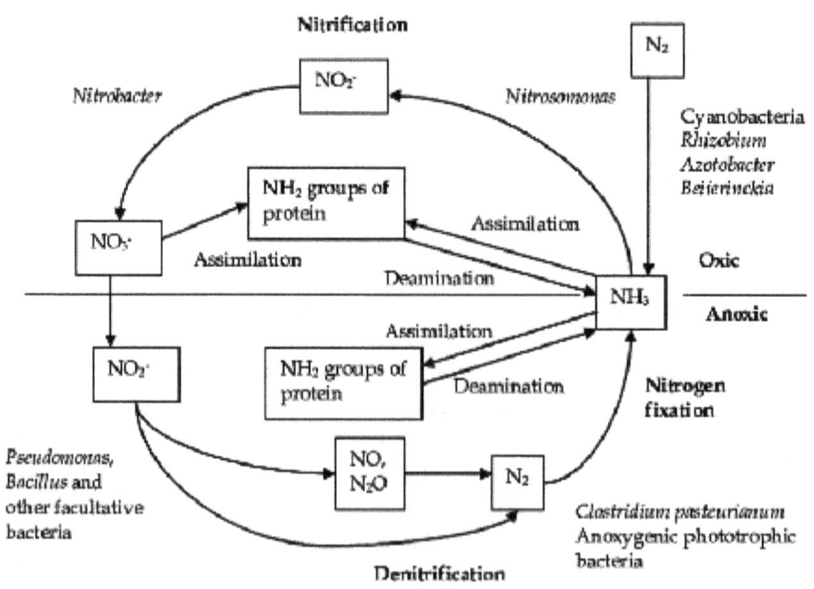

4. Biological nitrogen fixation

Biological nitrogen fixation is a component of nitrogen cycle which involves fixing up of atmospheric nitrogen by particular soil microorganisms. Nitrogen fixing ability has been restricted only to certain bacteria and few actinomycetes which belong to various groups and they are referred to as diazotrophs [7]. Diazotrophic microbes are ubiquitous to soil and are classified according to mode of nitrogen fixation to plants **Table 1**.

S. No.	Groups	Examples
1.	Free-living	*Azotobacter, Beijerinckia, Clostridium, Klebsiella, Anabaena, Nostoc*
2.	Symbiotic	*Rhizobium, Frankia, Anabaena azollae*
3.	Associative Symbiotic	*Azospirillum*

Table 1.
Groups of important diazotrophic organisms according to mode of nitrogen fixation.

The process of biological nitrogen fixation has been first documented in anaerobic bacterium *Clostridium pasteurianum* from which the enzyme nitrogenase has been isolated [8]. However today the organism has not been commercially used for the purpose. The nitrogen fixation is mediated by nitrogenase enzyme which reduces gaseous nitrogen to ammonia. All diazotrophs seemed to possess the enzyme and found to deliver quite similar mechanism of nitrogen fixation.

5. Important diazotrophs in commercial use

Rhizobium is the most studied bioinoculant which forms symbiotic association with legume plants. It was first shown by Boussingault that leguminous plant can fix atmosphere N_2 which Hellriegel and Wilfarth clarified that the process is done by bacteria residing in the roots of leguminous plants [9]. The purified bacterium was put into various examinations and now well-developed nitrogen fixing strains are available in various commercial production units.

This bioinoculant is specific for legume crops and forms nodules in the roots of the plants. It enriches the soil fertility also after harvesting of the crop. Hence it is the most preferred bioinoculant [10]. Other than root nodulating *Rhizobium* some of the strains found to nodulate stem known as *Azorhizobium* present in *Sesbania rostrata*. *Rhizobium* species are specific to legume crops because of nod factors they produce [11]. However some leguminous plants found to develop effective nodules on inoculation with the *Rhizobia* obtained from the nodules from other legume groups which is referred to as cross inoculation grouping **Table 2**.

1.	*Rhizobium leguminosarum*	CIG	Host it can nodulate
	bv. *viceae*	Pea	Peas, lentils, vicia
	bv. *phaseoli*	Bean	*Phaseolus* spp
	bv. *trifoli*	Clover	*Trifolium* spp
2.	*R. meliloti*	Alfalfa	Alfalfa, clover, fenugreek
3.	*R. loti*	Lotus	Trifoli, lupine
4.	*R. fredii*	Soybean	Soybean
5.	*R.* spp	Cowpea group	*Vigna, Arachis, Cajanus, Dolichus, Sesbania, Acacia, Prosopis*, green gram and blackgram
6.	*R.* spp	Chickpea group	Chickpea

Table 2.
Cross inoculation grouping of Rhizobium.

5.1 *Azospirillum*

Azospirillum is considered as very important diazotrophs as it form associative symbiotic relationship with the roots of graminaceous plants. It is generally recommended for rice crop [12]. The organism is microaerophillic, some are aerobic motile and gram negative in nature hence suits well for rice field conditions. It was first isolated by Beijernick and was named as *Sprillum lipoferum* later named as *Azospirillum*. In addition to nitrogen fixing ability, they also produce growth promoting substances such as IAA [13]. Some of the important species of *Azospirillum* has been listed below:

1. *A. brasilense*
2. *A. lipoferum*
3. *A. amazonense*
4. *A. halopraeferens*
5. *A. irkense*
6. *A. dobereinerae*
7. *A. largimobilis*

5.2 *Azotobacter*

Azotobacter are gram negative free living bacterium in the rhizosphere soil of many plant species, discovered by Beijernick. The bacterium is very well recognised diazotroph and fixes atmospheric nitrogen in its habitat. Owing to its versatile adaptability and nitrogen fixing ability, they are commercially used in agriculture for many crops and are known with a brand name azotobacterin. Some species of *Azotobacter* known to produce alginic acid, a compound used in medical industry and in food industry it is used as additive in ice creams and cakes [14]. Apart from its nitrogen fixing ability, it also synthesise many phytohormones such as auxins and helps in promoting growth of the plants [15]. They are involved in mobilising heavy metals in the soil thus used for bioremediation purposes as well. Many species of *Azotobacter* are pigment producers and found to degrade aromatic compounds in the agriculture lands.

5.3 Gluconoacterobacter diazotrophics

They are endotrophic bacterium which resides insides the stem of sugarcane as it prefers high sucrose and acid content for its survival. They have the ability of capturing atmospheric nitrogen and converting into ammonical form [16]. Moreov er they are known for stimulating plant growth by tolerant to acetic acid. The bacterium was first discovered in Brazil by scientists Vladimir A. Cavalcante and Johanna Dobereiner. They are originally known as *Acetobacter* belong to Acetobacteriaceae family and got the current name due to carbon source requirement. Besides nitrogen fixing ability they are known to synthesis indole-3-acetic acid which promo te the growth of the associated plant species [17]. Also reports suggest this bacterium controls pathogen especially *Xanthomonas albilineans* in sugarcane. Thus in recent years it is the most recommended bioinoculant for sugarcane.

Apart from these bacterial bioinoculants, cyanobacteria also fixes nitrogen which are referred as algal bioinoculants.

6. Algal bioinoculants

6.1 Algal biofertilizers

The potentiality of algal biofertilizers are realised long before by 1939, when WHO attributed the tropical rice natural fertility to green blue chlorophytic algae. Among algae, only blue green algae have biological nitrogen fixing ability due to the presence of heterocysts cells in them [18]. This bioinoculant is recommended only for rice crop and was proved to improve soil fertility by nitrogen fixation and organic matter enrichment after harvest. In some places, practice of culturing algae as dual crop along with rice has been done which found to inhibit small weed growth during cropping. Apart from this some of the algal species also promote growth by producing growth promoting substances [19].

The following list is some of the nitrogen fixing algal species:

a. Examples of unicellular nitrogen fixing algae: *Gloeothece*, *Gloeobacter*, *Synechococcus*, *Cyanothece*, *Gloeocapsa*, *Synechocystis*, *Chamaesiphon*, *Merismopedia*.

b. Filamentous non heterocystous forms of *Cyanobacteria*, *Oscillatoria*, *Spirulina*, *Arthrospira*, *Lyngbya*, *Microcoleus*, *Pseudanabaena*.

c. Filamentous heterocystous forms *Anabaena*, *Nostoc*, *Calothrix*, *Nodularia*, *Cylindrospermum*, *Scytonema*.

6.1.1 *Anabaena azollae*

Anabaena is a special type of algae which forms symbiotic association with free floating water fern *Azolla*. Water fern is bilobed in nature and algae resides in the roots of the fern. The common species of algae forming symbiotic association with *Azolla* are *A. microphylla*, *A. filiculoides*, *A. pinnata*, *A. caroliniana*, *A. nilotica*, *A. rubra* and *A. mexicana*. This algae takes shelter and carbon from the water fern and in turn fixes atmospheric nitrogen. They need sunlight and water for its multiplication and hence can be used for rice crop as dual crop. *Azolla* as dual crop in crop estimate to reduce nitrogen requirement by 20–25% [20].

6.2 Phosphate solubilizing and mobilizing bio inoculants

Next to nitrogen phosphorus is the key element for plant growth. Most of Indian soils contain significant amounts of inorganic form of phosphorus, but it is unavailable for plants as it is in the insoluble form. Hence it needs to be solubilised for plant use [21]. Moreover, phosphorus is also available in organic forms in soil which need to be mineralised for plant utilization. Thus mineralization and solubilisation of phosphorus in soil becomes important with respect to plant growth [22].

6.3 Phosphorus solubilising bioinoculants

The fate of phosphorus is that it forms apatite's with the salts present in the soil. In acid soil phosphorus will becomes Aluminium phosphates and Iron phosphates while in alkaline soils it becomes calcium phosphates or sodium phosphates and becomes unavailable to plants [23]. In order to make these form of phosphorus to available form some of the bioinoculants produces organic acids which convert them to soluble form like hypophosphites which can be taken by plants [24]. Examples of phosphorus solubilising Bacteria: *Bacillus megatherium* var. *phosphaticum*, *Bacillus megaterium* var. *phosphaticum*, *Bacillus subtilis*, *Bacillus circulans*, *Pseudomonas striata* Fungi: *Penicillium* sp., *Aspergillus awamori*.

The phosphorus transformations occurring in soil has been depicted below:

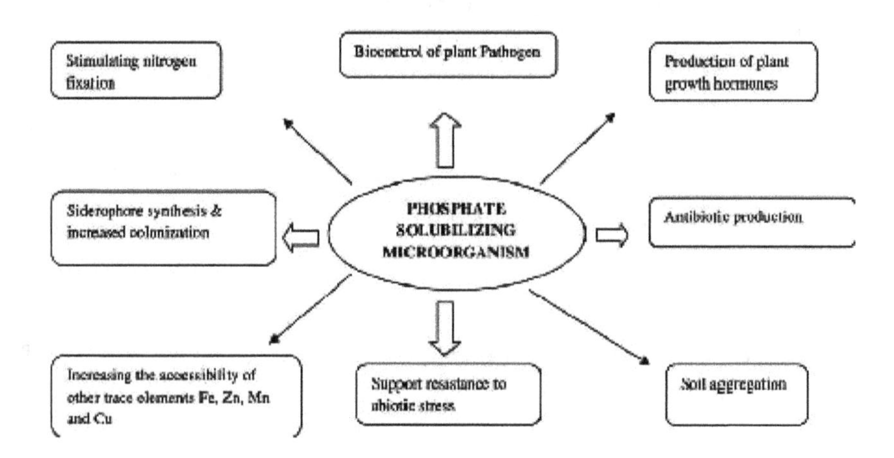

6.4 Phosphorus mobilising bioinoculants

Mycorrhiza is the special type of relationship between fungi and plants Existence of mycorrhizal fungi has been date back 450 million years ago [25]. The relationship is mutualistic in nature and the fungal members who enter into the relationship are members of Zygomycetes, Ascomycetes and Basidiomycetes [26]. Mycorrhizal fungi contribute According to the reports; roots of about 95% of all kinds of vascular plants are normally involved in symbiotic associations with mycorrhizae [27]. For angiosperms, gymnosperms, ferns and some mosses mycorrhizal association appears to be the norm. In the relationship fungus gets a supply of carbon from the associated plant and in turn plants gets lots of benefits from the fungus which is listed below:

- Fungi hyphae increases the root area hence produces more vigorous plants.
- Hyphae surrounding are thinner than roots, but longer than it hence absorb nutrients and water from deeper layers of soil. This helps the plants to tolerate drought.
- Mobilises phosphorus from distant places.
- Gives plants disease tolerance.
- Contribute to nutrient recycling due to production different enzymes.

Mycorrhizas are commonly divided into *ectomycorrhizas* and *endomycorrhizas* based on the mode of hyphal formation. The former do not penetrate the individual cells within the root, while the hyphae of later penetrate the cell wall and forms structures inside the cell membrane [28].

Ectomycorrhizas, are formed between the roots of around 10% of plant families, mostly woody plants including the birch, eucalyptus, oak, pine, and rose families, orchids, and fungi belonging to the *Basidiomycota*, *Ascomycota*, and *Zygomycota*. Thousands of ectomycorrhizal fungal species exist, hosted in over 200 genera [29]. In ectomycorrhizas the hyphae of the fungus do not penetrate the cells of plant roots. In ectomycorrhizae, the mycelium of the fungus forms a dense sheath over the surface of the root. These hyphae form a network in the apoplast, but do not penetrate the root cells. Ectomycorrhizae form a sheath and the fungus grows between the plant cells producing "Hartig net" [30].

One of the more important ectomycorrhizal fungi is *Pisolithus tinctorius*. Symbiosis begins when fungal spores germinate and emerging threadlike structures called hyphae, enter the epidermis of plant roots [31]. After colonization of the root, the fungus sends out a vast network of hyphae throughout the soil to form a greatly enhanced absorptive surface area. This results in improved nutrient acquisition and uptake by plant roots, particularly elemental phosphorus (P), zinc (Zn), manganese (Mn) and copper (Cu) and water. In return, the plant provides carbohydrates for the fungi [32].

6.5 Endomycorrhizae (AM fungi)

Endomycorrhizae form an association in which the hyphae penetrate and colonize epidermal and fleshy cortical cells of plant roots. The most common type of Endomycorrhizae is arbuscular endomycorrhizae (AM). Arbuscular mycorrhizae are characterized by the formation of unique structures such as arbuscles and vesicle by fungi within the plant root cortical cells [33]. Once the roots are colonized, individual hyphae extend from the root surface outward into the surrounding soil forming a vast hyphal network that absorbs nutrients and water that would otherwise be unavailable to the plant's root system. Endomycorrhizae can occur on most seed bearing plants, rain forest tree species, most agriculture crops and a vast variety of ornamental greenhouse crops.

Fungi forming AM associations include about 150 species belonging to genera *Gigaspora*, *Glomus*, *Sclerocystis*, *Acaulospora* and *Entrophospora*. Colonization of roots begins by the secretion of enzymes by arbuscular endomycorrhizae allowing hyphae to penetrate the epidermal and fleshy cortical cells of plant cells. Two to 3 days after colonizing the cell, the hyphae from structures within plant cells called arbuscules which resemble tiny trees and serve to facilitate the transfer of nutrients within the cortical cells.

Arbuscular endomycorrhizae provide the plant with certain mineral elements and water from the soil, and in turn, the plant provides sugars and other carbohydrates for the fungus. Between the cells, sac like structures, called vesicles may form midway or at the terminal ends of the hyphae. Vesicles contain lipids and serve primarily as storage organ for the fungus [34]. Vesicles can also serve as propagules that can colonize other parts of the plant root. Arbuscular endomycorrhizae hyphae also give rise to spores. Spores have very thick walls, which makes them very resistant to freezing and intense heat so they can survive for long periods of time. For this reason, spores are ideal for incorporating into growing media and for use as inoculants. It takes 2–6 weeks for arbuscular endomycorrhizae fungi to completely colonize plant roots and will remain with the plant throughout its life.

6.6 Plant growth promoting rhizobacteria (PGPR)

Other than these above mentioned bioinoculants some of the bacterium like *Pseudomonas*, *Bacillus thuringiensis* and fungi like *Trichoderma viride* are involved in the control of pests and diseases of plants by colonising the rhizosphere of many plants. They are referred to as biocontrol agents. Among the biocontrol agents *Pseudomonas* is largely used in seed treatments and soil application in large number of crops. They are known for siderophore production which chelates iron in the rhizosphere region thus creating iron deficiency for the pathogenic microbes. Apart from these report suggests that they are also involved in nitrogen fixation and nutrient transformations in soil. Some bacteria living in rhizosphere affect plant growth positively and some are detrimental. Rhizosphere bacteria that favor plant growth are termed as plant growth promoting rhizobacteria (PGPR). They improve plant growth directly by producing plant growth regulators such as auxins, gibberellins and cytokinins; by eliciting root metabolic activities and/or by supplying biologically fixed nitrogen. Consequently, germination, root development, nutrient and water uptake are improved. Other PGPR affect plant growth by indirect mechanisms such as biocontrol activity by suppression of bacterial, fungal and nematode pathogens. The mechanism of biocontrol include competition for colonization space and for nutrients, antibiosis, excretion of hydrogen cyanide and other volatile compounds, synthesis and absorption of siderophores, excretion of lytic enzymes (chitinases, glucanases) and systemic resistance. The well-known PGPR include *Azotobacter*, *Azospirillum*, *Azoarcus*, *Klebsiella*, *Bacillus*, *Pseudomonas*, *Arthrobacter*, *Enterobacter*, *Burkholderia*, *Serratia*, and *Rhizobium*.

7. Conclusion

In developing countries, the most important challenge is to produce sufficient food for the growing population from inelastic land area. Products of biological origin can be advantageously blended to replace a part of the energy-intensive inputs. It is in this context, biofertilizers can provide to the small and marginal farmers an economically viable lever for realizing the ultimate goal of increasing productivity. These microbes siphon out appreciable amounts of nitrogen from the atmospheri c reservoir, solubilise phosphorus and enrich the soil with this important but scarce nutrient. The crop-microbial-soil ecosystem can, therefore, be energized in sustainable agriculture with considerable ecological stability and environmental quality.

Author details

Murugaragavan Ramasamy*, T. Geetha and M. Yuvaraj
Adhiparasakthi Agricultural College, Vellore, Tamil Nadu, India

*Address all correspondence to: murugaragavanramasamy@gmail.com

References

[1] El A, Gawad AM, Hendawey MH, Farag HIA. Interaction between biofertilization and canola genotypes in relation to some biochemical constituents under Siwa Oasis conditions. Research Journal of Agriculture and Biological Sciences. 2009;**5**(1):82-96

[2] Bahadur AJ, Singh AK, Upadhyay Rai M. Effect of organic amendments and bio-fertilizers on growth, yield and quality attributes of Chinese cabbage (*Brassica pekinensis* Olsson). The Indian Journal of Agricultural Sciences. 2006;**76**(10):596-598

[3] Balachandar D, Nagarajan P. Dual inoculation of vesicular arbuscular mycorrhizae on *Rhizobium* in green gram. Legume Research. 1999;**22**(3):177-180

[4] Beg MA, Singh JK. Effects of biofertilizers and fertility levels on growth, yield and nutrient removal of green gram (*Vigna radiata*) under Kashmir conditions. The Indian Journal of Agricultural Sciences. 2009;**79**(5):388-390

[5] Begum MNSM. Response of soybean to *Rhizobium* inoculation and urea-M application under rainfed and irrigation condition [MS Thesis]. Bangladesh Agricultural University: Mymensingh; 1989

[6] Chandrasekar BR, Ambrose G, Jayabalan N. Influence of biofertilizers and nitrogen source level on the growth and yield of *Echinochloa frumentacea* (Roxb.) link. Journal of Agricultural Technology. 2005:222-233

[7] Chandrikapure KK, Sadawrte KT, Panchbhai DM, Shelke BD. Effect of bio-inoculants and graded doses of nitrogen on growth and flower yield of marigold (*Tagetes erects* L.). The Orissa Journal Of Horticulture. 2005;**28**(2):31-34

[8] Charitha Devi M, Reddy MN. Growth response of groundnut to VAM fungus and *Rhizobium* inoculation. Plant Pathology Bulletin. 2001;**10**:71-78

[9] Charitha Devi M, Reddy MN. Carbohydrate metabolism of groundnut (*Arachis hypogaea* L.) plants in relation to inoculation by VAM and *Rhizobium*. Legume Research. 2002;**25**(4):243-247

[10] Deshmukh DD, Dev DV. Effect of package of practices on nodulation, branching, nitrogen and crude protein content in groundnut. Legume Research. 2005;**28**(1):17-21

[11] Egamberdiyeva D. The effect of plant growth promoting bacteria on growth and nutrient uptake of maize in two different soils. Applied Soil Ecology. 2007;**36**:184-189

[12] El-Azouni IM, Hussein LD, Shaaban. The associative effect of VA mycorhizae with *Bradyrhizobium* as bio-fertilizers on growth and nutrient uptake of *Arachis hypogaea*. Research Journal of Agriculture and Biological Sciences. 2008;**4**(2):187-197

[13] Freiberg C, Fellay R, Broughton WJ, Rosenthal A, Perret X. Molecular basis of symbiosis between *Rhizobium* and legumes. Nature. 1997;**387**:394-401

[14] Gaballah MS, Gomaa AM. Performance of *faba* bean varieties grown under salinity stress and bio-fertilized with yeast. Journal of Applied Sciences. 2004;**4**(1):93-99

[15] Ganesan V, Mahadevan A. Effect of mycorrhizal inoculation of cassava, elephant foot yam and taro. The Journal of Root and Crops. 1994;**20**(1):1-14

[16] Jain LK, Singh P. Growth and nutrient uptake of chickpea (*Cicer arietinum* L.) as influenced by biofertilizers and phosphorous nutrition. Crop Research. 2003;**25**(3):410-413

[17] Jambotkar RK, Lakshman HC. Effect of AM fungus inoculation with additional phosphorous on the growth of *Brassica juncea* Linn. saplings. International Journal of Plant Sciences. 2008;**4**(1):52-54

[18] Kachroo D, Razdan R. Growth, nutrient uptake and yield of wheat (*Triticum aestivum*) as influenced by biofertilizers and nitrogen. Indian Journal of Agronomy. 2006;**51**(1):37-39

[19] Kale NY, Patil PL, Patil BC. A study on effect of *Rhizobium* and *Azotobacter* inoculation on nodulation, N-fixation and yield of gram. Indian Journal of Microbiology. 1982;**22**(3):203-205

[20] Lakshman HC. Interaction between VAM and Rhizobiumon *Petrocarpus marsupiuma* legume trophical tree. J. Nat. Con. Environ. 1998;**129**:75-81

[21] Neelamegam RK, Malarvizhi S, Christopher SG. Effect of biofertilizers on seed germination and early seedling growth of blackgram. Journal of Ecobiology. 2007;**20**(2):111-115

[22] Negi S, Singh RV, Dwivedi OK. Effect of biofertilizers, nutrient sources and lime on growth and yield of green pea. Legume Research. 2006;**29**(4):282-285

[23] Pannerselvam M, Thamizhiniyan P. Effect of AM fungi and diary effluent on the growth and photosynthetic pigments of *Phaseolus trilopus*. Plant Archives. 2008;**8**(1):171-173

[24] Parveen R, Kazi N, Lakshman HC. Combined inoculation of arbuscular mycorrhizal fungi and *Azotobacter* beneficial to *Proralea corylifolia* L. Asian Journal of Biological Sciences. 2008;**3**(1):11-14

[25] Patel PC, Patel JR, Sadhu AC. Response of forage sorghum (*Sorghum bicolor*) to bio-fertilizer and nitrogen levels. Indian Journal of Agronomy. 1992;**37**:466-469

[26] Prakash V, Sharma S, Kaushik S, Aggarwal A. Effect of soil sterilization on bio-inoculation activity in establishment of *Acacia catechu* wild. Phytomorphology. 2009;**59**(1 and 2): 51-56

[27] Rathore VP, Singh HP. Influence of vesicular arbuscular mycorrhizalfungi and phosphate on maize. Journal of the Indian Society of Soil Science. 1995;**43**:207-210

[28] Singh B, Pareek RJ. Studies on phosphorus and bioinoculants on biological nitrogen fixation, concentration, uptake quality and productivity of mungbean. Annals of Agricultural Research. 2003;**24**(3):537-541

[29] Singh HP. Response to inoculantion with *Bradyrhizobium*, vesicular arbuscular mycorrhiza and phosphate solublizing microbes on soybean in a Millisol. Indian Journal of Microbiology. 1994;**34**:27-37

[30] Singh A, Tripathi PN, Singh R. Effect of *Rhizobium* inoculation, nitrogen and phosphorus levels on growth, yield and quality of kharif cowpea (*Vigna ungiculata* H. Walp). Crop Research. 2007;**33**(1-3):71-73

[31] Singh LN, Devi YM, Singh AI. Effect of *Rhizobium* under different levels of nitrogen on nodulation and yield of broadbean (*faba* L.). Legume Research. 2005;**28**(2):99-102

[32] Tiwari DD, Katiyar NK, Nigam RC, Gupta BR. Yield, nutrient uptake

and quality characteristics of barley (*Hordeum vulgare* L.) as affected by graded levels of nitrogen and *Azotobacter* inoculation. Research on Crops. 2008;**9**(2):243-245

[33] Yadav KS, Suneja S, Sharma HR. Effect of dual inoculation of *Rhizobium* and *Azotobacter* in chickpea (*Cirar arietinum*). Environment and Ecology. 1994;**12**:865-868

[34] Zaidi A, Khan MDS, Amil MD. Interactive effect of rhizotropic microorganisms on yield and nutrient uptake of chickpea (*Cicer arietinum* L.). European Journal of Agronomy. 2003;**19**:15-21

4

Advancement of Nitrogen Fertilization on Tropical Environmental

Elizeu Monteiro Pereira Junior, Elaine Maria Silva Guedes Lobato, Beatriz Martineli Lima, Barbara Rodrigues Quadros, Allan Klynger da Silva Lobato, Izabelle Pereira Andrade and Letícia de Abreu Faria

Abstract

The nitrogen (N) fertilization synthetic or biological is primordial for food production worldwide. The consumption of N fertilizers in agricultural systems increased in exponential scale, mainly in developing countries. However, some negative points are associated to industrial N consumption; consequently the industry promoted ways to minimize N losses in production systems of tropical agriculture. Biological nitrogen fixation is a very important natural and sustainable process for the growth of leguminous plants, in which many micronutrients are involved, mainly as enzyme activators or prosthetic group. However, other mechanisms in the rhizosphere and molecular region still need to be clarified. Therefore, the aim of this chapter is to compile information about the historical and current affairs about the advances in N fertilization in tropical environments through a history from N fertilization world-wide, N balance in the main agricultural systems, introduction of alternative ways to avoid N losses, advances between BNF and micronutrients, as well as the effects of N absence in plant metabolisms. Biological nitrogen fixation is a very important natural process for the growth of leguminous plants, in addition many metallic nutrients, micronutrients, are involved in BNF metabolism, mainly as enzyme activators or prosthetic group. But other mechanisms in the rhizosphere and molecular region still need to be clarified.

Keywords: ammonia synthesis, biological N fixation, humic substances, N balance, volatilization

1. Introduction

Hellriegel and Wilfarth showed definitive evidence for N_2 fixation by microbes in legumes in 1886, but the industrial process to fertilizer production known as the Haber-Bosch was established just in 1906, which uses a catalytic agent at high pressure and high temperature [1].

Actually, the world population has now been increasingly relying on nitrogen (N) fertilizers in order to keep up with the demands of food and economic growth rates; on the other hand, less than 30% of synthetic fertilizers would actually be utilized; the unused chemicals sprayed on crops would be lost in the field and could subsequently cause serious environmental problems.

Urea is a popular N source in developing countries due to its advantages of a high N content, safety, and easy transportation [2]. However, the increase of pH and surface soil NH_4^+ concentrations resulting from urea hydrolysis can exacerbate NH_3 emission.

This causes low N use efficiency, especially in alkaline soils or soils with low sorption capacity, which limits the use of urea fertilizer in Europe [3]. In tropical areas, increasing the adoption of no-tillage systems also induces to high N losses from urea fertilization, in tropical soils, due to high temperatures and moisture; NH3 losses exceeding 40% of the surface-applied urea N have been reported, especially under no-till or perennial crops where plant residues are kept on the soil surface [4].

Nitrogen losses by NH_3 emission not only brings economic loss to farmers, but also detrimental effects to ecosystems and human health, while the biological nitrogen fixation (BNF) has the advantage of being environmentally friendly and therefore would be ideal for sustainable agriculture.

Enormous progress in almost all aspects of BNF has been made in the past century, especially in the recent two decades, in genetics and biochemistry, culminating in the determination of the crystallographic structures of both nitrogenase components and micronutrients metabolism.

These information collaborated to elucidate N assimilation routes in plants clarifying further its essentiality and allowing to infer that plants can be affected negatively in molecular even genetic level in N absence.

Therefore, the aim of this chapter is to compile information about the historical and current concerns about the advances in N fertilization in tropical environments through a history from N fertilization worldwide, N balance in the main agricultural systems, introduction of alternatives ways to avoid N losses, advances between BNF and micronutrients, as well as the effects of N absence in plant metabolism.

2. History of nitrogen fertilization on tropical environmental

Nitrogen is an essential element to all organisms, because it is part of protein, acids, and other organic compounds [5]. The importance of this nutrient for plants is already known since the 1660s; however, only at 1804 De Saussure received credits for N essentiality after observations of nitrate uptake from soil solution. In this same period, other researchers, as Liebig at 1840, fortified the idea of plants absorb N from atmosphere [6, 7].

Around 78% of the atmosphere gas is compound for N however in gaseous form chemically unavailable. In front of the increased demand by food production and need of N restitution after crop harvests, Fritz Haber at 1909 synthetizes the gaseous element to ammonia (NH_3) through a reaction with hydrogen and iron on high pressure and temperatures, which posteriorly was industrially developed by Carl Bosch in 1912–1913, resulting at the known Haber-Bosch process [8].

The N sources used on agricultural activities, even at the end of the eighteenth century, were from crop residues and animal manure modified or not through composting. The production and management of N fertilizers to increase crop yield, as well as corn [9, 10] and wheat [11] around the world [12] have begun at the Green Revolution of the nineteenth century, followed by ammonia synthesis in the beginning of the twentieth century and the increased need of high yield on agricultural areas [13].

World	2015	2016	2017	2018	2019	2020	Reference
Total capacity NH_3	174.781	181.228	185.222	186.804	186.920	188.310	[11]
Africa							
Total capacity NH_3	8.310	9.545	10.739	10.700	10.700	11.000	
Americas							
Total capacity NH_3	24.301	27.618	28.688	29.304	29.320	29.346	
Asia							
Total capacity NH_3	99.959	101.188	101.703	101.734	101.734	102.799	
Europe							
Total capacity NH_3	40.378	41.044	42.338	43.211	43.311	43.311	
Oceania							
Total capacity NH_3	1.833	1.833	1.854	1.854	1.854	1.854	

Table 1.
Estimative of supply capacity of N (NH_3) in continents (in thousand tons) of 2015–2020 (adapted of FAO [12]).

World	1960	1980	2000	Reference
Animal manure applied in soil	22%	16%	14%	[14]
Animal feces	56%	40%	40%	
Synthetic fertilizers	22%	44%	46%	
Africa				
Animal manure applied in soil	4%	4%	4%	
Animal manure left in pasture	91%	84%	84%	
Synthetic fertilizers	5%	12%	12%	
Americas				
Animal manure applied in soil	16%	13%	13%	
Animal manure left in pasture	60%	50%	47%	
Synthetic fertilizers	24%	37%	40%	
Asia				
Animal manure applied in soil	20%	13%	12%	
Animal manure left in pasture	61%	34%	30%	
Synthetic fertilizers	19%	53%	58%	
Europe				
Animal manure applied in soil	40%	28%	30%	
Animal manure left in pasture	27%	17%	17%	
Synthetic fertilizers	33%	55%	53%	
Oceania				
Animal manure applied in soil	2%	3%	4%	
Animal manure left in pasture	96%	91%	77%	
Synthetic fertilizers	2%	6%	19%	

The percentual represents averages from the 1960s, 1980s, and 2000s (adapted of FAO [12]).

Table 2.
Global cumulative of N fertilization from animal manure and fertilizers between 1961 and 2014.

Data from the FAO [14] estimated that the global capacity of N ammonia offer increases annually of 1.5% in average, with production of 174,781–188,310 thousands of tons of 2015–2020 (**Table 1**).

In addition, during this period, Africa, Oceania, Europe, and the Americas increased the capacity to 32.4, 1.1, 7.3, and 20.8%, respectively, however, stands out to Asia continent with the highest productive capacity estimated to 102,799 thousands of tons of N to 2020 (**Table 1**).

Estimates in global scale from FAOSTAT [15] show N inputs from animal manure increased from 66 to 113 million from 1961 to 2014, while N fertilizers applied in soils increased from 18 to 28 million of tons of N, respectively.

The use of N fertilizer at Europe continent increased 33% (about 5 million of tons of N), as a similar tendency observed in others regions (**Table 2**).

Brazil is one of the biggest fertilizer consumers in the world. The significant increase in fertilizer consumption occurred between 1988 and 2010 [16] as consequence of public policy implementation and Brazilian agriculture modernization.

Nitrogen had a higher growth consumption among the nutrients from NPK in the analyzed period, around 250%, from 814,952 to 2,854,189 tons; however, N fertilizers consumption was 12,211,855 ton from 2010 to 2013 and to around 15,469,549 tons from 2014 to 2017 [17, 18].

3. Nitrogen balance in the tropical agricultural systems

Nitrogen balance in the systems becomes a concern for tropical agriculture as a result of the high scale of N fertilizer production. Nutrient balance is a parameter that analyzes the relation between quantity of vegetable biomass produced and nutrient applied. Besides, nutrient balance is a tool with easy application and able to guide the management to efficient fertilization [19].

Nitrogen balance as a management technique accounts the nutrient exportation by crops, residual in soil and the N losses [20]; thus it is essential to a balanced fertilization strategy aiming to maximize the economic return and ensure the environmental quality.

The calculations of nutrient balance evaluation must account for the input and output of N because this nutrient can be distributed by soil, plant, and animal (**Table 3**). Between 95 and 100% of the total N input into soil is from the surface through rainfall or dust and aerosols, irrigation, runoff and groundwater, biological fixation by phototrophic and heterotrophic organisms, organic and inorganic fertilization, and seed reserves. Besides the plants exports, the N output occurs by erosion, leaching and drainage, ammonia volatilization, denitrification, and senescence plants [21, 22].

Brazilian crop exports 50% of N in harvested product mainly by the largest exportations of soybean (70%), corn (15%), sugarcane (8%), rice (2%), and wheat (2%) [17]. However, these N quantities have contribution from the N biologic fixation (NBF), mainly from soybean with 82% of the total N input in crops production.

Soybean occupied the largest area of agriculture in Brazil between 2013 and 2016 and also was responsible for the largest nutrient exportation, although N is not applied in this crop, it comprised 70% of the total N exported by all crops, while phosphorus and potassium reached 57.5 and 56.8%, respectively [23]. Analyze nutrient exportation nutrient exportation for area unity in this period was found out the largest nutrient exporters were soybean (181 kg ha^{-1}), tomato (159 kg ha^{-1}), and cotton (129 kg ha^{-1}).

Source	Amount	References
N input		[3]
Total N fertilization rates	A	
Total manure applied	B	
N symbiotic fixation	C	
Atmospheric deposition of N	D	
Irrigation water	E	
N input by seed in harvest	F	
N nonsymbiotic fixation	G	
Total N input	X = A + B + C + D + E + F + G	
N output		
N exportation in crops and/or biomass	I	
N losses by denitrification	J	
N losses by ammonia volatilization	K	
N losses by plants senescence	L	
Gaseous losses of N (except NH_3 volatilization)	M	
N losses by surface runoff	N	
N leaching	O	
N losses by soil erosion	P	
Total N output	Y=I + J + K + L + M + N + O + P	
Total N in soil		
Total N in beginning of the experiment	Q	
Total N in end of the experiment	R	
Total N changes in soil	Z = R-Q	
N balance		
During experiment performance	N_b = X-Y-Z	
N balance for year (kg N ha–1 yr–1)	N_b/Years of experiment	

Table 3.
N balance from total of inputs and outputs and N in soil in the beginning and final of agricultural experiments, modified from [3].

4. Ways to avoid N losses from agricultural systems

In agricultural systems there are losses in general; however, N losses are considered highly relevant [24, 25]. Nitrogen losses are a potential contaminant and can impact production cost. Nitrogen is a dynamic element in soil and can be lost to the atmosphere by denitrification and ammonia volatilization [24, 25].

Ammonia volatilization is a concerning problem because it represents high N losses in soil–plant system besides to be a threat for global environmental [26], while the N losses by denitrification in tropical areas are less significant in consequence of its restriction in the use of nitrate as fertilizer due its explosive potential [25–28].

Global agricultural production is responsible for 50% of N losses by ammonia volatilization meaning 37 tons of N for year; however, the losses can be higher according to the N source, application way, soil management, climate, soil temperature, and humidity [29–34].

	Rate of applied (kg N ha^{-1})	Mean % N volatilized	Location	Reference
Grassland soils	180	22.8	Argentina	[16–22]
	15–200	17.6	New Zealand	
	50	36.0	USA	
	30–150	26.7	UK	
	25	7.5	New Zealand	
Arable soils	50 150	55 30	Brazil Brazil	[11, 17, 23–29]
	120	77	Brazil	
	90	17.8	Denmark	
	200	30	India	
	60	7.9	Argentina	
	46	23	Australia	

Table 4.
Examples of ammonia volatilization due to urea application in different soils, modified from [20].

Urea is the most N source used in the world; however, also it has high susceptibility to be lost in agricultural systems [24, 25]. The high presence of urease enzyme in soil causes a rapid hydrolysis of urea and, consequently, ammonia losses to the atmosphere [35].

Variable quantities of ammonia lost to the atmosphere were related by urea use in agriculture [35–37] according to the exemplified in **Table 4**.

Urease is an extracellular enzyme naturally presents in soil, plants, and microorganisms acting as a catalyzer of urea in the hydrolysis process [30–32]. This chemical process induces excess of protons (H^+); consequently it rises pH in soil around the fertilizer granules of 6.5–8.8 or until 9.0 causing unbalance between ammonium (NH_4^+) and ammonia (NH_3) [33, 34].

During hydrolysis ammonium carbonate is formed, which is dissociated to produce ammonia ions and hydroxide; however, the relative concentration of ammonia and ammonium is determined by the pH in soil solution, and ammonia is favored under high pH condition according to equations [28].

$$NH_4^+ + OH^- \leftrightarrow NH_3 + H_2O \quad (1)$$

$$(NH_2)\ 2CO + 2H_2O \rightarrow (NH_4)\ 2CO_3 \rightarrow NH_4^+ + NH_3\uparrow + CO_2 + OH^- \quad (2)$$

Researches about urease inhibition in soil have begun over than 70 years ago, resulting in many compounds evaluated and patented as urease inhibitors [38]. Urease has a great effect on the soil-plant system through plant N efficiency, as well as being a versatile enzyme, presenting technological, biotechnological and transgenic applications [39].

Nitrogen losses can be avoided or reduced through organic or inorganic chemical compounds included in urea as an able technology to increase the efficiency of N fertilization at low cost [40–42]. Urea with urease inhibitor can cost around 30% higher than conventional urea [43].

The phosphorotriamides, hydroquinone, catechol, copper, boron, and zinc are the most evaluated additives as urease inhibitor [44]. There are more than

40 phosphorotriamides synthetized considered the most effective compound to urease inhibition because its composition comprises a functional group containing P=O or P=S bonded for at least one free amide (NH_2) to react with urease active sites and they are considered [45].

Urease inhibitor known as NBPt (N-(n-butyl) thiophosphoric triamide) has been the most used additives in Brazil, in which urea is the most used N source.

This additive is dissolved in a nonaqueous solvent to adding characteristics as (i) larger stability to NBPt molecule under temperature, humidity and transportation variances, and (ii) higher solubility; (iii) improves adherence of mix solvent + NBPt to urea granule, (iv) low toxicity, and inflammable potential; and (v) acts as buffer agent to keep alkaline pH similar to hydrolysis environment of urea in soil providing NBPt stability [43].

The largest of compounds used along with urea are low efficient when applied in soil [43]. NBPt aim is to retard the ammonia volatilization peak [46]. Generally, chemical compounds with similar structure as urea can be more efficient to retard the volatilization; thus, the bond sites and length of amide of phosphoryl triaside are similar to urea; however, there are no substrates for urease [45].

Recently, lab researches reported beneficial and/or synergic effects of the humic substances use with urea [47–49]; however, the action mechanism is still unknown [49]; also depending of humic substances, the results can be contradictory [50, 51], but there are hypotheses that urease enzymes reduce with the association of humic acid and urea [48]; besides it minimizes N losses, it can improve buffer effect in soil pH [52].

Urease inhibitor and humic substances with urea at adjusted pH (pH = 7) provided reduction of 50% from total N volatilization on a Latossolo Vermelho on sugar cane [53].

5. Interaction between biological N fixation (BNF) and micronutrients to higher plants

Biological N fixation (BNF) is an important process to global agricultural systems. This phenomenon was discovered in the mid of the nineteenth century by the German chemist Hermann Hellriegel (1831–1895); however, factors on root nodules were unknown, until the Dutch microbiologist and botanic Martinus Beijerinck (1851–1931) identifies microorganisms on root nodules able to realize chemical process to transform atmospherically N to ammonia allowing fixation and absorption by plants, proving the symbiosis between legumes and bacterial [54].

Fixation biological of N_2 (BNF) through the bacteria from genus *Bradyrhizobium* can supply N quantity necessary in legume crops as soybean, besides it is currently observed for many researchers as a clean technology contributing to replace mineral N fertilizers in legume crops [55].

Nitrogen fixation by bacteria already is well described [56]; however, currently studies are focused in nutrients involved in this metabolism, especially micronutrients [57, 58]. Among the micronutrients able to influence the BNF are boron, copper, zinc, cobalt, iron, nickel, manganese, and molybdenum, essential as structural components and enzyme activators in plants [56–59].

Iron is necessary to the production of cofactor FeMo that acts along with nitrogenase enzymes, which can affect significantly the BNF [60]. Excess or default of zinc and nickel can affect the established bacteria inside of the nodules and its symbiosis with plants [57].

There was an increase in BNF and N uptake as a result of the growth of nodules in number and mass with boron foliar application, and these results were attributed to the role of boron in the induction of nitrate assimilation by increasing protein synthesis by plant [58].

Manganese has direct role on many enzymatic processes on the BNF, including amide hydrolase enzyme which is directly dependent of Mn^{+2}, and it is responsible for ureide degradation being able to control the BNF under hydric deficiency [61].

Low copper affects the nodule formation and reduces the quantity of fixation bacteria; this element is essential for both bacteria and plants; however, its direct role on BNF is still unclear [59].

Molybdenum is an essential nutrient to BNF taking part on nitrate reductase with the reduction of nitrate (NO_3^-) to nitrite (NO_2^-) and on the nitrogenase process in conversion of dinitrogen (N_2) to ammonia (NH_3) by fixation bacteria. The low quantity demand of molybdenum allows its application on soil and foliar or even by seeds treatments, which is a form of quality aggregation to the seeds by affecting positively on germination [62].

Cobalt is a component of cobalamin and leghemoglobin synthesis, which is controlling their levels on nodules and avoiding nitrogenase enzyme inactivation; thus this element can be considered essential to N_2 fixation [63].

Nickel can affect directly the presence and quantity of fixation microorganisms because it is a hydrogenase component (Ni-Fe), which can recycle H_2 that is

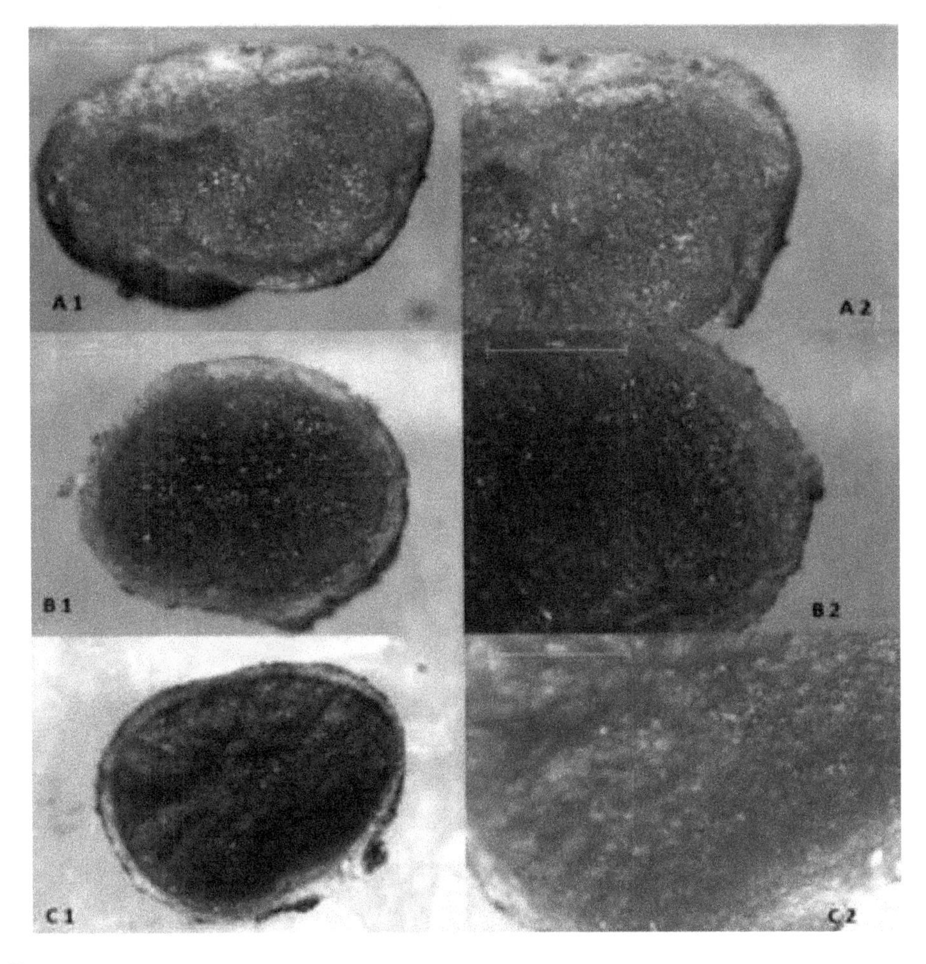

Figure 1.
Root nodules from legume. A1, longitudinal section; A2, approximated image on nodules developed with no Ni; B1,longitudinal section; B2, approximated image on nodules developed with 0.5 g dm^{-3} of Ni; C1, longitudinal section; C2, approximated image on nodules developed with 10 g dm^{-3} of Ni [64].

generated from N reduction and could affect positively or negatively the legume metabolism [64]. Nickel balance on BNF can be seen on fixation nodules where in its absence causes large formation of internal cells according to **Figure 1**.

6. Recent reports about N absence on plants metabolism

Even though the essentiality had been established for N at higher plants, there are still remained doubts about how the N absence can affect the metabolism. Recently, by modern techniques and sensible equipment, it was possible to determine clearly as N absence affects plant metabolism and production.

The N deficiency exposure of *Olea europaea* plants was described as a significant decrease on chlorophyll a and net photosynthetic rate (**Figure 2**). Photosynthesis is a process that involves light absorption by the photosynthetic pigments present in

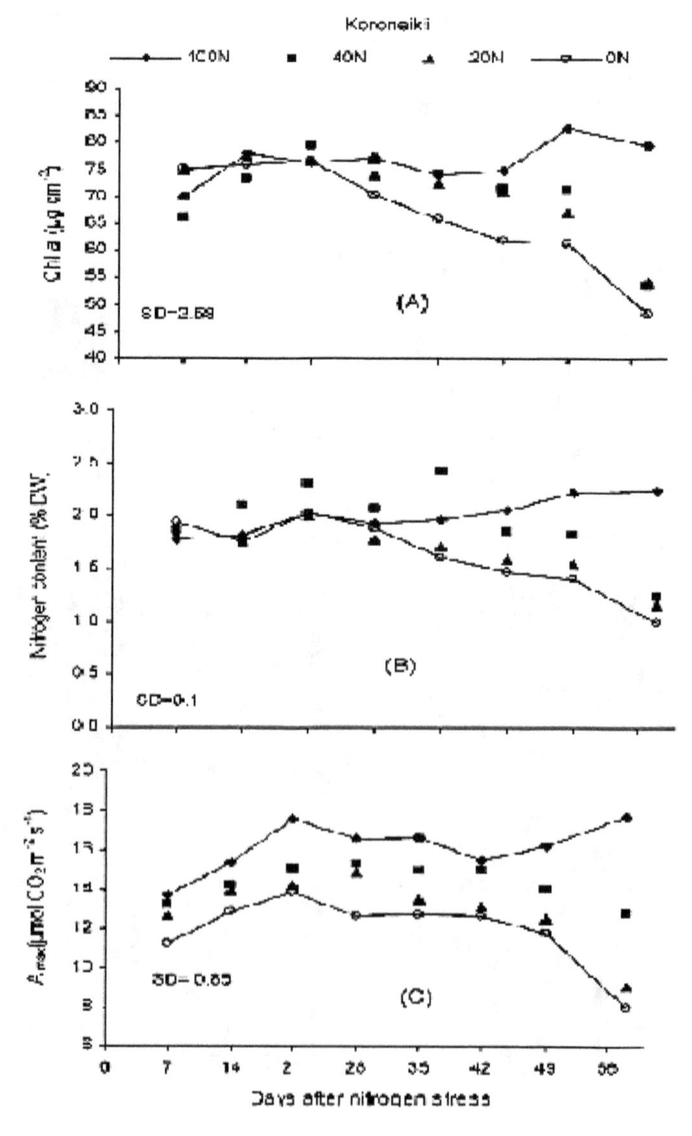

Figure 2.
Chlorophyll a (Chl a), nitrogen content, and net photosynthetic rate (Amax) in Olea europaea *plants exposed to nitrogen deficiency [65].*

light-harvesting complexes, being crucial for plant development and largely dependent on the leaf N content, because N composes the chlorophyll molecules [65].

The effects of N deficiency in the leaves of *Oryza sativa* seedlings were verified that the fluorescence parameters were negatively modulated in N-deficient plants [66]. While **Figure 2** presents few modifications until the fifth day in N-deficient plants, when compared with control plants, however as nitrogen deficiency continued, chlorophyll fluorescence of N-deficient plants was significantly impacted, in comparison with control plants.

The decrease in the ratio Fv/Fm of plants under water deficit indicates reduction in the photochemical activity, leading to the inhibition of the photosynthetic rate and the generation of reactive oxygen radicals in the chloroplast, causing damages to PSII components. Additionally, the decline in ETR values of plants under water deficit is due to the deficiency of plastoquinone (PQ) used in oxidation-reduction reactions.

7. Concluding remarks

Nitrogen fertilizer consumption follows the increasing demand by food, fiber, and energy production. The quantification of nitrogen inputs and outputs on agricultural system has been a useful and efficient tool to the evaluation of managements, mainly to the tropical agricultural.

Biological fixation is an important nitrogen input to productive systems comprising benefits in economic and environmental concerns, mainly for tropical agriculture; however, the narrow relation among this process and micronutrients and its metabolic routes still needs to be clarified.

Advances of the N fertilization on tropical environment reported at this chapter are focused mainly in an attempt to reduce ammonia volatilization from urea in consequence of its largest use as N source.

Among urease inhibitors used in tropical agriculture, NBPt has been highlighted; however, humic substances have been shown as a future alternative to reduce ammonia volatilization that still requires knowledge about its origin, molecular composition, and environmental questions.

Author details

Elizeu Monteiro Pereira Junior, Elaine Maria Silva Guedes Lobato*,
Beatriz Martineli Lima, Barbara Rodrigues Quadros, Allan Klynger da Silva Lobato,
Izabelle Pereira Andrade and Letícia de Abreu Faria
Federal University of Rural Amazônia (UFRA), Paragominas, Brazil

*Address all correspondence to: elaine.guedes@ufra.edu.br

References

[1] Cheng Q. Perspectives in biological nitrogen fixation research. Journal of Integrative Plant Biology. 2008;**50**(7): 784-796. DOI: 10.1111/J.1744-7909. 2008.00700.x

[2] Glibert PM, Harrison J, Heil C, Seitzinger S. Escalating worldwide use of urea – A global change contributing to coastal eutrophication. Biogeochemistry. 2006;**77**:441-463. DOI: 10.1007/s10533-005-3070-5

[3] Sommer SG, Schjoerring JK, Denmead OT. Ammonia emission from mineral fertilizers and fertilized crops. Advances in Agronomy. 2004;1:69-84. DOI: 10.1016/S0065-2113(03)82008-4

[4] Cantarella H, Trivelin PCO, Contin TLM, Dias FLF, Rossetto R, Marcelino R, et al. Ammonia volatilization from urease inhibitor treated urea applied to sugarcane trash blankets. Science in Agriculture. 2008;**65**:397-401

[5] McKee HS. Nitrogen Metabolism in Plants. EUA: Oxford University Press; 1962. pp. 1-18

[6] Russel EW. Soil Conditions and Plant Growth. 9th ed. New York: Wiley; 1973. p. 23

[7] Erisman JW, Galloway J, Klimont Z, Winiwarter W. How a century of ammonia synthesis changed the world. Nature Geoscience. **2**:163-165. DOI: 10.1038/ngeo325

[8] Standage T. Uma história comestível da humanidade. Rio de Janeiro: Zahar; 2010 276p

[9] Szulc P, Waligora H, Michalski T, Zajac-Rybus M, Olejarski P. Efficiency of nitrogen fertilization based on the fertilizer application method and type of maize cultivar (Zea mays L.). Plant, Soil and Environment. 2016;**62**:135-142. DOI: 10.17221/654/2015-PSE

[10] Litke L, Gaile Z, Ruza A. Nitrogen fertilizer influence on winter wheat yield and yield components depending on soil tillage and forecrop. Research for Rural Development. 2017;**1**:9-12. DOI: 10.22616/rrd.23.2017.049

[11] FAO. World fertilizer trends and outlook to 2020. Rome; 2017

[12] Lucas FT, Borges BNMN, Coutinho ELM. Nitrogen fertilizer management for Maize production under tropical climate. Agronomy Journal. 2018;**111**:2031-2037. DOI: 10.2134/agronj2018.10.0665

[13] Cao P, Lu C, Yu Z. Historical nitrogen fertilizer use in agricultural ecosystems of the contiguous United States during 1850-2015: Application rate, timing, and fertilizer types. Earth System Science Data. 2018; **10**:969-984. DOI: 10.5194/essd-10-969-2018

[14] FAO. Nitrogen inputs to agricultural soils from livestock manure – New statistics. Rome; 2018

[15] Da Cunha JF, Casarin V, Prochnow LI. Balanço de Nutrientes na Agricultura Brasileira no período de 1988 a 2010. Informações Agronômicas n° 130; 2010

[16] Da Cunha JF, Francisco EAB, Casarin V, Prochnow LI. Balanço de Nutrientes na Agricultura Brasileira – 2009 a 2012. Informações Agronômicas n° 145; 2014

[17] Da Cunha JF, Francisco EAB, Prochnow LI. Balanço de Nutrientes na Agricultura Brasileira no período de 2013 a 2016. Informações Agronômicas n° 162; 2018

[18] CNPASA. Balanço de nutrientes em sistemas agrícolas: importância do manejo de nutrientes em safras de

quebra de produtividade. Informativo técnico do Núcleo de Sistemas Agrícolas da Embrapa Pesca e Aquicultura n° 14; 2016

[19] Reetz HF. In: Lopes AS, editor. Fertilizantes e o seu uso eficiente. São Paulo: ANDA; 2017. 178p

[20] Sainju UM. Determination of nitrogen balance in agroecossystems. MethodsX. 2017;**4**:199-208

[21] Wetsellar R, Ganry F. In: Dommergues YR, Diem HG, editors. Microbiology of Tropical Soils and Plant Productivity. The Hague/Boston/London: Junk Publishers; 1982

[22] Guareschi FR, Boddey RM, Alves JR, Sarkis LF, Martins MR, Jantalia CP, et al. Balanço de nitrogênio, fósfoto e potássio na agricultura da América Latina e o Caribe. Revista Terra Latinoamericana. 2019;**37**:105-119

[23] De Datta SK. Principles and Practices of Rice Production. New York: John Wiley; 1981. 618 p

[24] Cantarella H, Mattos Júnior D, Quaggio JA, Rigolin AT. Fruit yield of Valencia sweet orange fertilized with different N sources and the loss of applied N. Nutrient Cycling in Agroecosystems. 2003;**67**:215-223

[25] Malavolta E. Manual de Nutrição Mineral de Plantas. Agronômica Ceres: São Paulo; 2006. 638 p

[26] Cantarella H. Nitrogênio. In: Novais RF, Alvarez VVH, Barros NF, Fontes RLF, Cantarutti RB, Neves JCL, editors. Fertilidade do Solo. Viçosa: SBCS; 2007. pp. 422-423

[27] Vitti AC, Trivelin PCO, Gava GJC, Franco HCJ, Bologna IR, Faroni CE. Produtividade da cana-de-açúcar relacionada à localização de adubos nitrogenados aplicados sobre os resíduos culturais em canavial sem queima. Revista Brasileira de Ciência do Solo. 2007;**31**:491-498

[28] Cameron KC, Moir HJD. Nitrogen losses from the soil/plant system: A review. The Annals of Applied Biology. 2013;**162**:145-173

[29] Martens DA, Bremmer JM. Soil properties affecting volatilization of ammonia from soils treated with urea. Communications in Soil Science and Plant Analysis. 1989;**20**:1645-1657

[30] Watson CA. The influence of soil properties on the effectiveness of phenylphosphorodiamidate (PPD) in reducing ammonia volatilization from surface applied urea. Nutrient Cycling in Agroecosystems. 1990;**24**:1-10

[31] Bussink DW, Oenema O. Ammonia volatilization from dairy farming systems in temperate areas: A review. Nutrient Cycling in Agroecosystems. 1998;**51**:19-33

[32] Bishop P, Manning M. Urea Volatilization: The Risk Management and Mitigation Strategies. Palmerston North, New Zealand: Fertilizer and Lime Research Centre, Massey University; 2010

[33] Sutton MA, Bleeker A, Howard CM, Bekunda M, Grizzetti B, De Vries W, et al. Our Nutrient World: The Challenge to Produce More Food and Energy with Less Pollution. Edinburgh: Centre for Ecology and Hidrology; 2013

[34] Stafanato JB, Goulart RS, Zonta E, Lima E, Mazur N, Pereira CG, et al. Volatilização de amônia oriunda de ureia pastilhada com micronutrientes em ambiente controlado. Revista Brasileira Ciência do Solo. 2013;**37**:726-732

[35] Cabezas WARL, Souza MA. Volatilização de amônia, lixiviação de

nitrogênio e produtividade de milho em resposta à aplicação de misturas de ureia com sulfato de amônio ou com gesso agrícola. Revista Brasileira Ciência do Solo. 2008;**32**:2331-2342

[36] Zaman M, Saggar S, Blenner-hassett JD, Singh J. Effect of urease and nitrification inhibitors on N transformation, gaseous emissions of ammonia and nitrous oxide, pasture yield and N uptake in grazed pasture system. Soil Biology and Biochemistry. 2009;**41**:1270-1280

[37] Nascimento CAC, Vitti GC, Faria LA, Luz PHC, Mendes FL. Ammonia volatilization from coated urea forms. Revista Brasileira Ciência do Solo. 2013;**37**:1057-1063

[38] Conrad JP. Catalytic activity causing the hydrolysis of urea in soil as influenced by several agronomic factors. Soil Science Society of America Journal. 1940;**5**:238-241

[39] Kappaun K, Piovesan A, Celia R, Carlini R, Ligabue-Braun R. Ureases: Historical aspects, catalytic and non-catalytic properties – A review. Journal of Advanced Research. 2008;**13**:3-17

[40] Trenkel ME. Slow and Controlled-Release and Stabilized Fertilizers: An Option for Enhancing Nutrient Use Efficiency in Agriculture. Paris: Internacional Fertilizer Industry Association; 2010. 167 p

[41] Azeem B, Kushaari K, Man ZB, Basit A, Thanh TH. Review on materials and methods to produce controlled release coated urea fertilizer. Journal of Controlled Release. 2014;**181**:11-21

[42] Timilsena YP, Adhikari R, Casey P, Muster T, Gill H, Adhikari B. Enhanced efficiency fertilizers: A review of formulation and nutrient release patterns. Journal of the Science of Food and Agriculture. 2014;**95**:1131-1142

[43] Guelfi D. Fertilizantes Nitrogenados Estabilizados de liberação lenta ou controlada. Informações Agronômicas n°. 157; 2017

[44] Bremner JM, Douglas LA. Inhibition of urease activity in soils. Soil Biology and Biochemistry. 1971;**3**:297-307

[45] Dominguez MJ, SanMartin C, Font M, Palop J, San Francisco J, Urrutia O, et al. Design synthesis and biological evaluation of phosphoramide derivatives as ureases inhibitors. Journal of Agricultural and Food Chemistry. 2008;**56**:3721-3731

[46] Watson CJ, Akhonzada NA, Hamilton JTG, Matthews DI. Rate and mode of application of the urease inhibitor N-(n-butyl) thiophosphoric triamide on ammonia volatilization from surface-applied urea. Soil Use and Management. 2008;**24**:246-253

[47] Ahmed OH, Aminuddin H, Husni MHA. Reducing ammonia loss from urea and improving soil-exchangeable ammonium retention through mixing triple superphosphate, humic acid and zeolite. Soil Use and Management. 2006;**22**:315-319

[48] Dong L, Kreylos AL, Yang J, Yuana H, Scowb KM. Humic acids buffer the effects of urea on soil ammonia oxidizers and potential nitrification. Soil Biology and Biochemistry. 2009;**4**:1612-1621

[49] Kasim S, Ahmed OH, Majid NMA, Yusop MK, Jalloh MB. Reduction of ammonia loss by mixing urea with liquid humic and fulvic acids isolated from tropical peat soil. American Journal Agricultural Biology Science. 2009;**4**:18-23

[50] Canellas LP, Piccolo A, Dobbss LB, Spaccini R, Olivares FL, Zandonadi DB, et al. Chemical composition and bioactivity properties of size fractions

separated from a vermicompost humic acid. Chemosphere. 2010;**78**:457-466

[51] Rose MT, Patti AF, Little KR, Brown AL, Jackson WR, Cavagnaro TR. A meta-analysis and review of plant-growth response to humic substances: Practical implications for agriculture. Advances in Agronomy. 2014;**124**:37-89

[52] Pertusatti J, Prado AGS. Buffer capacity of humic acid: Thermodynamic approach. Journal of Colloid and Interface Science. 2007;**314**:484-489

[53] Leite JM. Eficiênci agronômica da adubação nitrogenada associada à aplicação de substâncias húmicas em cana-de-açúcar [Tese]. Piracicaba: Escola Superior de Agricultura Luiz de Queiroz; 2016

[54] Hirsh AN. Brief History of the Discovery of Nitrogen-Fixing Organism. Los Angeles: University of California; 2009

[55] Campo RJ, Hungria M. Importância dos Micronutrientes na Fixação Biológica do N2. Informações Agronômicas nº 98; 2002

[56] Rodrigues ASP. Aspectos que interferem na nodulação e fixação biológica de Nitrogênio por Bradyrhizobium na Cultura da Soja. Primavera Leste, Mato Grosso: Universidade de Cuiabá; 2017

[57] Kryvoruchko IS. Zn-use efficiency for optimization of symbiotic nitrogen fixation in chickpea (*Cicer arietinum* L.). Turkish Journal of Botany. 2017;**2**: 669-972. DOI: 10.3906/bot-1610-6

[58] Bellaloui N, Mengistu A, Kassem MA, Abel CA, Zobiole LHS. Role of boron nutrient in nodules growth and nitrogen fixation in soybean genotypes under Wates stress conditions. In: Advances in Biology and Ecology of Nitrogen Fixation. Editor. Takuji Ohyama, Zobiole LHS: Intech Open; 2006. p. 155-159. DOI: 10.5772/56994

[59] Weisany W, Raei Y, Allahverdipoor KH. Role of some of mineral nutrients in biological nitrogen fixation. Bulletin of Environment, Pharmacology and Life Sciences. 2013;**2**:77-84

[60] Dynarski KA, Houlton BZ. Nutrient limitation of terrestrial free-living nitrogen fixation. The New Phytologist. 2018;**217**(3):1050-1051. DOI: 10.1111/nph.14905

[61] De Souza LGM. Otimização da Fixação Biológica de Nitrogênio na soja em função da reinoculação em cobertura sob Plantio Direto. Ilha Solteira: Universidade Estadual Paulista; 2015

[62] Dias JAC. Enriquecimento de sementes de ervilha com molibdênio, fixação simbiótica de nitrogênio, produção e qualidade de sementes. Ilha Solteira: Universidade Estadual Paulista; 2017

[63] Galdino PLF. Aplicação de Cobalto e Molibdênio no crescimento vegetativo da Soja. Sinop, Mato Grosso do Sul: Universidade Federal do Mato Grosso do Sul; 2018

[64] Macedo FG, Bresolin JD, Santos EF, Furlan F, Lopes da Silva WT, Pollaco JC, et al. Nickel availability in soil as influenced by liming and its role in soybean nitrogen metabolism. Frontiers in Plant Science. 2016;**8**(7):1358. DOI: 10.3389/fpls.2016.01358

[65] Boussadia O, Steppe K, Zgallai H, El Hadj SB, Braham M, Lemeur R, et al. Effects of nitrogen deficiency on leaf photosynthesis, carbohydrate status and biomass production in two olive cultivars 'Meski' and 'Koroneiki'. Scientia Horticulturae. 2010;**123**(3):336-342

[66] Huang ZA, Jiang DA, Yang Y, Sun JW, Jin SH. Effects of nitrogen deficiency on gas exchange, chlorophyll fluorescence, and antioxidant enzymes in leaves of Rice plants. Photosynthetica. 2004;**42**:357-364. DOI: 10.1023/B:PHOT.0000046153.08935.4c

5

Nitrogen Fertilization in Blackberry

Ivan dos Santos Pereira, Adilson Luis Bamberg, Carlos Augusto Posser Silveira, Luis Eduardo Corrêa Antunes and Rogério Oliveira de Sousa

Abstract

Nutrition studies for blackberry crop are scarce worldwide. This chapter presents several aspects of nitrogen (N) in blackberry (*Rubus* spp.) nutrition. Soil characteristics that can influence nitrogen fertilization are the large discrepancies in the rates recommended in the literature, forms and times of application, sources of nitrogen, differences between cultivars and the main symptoms of N deficiency. The impact of moderate and severe nitrogen deficiency on vegetative growth and yield of 'Tupy' blackberry is also presented. In addition, a nitrogen fertilization recommendation system is proposed, based on the organic matter content of the soil, the age of the plants, and the expected productivity of the cultivars.

Keywords: *Rubus* spp., nutritional requirements, cultivar differences, nutritional deficiency, soil organic matter, fertilization recommendation

1. Introduction

Nitrogen (N) is the mineral element that plants generally need in greater quantity, since they serve to form components of plant cells, such as amino acids and nucleic acids, besides participating in the chlorophyll molecule [1]. N deficiency rapidly reduces plant growth, as it causes reduction of cell division and expansion, leaf area, and photosynthesis [2].

In blackberry (*Rubus* spp.), N is the most abundant element and plays a major role in its growth, development, and productivity [3–6]. The optimum leaf content required for a satisfactory performance of blackberry varies from 2.2 to 3.0% of the dry matter of the leaves [7, 8].

The need for N supply may vary according to soil organic matter (SOM) content, yield, growth habit, age, and cultivar [8, 9]. The N rates recommended in the literature vary widely, mainly due to differences between cultivars and soil characteristics, but another important factor is the age of the plants. In the first years, the productive capacity of the plants is smaller, and therefore the demand for nitrogen is also lower. High N rates in the first 2 years can reduce fruit quality and increase disease incidence. On the other hand, low rates from the third year make it difficult to obtain high yield.

Nitrogen fertilization provides immediate effect (same season) and residual (next season). The immediate effect is mainly on the productive capacity of the floricans and the size of the fruits. Already in the following season, the greatest influence is on the growth and formation of primocanes and floral buds [10].

Considering the N importance to blackberry, this chapter aims to review the dynamics of nitrogen fertilization and present a recommendation proposal.

2. Soil characteristics

The recommended soil pH for blackberries is 5.5 to 6.5 [7, 8, 11]. The availability of N from fertilizers mainly depends on the fertilizer source [12] and the method of application [13]. Soil pH affects nitrification, with ammonium-N sources converted more rapidly to nitrate-N at a pH of 6.0 than at 5.5 [8].

Another issue that must be observed in relation to pH is the presence of aluminum (Al^{3+}) in the soil. In these cases it is essential that the pH is raised to values close to 6.0 through liming, avoiding problems with phytotoxicity by aluminum.

The ideal soil organic matter content is around 2.0 to 4.5% [3, 14, 6]. However, what is observed is that the blackberry is a rustic species, which can be cultivated in soils with a wide range of SOM levels. The difference is that in soils with low levels, less than 2%, there is a need for special care in relation to nitrogen fertilization. On the other hand, when the content of SOM is high, more than 4.5%, the vigor of the plants is driven in a way that increases the frequency and intensity of pruning, as well as decreases the need for N fertilization.

Some studies indicate a lack of effect of fertilization on yield in soils with a SOM content of 3.9%, while in soil with a SOM content of 1.1%, there was a linear increase of the yield in response to increasing rates of N [4, 15]. These results suggest an important influence of SOM on the response of blackberry to fertilization with N.

3. Nitrogen fertilizer rate

There are significant variations in the nitrogen rates recommended by the literature, with values ranging between 0.0 and 200 kg ha^{-1}. Such differences may be related to factors such as soil characteristics, climate, and genotypes [3, 5].

Some authors recommend the application of 34–56 kg ha^{-1} in the first year, regardless of growth habit, and in the following years, the dose would be from 56 to 78 kg ha^{-1} for training and semi-erect cultivars and from 56 to 90 kg ha^{-1} for cultivars of erect habit [8].

Other indications for semi-erect and erect cultivars are 25–45 kg ha^{-1} at the establishment and 45–70 kg ha^{-1} in subsequent years [16, 17]. The recommendations for training cultivars are between 25 and 45 kg ha^{-1} in the year of establishment and between 45 and 60 kg ha^{-1} in the following seasons [17].

Considering that Tupy and Xavante has semi-erect and erect growth habit, respectively, there was an inversion of the logic indicated in the recommendations of the literature. In this case, probably the greatest nutritional need of the cultivar Tupy may come as a result of its greater yield and export of nutrients, about 40% higher than that obtained with Xavante.

Another feature that diverges between these two genotypes is the presence of thorns. There is a need for new studies in order to verify if there is a possible relationship of this characteristic with the nutritional need.

The amount of nutrients applied in blackberry cultivation can also vary with the age of the plants. The application of N in the year of establishment of the crop is controversial; there are authors who do not recommend the application, due to the risk of damages to the vegetative buds and others that, although emphasizing that the application is less necessary than in the subsequent years, suggest the application of up to 56 kg ha^{-1} [8, 14]. On the other hand, in a study carried out in the south of Brazil, in an area with a 1.1% SOM content, the maximum productive efficiency rate was 109 kg ha^{-1} in the first harvest after planting and increased up to 155 kg ha^{-1} at the third crop [4]. Thus, these results indicate that there is a need for differential fertilization strategy, depending on the age of the plants.

4. Nitrogen fertilizer application

The N application is usually done in granular form while planting directly in the row but can be carried out during crop development, via fertigation or foliar fertilization [14, 17]. In a study comparing the application methods, it was concluded that the granular fertilization in the spring had a better result than the drip fertigation, mainly due to the higher leaching observed in the drip method [13]. Although it can be used, foliar fertilization has not presented satisfactory results [8], which is justified by the high demand that the plant has in relation to this nutrient.

About the N sources, blackberry responds satisfactorily to several, but ammonium nitrate, urea, and ammonium sulfate are the most commonly used sources [3, 17]. In Brazil, the ammonium sulfate is currently recommended, as blackberries also demand important sulfur amount (about 3 kg per ton of fruit and pruning material removed) [7, 14].

However, research results indicate that after 3 years of using ammonium sulfate as N source, there is a significant reduction of pH (**Figure 1a**) and increase of Al^{3+} content (**Figure 1b**) in the soil, which may have a negative impact on the productivity (**Figure 1c**). In this specific case, when the soil pH was lower than 4.7 and the Al^{3+} content was higher than 0.66 cmol$_c$ dm^{-3}, blackberry production decreased. However, further studies on different soil types should be carried out in order to confirm this relationship. Therefore, when using ammonium sulfate as N source, the pH and Al^{3+} content of the soil should be monitored.

As for the time of application, N should be applied in the spring and after harvest [7, 14, 18]. Fertilization carried out at the end of winter or early spring aims to provide fruit production and the growth of primocanes, new stems that will be responsible for fruit production in the following year.

N has an important role in the formation of "primocane" numbers since it stimulates budding of crown buds, thus impacting the number of stems and the yield of the next season [3]. Maintaining the proper number of stems over the years is important. There is a positive correlation between the number of stems and the

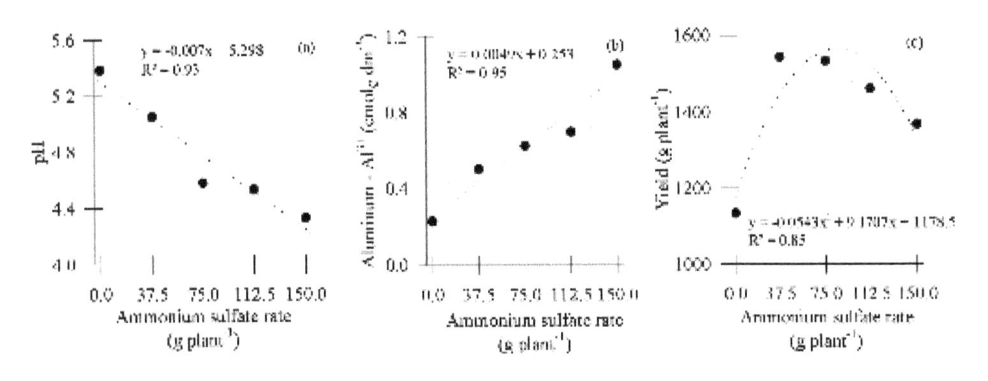

Figure 1.
Influence of ammonium sulfate rates on pH (a), aluminum (Al^{3+}) concentration (b), and blackberry 'Tupy' productivity (c) [9].

yield of the blackberry, that is, the larger the number of stems (up to 12 stems m $^{-1}$), the greater the productivity [19].

On the other hand, postharvest fertilization, usually performed after post-harvest pruning, has the function of stimulating the development of primocanes, inducing the formation of vigorous stems and thus capable of supporting high yields and larger fruits in the next season. In blackberry cultivation, there is a significant correlation between stem diameter and fruit size, and more vigorous stems have potential for larger fruit production [19, 20].

5. Nutritional differences between cultivars

It has been observed in the literature that blackberry cultivars present important differences in relation to their nutritional requirements [4, 5, 8, 17]. Some authors recommend fertilization for groups of cultivars with the same growth habit [8, 17]. In addition, there are research results that also indicate differences between cultivars with and without thorns [5, 21]. But, in general, the greatest difference between demands of each cultivar is actually related to their productive capacity, that is, more productive cultivars export larger quantities of N and therefore also demand higher rates of fertilizer. For this reason, the tendency is that fertilization recommendations incorporate the expected productivity as a criterion to define the most adequate N rate.

6. Nitrogen deficiency

Compared to blackberry plants with no N deficiency (**Figure 2a**), N deficiency in leaves is characterized by foliar chlorosis (**Figure 2b**), and in severe deficiency situations, reddish patches may appear distributed throughout the leaf blade (**Figure 2c**). In addition to the leaves, the stems may also exhibit reddish pigmentation, and the greater the deficiency, the greater the intensity of the red (**Figure 2e** and **f**), being this type of pigmentation originated from the anthocyanin accumulation [21]. N deficiency appears on old leaves and progresses to the younger ones. This is due to the fact that N is easily translocated and redistributed within the plant [21]. Therefore, when the nutrient supply in the roots is insufficient, the nutrient of the older leaves is mobilized for the younger ones.

The vegetative growth is the major aspect affected by the deficiency of N, being the foliar chlorosis the first visible symptom. This happens because N-deficient plants present a lower chlorophyll leaf concentration, which can be verified by the lower SPAD (soil plant analysis development) index in leaves of nitrogen-deficient plants (**Figure 3a**), which is an indirect measure of the foliar chlorophyll content. It is noted that the SPAD index was not able to identify differences between moderate and severe deficiency. However, the method clearly identified N deficiency even in a moderate situation. In this way, it is possible that through calibration studies, the SPAD index can be used as a rapid method to evaluate the N leaf content in blackberry plants.

In general, when foliar chlorosis is identified, vegetative growth has generally been compromised. Among the main growth parameters that may indicate N deficiency problems in the blackberry are the length of internodes and the length of the stems. In **Figure 3b**, it is possible to observe the average reduction of 15% in the length of the internodes in plants with moderate N deficiency. However, the length of the stems is further reduced. In plants with moderate deficiency, the reduction in stem length was 35%, whereas in plants with severe deficiency, the reduction was 52% (**Figure 3c**).

N deficiency also causes a reduction in the number and in the mass of black-berry fruits. **Figure 3c** shows a reduction of 63% in the number of fruits produced

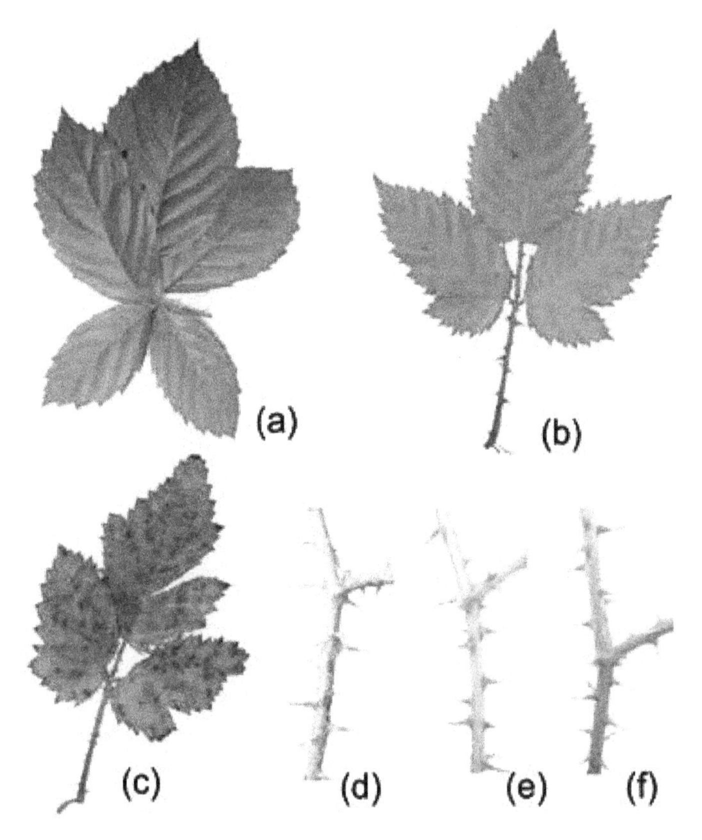

Figure 2.
Leaf superior surface with normal aspect, no deficiency symptoms (a), with moderate (b) and severe nitrogen deficiency (c) in Tupy cultivar. Besides stems without deficiency symptoms (d) and with moderate (e) and severe nitrogen deficiency (f) [9].

in blackberry plants with moderate N deficiency and 65% in plants with severe deficiency. Regarding the fruit mass, the reduction was 52% in plants with moderate deficiency and 60% when the deficiency was severe (**Figure 3d**).

The increase in fruit size is the main effect of nitrogen fertilization on the fruit quality of blackberries [22]. In general, nitrogen fertilization induces the formation of vigorous stems which, in turn, provide the formation of larger fruits [20]. In terms of sensory quality, the effects are varied. Nitrogen fertilization performed at the recommended amount increases soluble solids concentration but has little effect on attributes such as pH, acidity, sugar/acid ratio, and firmness [23, 24]. On the other hand, the excess of N can cause reduction of soluble solids, increase acidity and pH, and also decrease fruit firmness.

The accumulation of negative effects caused by N deficiency on vegetative growth and fruit formation in blackberry plants had a catastrophic impact on plant productivity. It was observed a reduction of 71% in the productivity of plants with moderate deficiency and 75% in plants with severe deficiency (**Figure 3f**).

The results presented in **Figure 3** demonstrate that in many aspects there are no significant differences in vegetative growth or fruit yield between plants with moderate and severe N deficiency. This shows that even when N deficiency is moderate, the crop yield is extremely impacted.

Excess of N is characterized by excessive plant vigor, long internodes, thinner stems, dark green leaves, low yield, low quality of fruits, less conservation potential, and greater risk of diseases [8, 14, 21]. In addition, very high rates of N may result in a decrease in the foliar content of manganese (Mn), potassium (K),

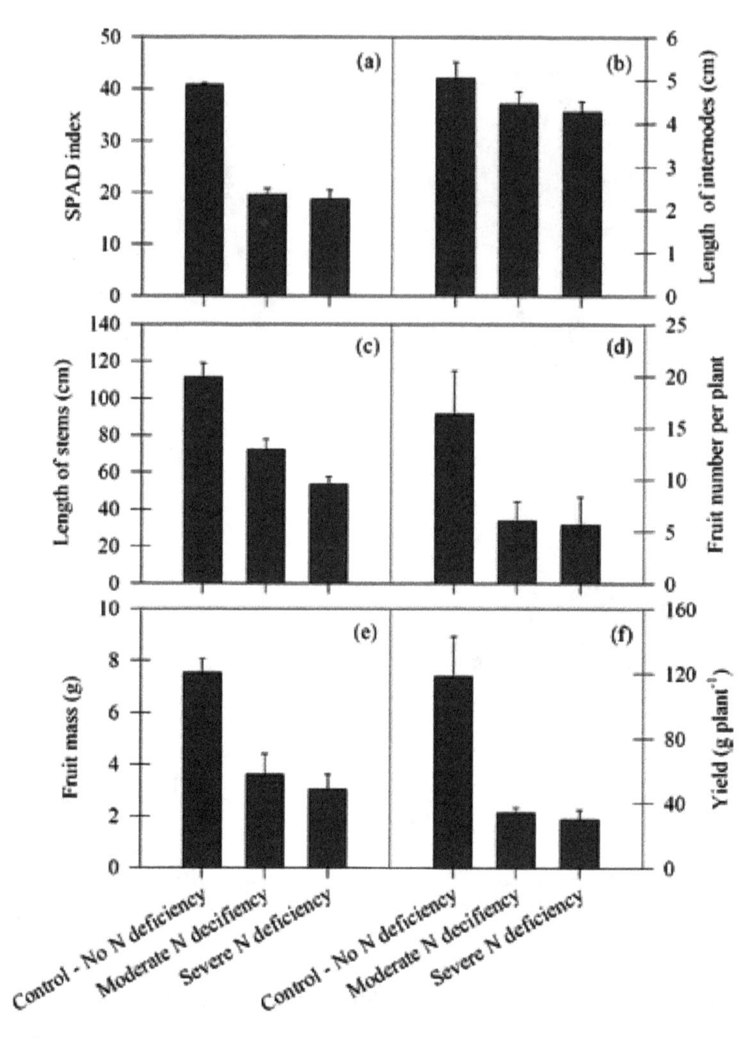

Figure 3.
Effects of different levels of nitrogen (N) deficiency on the SPAD index (a), length of internodes (b), length of stems (c), fruit number per plant (d), fruit mass (e) and yield (f) of 'Tupy' blackberry. Deficiency levels: control or no N deficiency; moderate deficiency, and severe deficiency.

calcium (Ca), and magnesium (Mg) and increase of copper (Cu). On the other hand, the reduction of fertilization with N reduces the iron content (Fe) and may reduce production due to nutritional imbalance [4, 25]. The reduction of other nutrients can be attributed to two main factors, competition for the same binding sites and the dilution effect, since N stimulates vegetative growth [4].

7. Nitrogen recommendation fertilization

The fertilization recommendation is divided into **fertilization of preplanting** and **production fertilization** [7]. The preplanting fertilization aims to correct soil problems, as in the case of N, low levels of organic material, and must be carried out during soil preparation, before planting of seedlings, while the production fertilization has the objective to restore to the soil the quantities of N exported by the production of fruits (considers the expectation of production) and natural losses, as, for example, by leaching.

7.1 Preplanting fertilization

Nitrogen fertilization of preplanting is recommended in soils with organic matter content of less than 2.5%. In these cases, it is recommended to apply 40 kg ha^{-1} of N, and it is preferable to use organic sources, such as compost from manure.

7.2 Production fertilization

Production fertilization begins in the year following to planting of the seedlings and takes into account soil organic matter content, plant age, and production expectation (**Figure 4**).

The N rate applied in the spring should be divided into three times: the first application should be made at the beginning of the budding; the second, 30 days

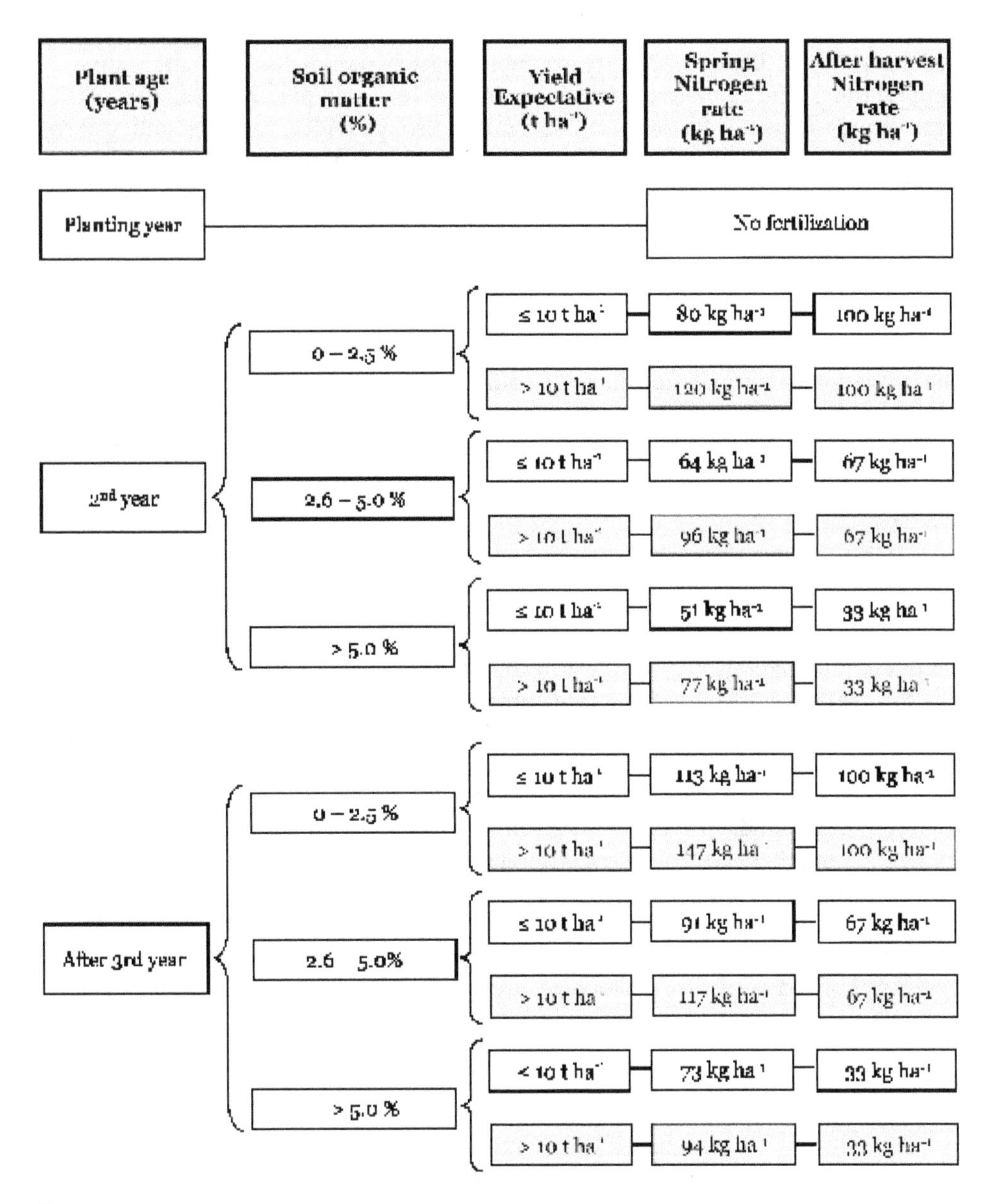

Figure 4.
Recommendation of nitrogen production fertilization.

Class	Nitrogen content (%)
Insufficient	<2.20
Normal	2.30–3.00
Excessive	>3.00

Table 1.
Tissue N concentration from sampling after harvesting the primocane's leaves [8].

after; and the third, at 60 days after the first application. The application postharvest can be applied in a single rate.

The N may be applied in the form of urea (45% of N), calcium nitrate (14% of N), ammonium nitrate (32% of N), or ammonium sulfate (20% of N). Ammonium sulfate is preferably recommended because the blackberries require sulfur. However, the consecutive use of ammonium sulfate can significantly reduce soil pH, being necessary to monitor this parameter. An alternative to supply the sulfur demand with little impact on soil pH may be the use of agricultural gypsum ($CaSO_4.2H_2O$), which also provides calcium and can be applied at a rate of 34–45 kg ha^{-1} to supply possible S deficiencies of blackberries [8].The N source application should be performed at the soil surface, along the planting row, approximately 15 cm away from crown of plants.

7.3 Leaf N concentrations

In contrast to other species, in blackberry, leaf analysis is applied to evaluate the nutritional status after harvest, aiming the elaboration of a fertilization program for the next productive cycle (**Table 1**). For this reason, sampling is performed after harvesting the primocane's leaves. Sampling is performed on the 6th completely expanded leave from the stem apex.

8. Conclusion

Nutrition studies for blackberry crop are scarce worldwide. However, as presented in this chapter, nitrogen fertilization is one aspect of crop management that has the greatest impact on vegetative growth, fruit yield, and quality. It has been demonstrated in this chapter that even under conditions of moderate N deficiency, there can be significant impacts on blackberry yield. Nitrogen-deficient plants tend to form weak, low-yielding stems. In this way, the application of adequate rates for each type of soil and cultivar is fundamental for satisfactory productivities.

In addition, there are important differences in the demand of N between cultivars, probably due to productive capacity, that is, cultivars that provide higher yields also export larger amounts of N and therefore need larger amount of nitrogen fertilizers.

In this way, a nitrogen fertilization program was proposed, based on SOM (indicator of N availability), plant age, and production expectation.

Acknowledgements

The authors are grateful to the Brazilian Agricultural Research Corporation (EMBRAPA) and the Soil and Water Management and Conservation Post-Graduate Program at the Federal University of Pelotas (UFPel) for the financial and infrastructure support.

Author details

Ivan dos Santos Pereira[1*], Adilson Luis Bamberg[2], Carlos Augusto Posser Silveira[2], Luis Eduardo Corrêa Antunes[2] and Rogério Oliveira de Sousa[1]

1 Federal University of Pelotas (UFPel), Pelotas, Brazil

2 Brazilian Agricultural Research Corporation (EMBRAPA), Pelotas, Brazil

*Address all correspondence to: ivanspereira@gmail.com

References

[1] Taiz L, Zeiger E. Fisiologia Vegetal. 5th ed. Artmed: Porto Alegre; 2013. 820 p

[2] Chapin FS. The mineral nutrition of wild plants. Annual Review of Ecology and Systematics. 1980;**11**:233-260

[3] Grandall PC. Bramble Production: The Management and Marketing of Raspberries and Blackberries. 1st ed. Binghamton, N.Y.: CRC Press; 1995. 172 p

[4] Pereira IS, Picolotto L, Messias RS, Potes ML, Antunes LEC. Adubação nitrogenada e características agronômicas em amoreira-preta. Pesquisa Agropecuária Brasileira. 2013;**48**:373-380. DOI: 10.1590/S0100-204X2013000400004

[5] Pereira IS, Silveira CAP, Picolotto L, Schneider FC, Gonçalves MA, Vignolo GK, et al. Constituição química e exportação de nutrientes da amoreira-preta. Revista Congrega URCAMP. 2013;**9**:1-10

[6] Castaño CA, Morales CS, Obando FH. Evaluación de las deficiencias nutricionales em el cultivo de la mora (*Rubus glaucus*) en condiciones controladas para bosque montano bajo. Agronomía. 2008;**16**:75-88

[7] CQFS. Comissão de Química e Fertilidade do Solo—RS/SC. Manual de calagem e adubação para os estados do Rio Grande Sul e Santa Catarina. 11ª ed. SBCS-NRS: Porto Alegre; 2016. 376 p

[8] Hart J, Strik B, Rempel H. Caneberries. In: Nutrient Management. Oregon State University; January 2006. 8 p

[9] Pereira IS, Nava G, Picolotto L, Silveira CAP, Vignolo GK, Gonçalves MA, et al. Exigência nutricional e adubação da amoreirapreta: Nutritional requirements and blackberry fertilization. Revista de Ciências Agrárias. 2015;**58**:96-104. DOI: 10.4322/rca.1755

[10] Lawson HM, Waister PD. The response to nitrogen of a raspberry plantation under contrasting systems of management for a weed and sucker control. Horticulture Research. 1972;**12**:43-55

[11] Bushway L, Pritts M, Handley D, editors. Raspberry & Blackberry Production Guide for the Northeast, Midwest, and Eastern Canada. Plant and Life Sciences Publishing Coop.; 2008. 157 p. Ext. NRAES-35

[12] Gutser RT, Ebertseder A, Weber M, Schraml, Schmidhalter U. Short-term and residual availability of nitrogen after long-term application of organic fertilizers on arable land. Journal of Plant Nutrition and Soil Science. 2005;**168**:439-445

[13] Kowalenko CG, Keng JCW, Freeman JA. Comparison of nitrogen application via a trickle irrigation system with surface banding of granular fertilizer on red raspberry. Canadian Journal of Plant Science. 2000;**80**:363-371. DOI: 10.4141/P99-094

[14] Freire CJS. Nutrição e adubação. In: Cultivo da amoreira-preta. In: Antunes LEC, Raseira MCB, editors. Sistemas de Produção. Pelotas; 2007. pp. 45-54

[15] Pereira IS. Adubação de pré-plantio no crescimento, produção e qualidade da amoreira-preta (*Rubus* sp.) [Dissertação]. Pelotas: Universidade Federal de Pelotas; 2008

[16] Pritts M, Handley D. Bramble Production. Ithaca, NY: Northeast Regional Agricultural Engineering Service Pub. 35; 1989

[17] Strik BC. A review of nitrogen nutrition of Rubus. Acta Horticulturae. 2008;(777):403-410

[18] Dickerson GW, editor. Blackberry Production in New Mexico. Cooperative Extension Service College of Agriculture and Home Economics, New Mexico State University (NMSU) and the U.S. Department of Agriculture Cooperating, Guide H-325; 2000. 8 p

[19] Pereira IS, Antunes LEC, Silveira CAP, Messias RS, Gardin JPP, Schneider FC, et al. Caracterização agronômica da amoreira-preta cultivada no sul do estado do Paraná. Pelotas: Embrapa Clima Temperado; 2009. 19 p

[20] Eyduran SP, Eyduran E, Agaoglu YS. Estimation of fruit weight by cane traits for eight American blackberries (*Rubus fruticosus* L.) cultivars. African Journal of Biotechnology. 2008;7:3031-3038

[21] Antunes LEC, Pereira IS, Picolotto L, Vignolo GK, Gonçalves MA. Produção de amoreira-preta no Brasil. Revista Brasileira de Fruticultura. 2014;**36**:100-111. DOI: 10.1590/0100-2945-450/13

[22] Pereira IS, Antunes LEC, Messias RS, Silveira CAP, Vignolo GK. Avaliações da subtração dos elementos N, P e K sobre a produção e qualidade de frutos de amoreira-preta. In: Comunicado Técnico 275. Pelotas: Embrapa Clima Temperado; 2012. 6 p

[23] Nelson E, Martin LW. The relationship of soil-applied N and K to yield and quality of 'Thornless Evergreen' blackberry. HortScience. 1986;**21**:1153-1154

[24] Alleyne V, Clark JR. Fruit composition of 'Arapaho' blackberry following nitrogen fertilization. HortScience. 1997;**32**:282-283

[25] Spiers JM, Braswell JH. Influence of N, P, K, Ca, and Mg rates on leaf macronutrient concentration of 'Navaho' blackberry. Acta Horticulturae. 2002;(585):659-663

6

Phylogenomic Review of Root Nitrogen-Fixing Symbiont Population Nodulating Northwestern African Wild Legumes

Mokhtar Rejili, Mohamed Ali BenAbderrahim and Mohamed Mars

Abstract

The present review discusses the phylogenomic diversity of root nitrogen-fixing bacteria associated to wild legumes under North African soils. The genus *Ensifer* is a dominant rhizobium lineage nodulating the majority of the wild legumes, followed by the genus *Rhizobium* and *Mesorhizobium.* In addition, to the known rhizobial genera, two new *Microvirga* and *Phyllobacterium* genera were described as real nodulating and nitrogen-fixing microsymbiotes from *Lupinus* spp. The promising rhizobia related to nitrogen fixation efficiency in association with some legumes are shared. Phylogenetic studies are contributing greatly to our knowledge of relationships on both sides of the plant-bacteria nodulation symbiosis. Multiple origins of nodulation (perhaps even within the legume family) appear likely. However, all nodulating flowering plants are more closely related than previously suspected, suggesting that the predisposition to nodulate might have arisen only once. The origins of nodulation, and the extent to which developmental programs are conserved in nodules, remain unclear, but an improved understanding of the relationships between nodulin genes is providing some clues.

Keywords: rhizobia, North Africa, symbiosis, legumes, phylogenomic

1. Introduction

Africa has a vast array of indigenous legumes, ranging from large rain forest trees to small annual herbs [1]. However, in recent years, there has been a tendency in agriculture and forestry to use exotic species for crops and wood. As has been pointed out several times over nearly 30 years, most recently [2], by the US National Academy of Sciences, this ignores the potential of the native species, which are arguably better adapted to their environment. For this review, the nodulated indigenous legume genera in Northwestern Africa with known uses have been selected to illustrate the problems and potential for their better exploitation.

The wild legume flora in Northwestern Africa is rich, with great specific and infraspecific diversity [3]. The overgrazing and expansion of agriculture has gradually led to the regression and extinction of many pastoral and forage species. In addition, desertification causes disturbance of plant-microbe symbioses, which are a critical ecological factor in helping further plant growth in degraded ecosystems [4]. In this context, the establishment of indigenous pastoral legume species associated with their appropriate symbiotic bacterial partners may be of increased value for success in soil fertility restoration. Biological N2 fixation (BNF) is the major way for N input into desert ecosystems. Rhizobium-legume symbioses represent the major mechanism of BNF in arid lands, compared with the N_2-fixing heterotrophs and associative bacteria [5, 6] and actinorhizal plants [7, 8]. Deficiency in mineral N often limits plant growth, and so symbiotic relationships have evolved between plants and a variety of N2-fixing organisms [9]. The symbiotically fixed N_2 by the association between rhizobium species and the legumes represents a renewable source of N for agriculture. Values estimated for various legume crops and pasture species are often impressive [10]. In addition to crop legumes, the nodulated wild (herb and tree) legumes have potential for nitrogen fixation and reforestation and to control soil erosion [11]. It has been reported that a novel, suitable wild legume-rhizobia associations are useful in providing a vegetational cover in degraded lands [12].

Considering the major ecological importance of many wild legumes such as *Retama* sp., *Acacia* sp., *Lotus* sp., *Lupinus* sp., *Medicago* sp., etc. in Northwest Africa by their important role in soil fertility maintenance, coverage, and dune stabilization, the present chapter proposes to review the phylogenomic diversity of root nitrogen-fixing symbiont population nodulating Northwestern African wild legumes listed in the bibliography, some of which are common and play important ecological and pastoral roles, but others are rare and endangered. As well as the host legumes, the nodule endosymbionts also vary widely in Africa and include newly described members of both α and β branches of the Proteobacteria, now often referred to as α- or β-rhizobia, even though they do not have "rhizobium" as part of their generic names [13].

Therefore, understanding the nature of indigenous populations of rhizobia-nodulating wild legumes is of considerable agricultural significance. It is also of interest to identify a wider variety of bacterial strains in a bid to define new strains for the production of inoculants for smallholder farms.

2. Genetics and functional genomics of legume nodulation

The interaction between rhizobia and legumes in root nodules is an essential element in sustainable agriculture, as this symbiotic association is able to enhance biological fixation of atmospheric nitrogen (N_2) and is also a paradigm in plant-microbe signaling [14–16]. The knowledge of the whole genome would allow the specific features of each rhizobium to be identified. The prominent feature of this group of bacteria is their molecular dialog with plant hosts, an interaction that is enabled by the presence of a series of symbiotic genes encoding for the synthesis and export of signals triggering organogenetic and physiological responses in the plant [17, 18]. In recent years, significant progress has been made in resolving the complex exchange of signals responsible for nodulation through genome assembly, mutational and expression analysis, and proteome characterization of legumes [14, 19, 20] and rhizobia [15, 21–23].

3. Phylogenomic of wild legume root nitrogen-fixing symbionts

The known diversity of rhizobia increases annually and is the subject of several reviews, the most recent and comprehensive being that of [24]. It is not our intention to revisit this subject nor the genetic basis of nodulation [25, 26], the horizontal

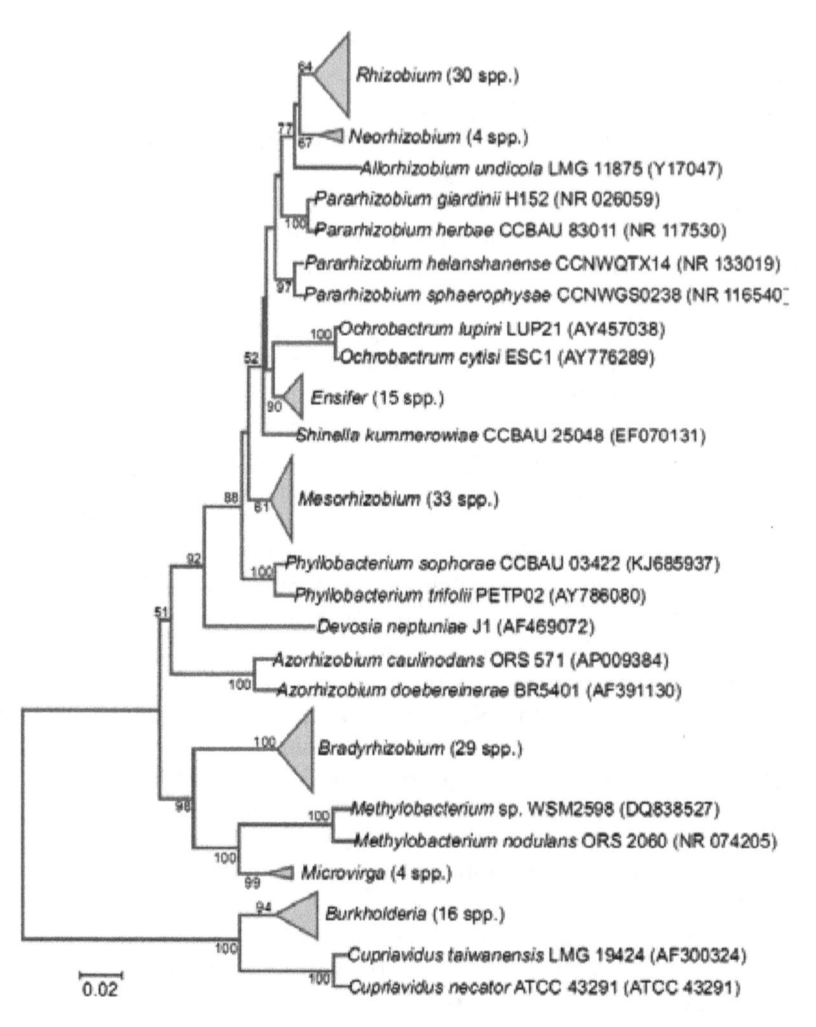

Figure 1.
Phylogenetic tree showing the relationships of currently described genera and species of alpha- and beta-rhizobia, based on aligned sequences of the 16S rRNA gene (1341-bp internal region) (adapted from [28]).

transfer of symbiosis-related genes [23], or the symbiovar concept [27] but instead to attempt to link, where possible, rhizobial genotypes with their geographical locations and/or legume tribes/genera. At the time of writing, rhizobia consist of a diverse range of genera in the Alphaproteobacterial and Betaproteobacterial classes and are termed "alpha-rhizobia" and "beta-rhizobia," respectively (**Figure 1**).

3.1 The genus *Bradyrhizobium* (*Bradyrhizobiaceae*)

The *Bradyrhizobium* genus was described by Jordan in 1982 [29]. It currently consists of nine rhizobia species.

For the *Loteae* tribe, previous studies found that *Lotus palustris* and *L. purpureus* species from Algeria were nodulated by *Bradyrhizobium lupini*, and *L. pedunculatus* by *B. japonicum* [30]. However, *L. creticus* ssp. *maritimus* is nodulated by both [30]. At Tunisia, *L. roudairei* microsymbiont is closely related to *B. japonicum* [31]. For the *Acacieae* tribe, two studies reported that rhizobial strains associ-ated to the *Acacia saligna*, an Australian introduced species, belonged to the genus *Bradyrhizobium* genus under Algerian and Moroccan soils [32–34]. For the *Genisteae* tribe, it has been noticed that *Bradyrhizobium* is the dominant genus of symbiotic nitrogen-fixing bacteria associated with *Retama* species in North Africa: *Retama monosperma*, *R. raetam*, and *R. sphaerocarpa* (Algeria: [35, 36]; Morocco: [37]). Recently, the novel *B. retamae* species, in which groups with *B. elkanii* and *B. pachyrhizi* and related *B. lablabi* and *B. jicamae* type strains are included in *Bradyrhizobium* group II [38], has been isolated from *R. sphaerocarpa* and *R. monosperma* in Morocco [37]. For the genus *Cytisus*, two studies reported that *Cytisus villosus* is nodulated by *B. cytisi* sp. nov. and *B. rifense* sp. nov. in Morocco [39, 40] and by genetically diverse *Bradyrhizobium* strains in Algeria belonging to *B. japonicum* and *B. canariense* and to new lineage within the *Bradyrhizobium* genus [41]. Fiftytwo strains isolated from root nodules of the Moroccan shrubby legume *Cytisus triflorus* were genetically characterized, and results showed that it is nodulated by *Bradyrhizobium* strains, with 99% homology with *Bradyrhizobium* genosp. AD [42]. For the genus *Lupinus*, some endosymbiotic bacteria of *L. luteus and L. micranthus* from Tunisia and Algeria belonged to *B. lupini*, *B. canariense*, *B. valentinum*, *B. cytisi/B. rifense*, *B. japonicum*, *B. elkanii*, and *B. retamae* [43–45].

3.2 The genus *Mesorhizobium* (*Phyllobactericeae*)

The genus *Mesorhizobium* was described by Jarvis et al. [46]. Several *Rhizobium* species were transferred to this genus. It currently consists of 21 rhizobia species.

For subtribe *Astragalinae* (Coluteinae Clade), Guerrouj et al. [37] reported that rhizobial symbiont of *Astragalus gombiformis* in Eastern Morocco is closely related to *M. camelthorni*. A polyphasic approach analysis indicated that bacterial strains isolated from the pasture legume *Biserrula pelecinus* growing in Morocco belong to the genus *Mesorhizobium*. At Tunisia, Mahdhi et al. [47] showed that five strains isolated from *Astragalus corrugatus* were phylogenetically related to *M. temperatum* and to *Mesorhizobium* sp. From the tribe *Galegeae* (subtribe *Coluteinae*), Ourarhi et al. [48] reported that *Colutea arborescens* is nodulated by diverse rhizobia in Eastern Morocco, among them, the genus *Mesorhizobium*. For the *Loteae* tribe, *M. alhagi* as well as *M. temperatum* were isolated, at Tunisia, from *Lotus creticus* [49–51]. Zakhia et al. [31] reported that *Lotus argenteus* microsymbiotes are closely related to *M. mediterraneum* in the infra-arid zone of Tunisia. Roba et al. [52] reported that *M. delmotii* and *M. prunaredense* are two new rhizobial species nodulating *Anthyllis vulneraria* growing on Tunisian soils. From the *Acacieae* tribe, Boukhatem et al. [33] reported that rhizobial strains associated to the *Acacia saligna*, an Australian introduced species, to *A. ehrenbergiana* and *F. albida* belonged to *M. mediterraneum* under Algerian soils. From the *Genisteae* tribe, the genetic diversity of *Genista saharae* microsymbionts in the Algerian Sahara reported that they belonged to *M. camelthorni* [53]. For the *Mimoseae* tribe, root-nodulating bacteria associated to *Prosopis farcta* growing in the arid regions of Tunisia were assigned to the genus *Mesorhizobium* [54]. From the *Hedysareae* tribe, Zakhia et al. [31] reported that one strain isolated from *Ebenus pinnata* root nodules is closely related to *M. ciceri* in the infra-arid zone of Tunisia.

3.3 The genus *Rhizobium* (*Rhizobiaceae*)

The genus *Rhizobium* was the first named (from Latin meaning "root living"), and for many years this was a "catch all" genus for all rhizobia. Some species were later moved in to new genera based on phylogenetic analyses [55]. It currently consists of 49 rhizobial species.

For *Galegae* tribe, Zakhia et al. [31] reported that rhizobial symbionts of *Astragalus gombiformis*, *A. armatus*, and *A. cruciatus* are closely related to *Rhizobium mongolense*, *R. leguminosarum*, and *R. galegae*, in the infra-arid zone of Tunisia. From

Genisteae tribe, it was shown that strains from Tunisia nodulating *Argyrolobium uniflorum* are closely affiliated to *R. giardinii*, *Calicotome villosa* to *R. mongolense*, and *Genista microcephala* to *R. mongolense* and *R. leguminosarum* [31]. Mahdi et al. [56–58] reported that strains nodulating *Genista saharae* and *Retama retam* are members of the genus *Rhizobium*. Nonetheless, there are reports indicating that members of the genus *Rhizobium* nodulate *Adenocarpus decorticans* and *Cytisus arboreus* at Morocco [59]. For the *Loteae* tribe, *R. leguminosarum* and *R. mongolense* were isolated, at Tunisia, from *Anthyllis henoniana*, *R. leguminosarum* from *Coronilla scorpioides*, and *R. mongolense* from *Lotus creticus* [31]. Rejili et al. [51] reported that *Lotus creticus* microsymbiotes are closely related to *R. huautlense* in the arid areas of Tunisia. Bacterial strains isolated from root nodules of *Scorpiurus muricatus* sampled from different regions of western Algeria are affiliated to *R. vignae*, *R. radiobacter*, and *R. leguminosarum* [60]. For the *Trifolieae* tribe, *R. galegae* species was isolated, in Tunisia and Algeria, from *Medicago marima* and *M. truncatula* [31]. In Algeria, Merabet et al. [61] reported that *Medicago ciliaris* and *M. polymorpha* are nodulated by *Rhizobium* sp. Similarly, genetic diversity of rhizobia from annual *Medicago orbicularis* showed that they are affiliated to *Rhizobium tropici* [62]. For the *Vicieae* tribe, *R. leguminosarum* species was isolated from *Lathyrus numidicus* [31]. Mahdhi et al. [63] reported that *Vicia sativa* isolates from Tunisia had 16S rDNA type identical to that of the reference *R. leguminosarum*. From *Acacieae*, Boukhatem et al. [33] reported that bacteria-nodulating *Acacia saligna* and *A. seyal* under Algerian soils are affiliated to the *R. tropici* clade and *R. sullae* clade, respectively. On the other hand, the same study mentioned that five bacterial isolates, all from *A. saligna*, formed a separate clade in the vicinity of the *R. galegae-R. huautlense-R. loessense* branch [33]. The same authors showed that the *R. leguminosarum* reference strain was represented by five *A. karroo* isolates and five *A. seyal* isolates [33]. At Tunisia, the genetic diversity of root nodule bacteria associated to *Hedysarum coronarium* (sulla), from *Hedysareae* tribe, showed that they are closely related to *R. sullae* [64]. Similarly, Ezzakkioui et al. [65] indicated that the strains from the Moroccan *Hedysarum flexuosum* legume had 99.75–100% identity with *R. sullae*.

3.4 The genus *Ensifer (Sinorhizobium) (Rhizobiaceae)*

The genera *Sinorhizobium* and *Ensifer* were recently recognized as forming a single phylogenetic clade [66, 67] and are now united, and all species of the genus *Sinorhizobium* have been transferred to the genus *Ensifer*, in line with rule 38 of the Bacteriological Code [68, 69]. The genus currently consists of 17 species.

Bacteria belonging to *Ensifer* genus are widely distributed in arid regions of Tunisia. From the *Loteae* tribe, *E. meliloti* and *E. numidicus* were isolated, at Tunisia, from *Lotus creticus* [49–51, 69] and *Rhizobium* sp. from *Hippocrepis areolata* [47]. From the *Acacieae* tribe, genetic characterization of rhizobial bacteria-nodulating *Acacia tortilis* subsp. *raddiana*, *A. gummifera*, *A. cyanophylla*, *A. karroo*, *A. ehrenbergiana,* and *A. horrida* in Tunisia, Algeria, and Morocco reported that they belonged to the species *E. meliloti*, *E. garamanticus*, and *E. numidicus* and *Ensifer* sp. [31, 33, 70–72, 73]. At Algeria, isolates from four different host species, namely, *A. karroo*, *A. ehrenbergiana*, *A. saligna*, and *A. tortilis*, were closely related to *E. fredii*, *E. terangae*, and *E. kostiense* reference strains [33]. For the *Mimoseae* tribe, 40 isolates associated to *Prosopis farcta* growing in the arid regions of Tunisia belonged to *E. meliloti*, *E. xinjiangense/E. fredii*, and *E. numidicus* species [54]. For the *Trifolieae* tribe, strains nodulating different *Medicago* species in Tunisia, Algeria, and Morocco such as *M. sativa*, *M. arborea*, *M. truncatula*, *M. ciliaris*, *M. laciniata*, *M. polymorpha*, *Medicago arabica*, *M. marima*, *Medicago littoralis*, and *M. scutella* are associated to *E. meliloti*, *E. medicae* , or *E. garamanticus* [31, 61, 62, 69, 74–79]. Similarly, *Ononis natrix*

Subfamily tribe	Genus	Species	Symbiont	Geographic origin
Mimosoideae				
Acacieae	*Acacia*	*A. cyanophylla*	*E. meliloti*, *E. fredii*, *Ensifer* sp.	Tunisia, Morocco
		A. gummifera	*E. meliloti*, *E. garamanticus*, *E. numidicus*, *Ensifer* sp.	Tunisia, Morocco
		A. horrida	*E. meliloti*, *E. garamanticus*, *E. numidicus*, *Ensifer* sp.	Tunisia, Morocco
		A. tortilis raddiana	*E. meliloti*, *E. garamanticus*, *E. numidicus*, *Ensifer* sp.	Tunisia, Morocco
		A. saligna	*Bradyrhizobium* sp., *Mesorhizobium* sp., *Rhizobium* sp., *Ensifer* sp.	Algeria, Morocco
		A. ehrenbergiana	*Mesorhizobium* sp., *Ensifer* sp.	Algeria
		A. karroo	*Rhizobium* sp., *Ensifer* sp.	Algeria
		A. nilotica	*Rhizobium* sp.	Algeria
		A. seyal	*Rhizobium* sp.	Algeria
		F. albida	*Mesorhizobium* sp.	Algeria
Mimosae	*Prosopis*	*P. farcta*	*Mesorhizobium* sp., *E. meliloti*, *E. xinjiangense*, *E. fredii*, *E. numidicus*	Tunisia
Papilionoideae				
Galegae	*Astragalus*	*A. armatus*	*R. mongolense*, *R. leguminosarum*, *R. galegae*	Tunisia
		A. cruciatus	*R. mongolense*, *R. leguminosarum*, *R. galegae*	Tunisia
		A. corrugatus	*M. temperatum*, *Mesorhizobium*	Tunisia
		A. gombiformis	*M. camelthorni*, *R. mongolense*, *R. leguminosarum*, *R. galegae*	Morocco, Tunisia
	Biserrula	*B. pelecinus*	*Mesorhizobium*	Morocco
	Colutea	*C. arborescens*	*Mesorhizobium*	Morocco
Genisteae	*Argyrolobium*	*A. uniflorum*	*R. giardinii*	Tunisia
	Adenocarpus	*A. decorticans*	*Rhizobium*	Morocco
	Calicotome	*C. villosa*	*R. mongolense*	Morocco
	Cytisus	*C. arboreus*	*Bradyrhizobium* sp.	Morocco
		C. triflorus	*Bradyrhizobium*	Morocco
		C. villosus	*B. cytisi*, *B. rifense*, *B. japonicum*, *B. canariense*	Algeria, Morocco
	Lupinus	*L. luteus*	*B. lupini*, *B. canariense*, *B. valentinum*, *B. cytisi*, *B. rifense*, *B. japonicum*, *B. elkanii*, *B. retamae*, *Microvirga*	Algeria, Tunisia

Subfamily tribe	Genus	Species	Symbiont	Geographic origin
		L. micranthus	*B. lupini*, *B. canariense*, *B. valentinum*, *B. cytisi*, *B. rifense*, *B. japonicum*, *B. elkanii*, *B. retamae*, *Microvirga*, *Phyllobacterium*	Algeria, Tunisia
	Genista	*G. microcephala*	*R. mongolense*, *R. leguminosarum*, *Rhizobium*	Tunisia
		G. saharae	*M. camelthorni*	Algeria
	Retama	*R. monosperma*	*B. retamae*	Algeria, Morocco
		R. raetam	*B. retamae*, *Rhizobium*	Algeria, Tunisia, Morocco
		R. sphaerocarpa	*B. retamae*	Algeria, Morocco
Hedysareae	*Hedysarum*	*H. carnosum*	*E. meliloti*	Tunisia
		H. flexuosum	*R. sullae*	Morocco
		H. coronarium	*R. sullae*	Tunisia
		H. spinosissimum	*E. meliloti*	Tunisia
	Ebenus	*E. pinnata*	*M. ciceri*	Tunisia
Loteae	*Anthyllis*	*A. henoniana*	*R. leguminosarum*, *R. mongolense*	Tunisia
		A. vulneraria	*M. delmotii*, *M. prunaredense*	Tunisia
	Coronilla	*C. scorpioides*	*R. leguminosarum*	Tunisia
	Hippocrepis	*H. areolata*	*Rhizobium*	Tunisia
		H. bicontorta	*E. meliloti*	Tunisia
	Lotus	*L. argenteus*	*M. mediterraneum*	Tunisia
		L. creticus	*B. lupini*, *B. japonicum*, *M. alhagi*, *M. temperatum*, *R. mongolense*, *R. huautlense*, *E. meliloti*, *E. numidicus*	Algeria, Tunisia
		L. palustris	*B. lupini*	Algeria
		L. pedunculatus	*B. japonicum*	Algeria
		L. purpureus	*B. lupini*	Algeria
		L. pusillus	*M. alhagi*, *M. temperatum*, *E. meliloti*	Tunisia
		L. roudairei	*B. japonicum*	Tunisia
	Scorpiurus	*S. muricatus*	*R. vignae*, *R. radiobacter*, *R. leguminosarum*	Algeria
Trifolieae	*Medicago*	*M. arabica*	*E. meliloti*, *E. medicae*, and *E. garamanticus*	Morocco
		M. arborea	*E. meliloti*, *E. medicae*, and *E. garamanticus*	Morocco

Subfamily tribe	Genus	Species	Symbiont	Geographic origin
		M. ciliaris	*Rhizobium*, *E. meliloti*, *E. medicae*, and *E. garamanticus*	Algeria
		M. marima	*R. galegae*	Algeria, Tunisia
		M. laciniata	*E. meliloti*, *E. medicae*, and *E. garamanticus*	Tunisia
		M. littoralis	*E. meliloti*, *E. medicae*, and *E. garamanticus*	Tunisia
		M. orbicularis	*R. tropici*	Tunisia, Algeria, Morocco
		M. polymorpha	*Rhizobium*, *E. meliloti*	Tunisia, Algeria, Morocco
		M. sativa	*E. meliloti*, *E. medicae*	Tunisia, Algeria, Morocco
		M. scutella	*E. meliloti*	Algeria, Tunisia
		M. truncatula	*R. galegae*, *E. meliloti*, *E. medicae*	Tunisia, Algeria, Morocco
	Melilotus	*M. indicus*	*E. meliloti*	Algeria
	Ononis	*O. natrix* ssp. *filifolia*	*E. meliloti*	Tunisia
	Trigonella	*T. maritima*	*E. meliloti*	Tunisia
Vicieae	*Lathyrus*	*L. numidicus*	*R. leguminosarum*	Tunisia
	Vicia	*V. sativa*	*R. leguminosarum*	Tunisia

Table 1.
Recapitulative results of root nodule symbionts from Northwestern African wild legumes.

and *Trigonella maritima* are nodulated by *E. meliloti* [31, 63]. Nodule rhizobia of *Melilotus indicus* growing in the Algerian Sahara are affiliated to *E. meliloti* [78]. *E. meliloti* and *E. numidicus* strains were isolated from the Genisteae tribe such as *Argyrolobium uniflorum*, *Retama raetam*, and *Genista saharae* [56–58, 69]. For *Galegae* tribe, Mahdhi et al. [47] reported that rhizobial symbionts of *Astragalus corrugatus* are closely related to *E. meliloti* under Tunisian soils. From the *Hedysareae* tribe, Mahdhi et al. [35] reported that strains isolated from *Hedysarum spinosissimum* root nodules are closely related to *E. meliloti* in the infra-arid zone of Tunisia.

3.5 The genus *Neorhizobium* (*Rhizobiaceae*)

The genus *Neorhizobium* was proposed by Mousavi et al. [80] as an alternative to solve the issue of grouping the members of this genus with *Agrobacterium* and *Rhizobium* genera. The genetic diversity of the Algerian legume *Genista saharae* isolates was assessed, and results reported that they are affiliated to *Neorhizobium alkalisoli*, *N. galegae*, and *N. huautlense* [53]. Several studies reported that *N. galegae* is isolated from different legumes in Tunisia such as *Astragalus sp.* [31, 54], *Argyrolobium uniflorum* [31], *Anthyllis henoniana* [31], *Lotus creticus* [31, 50], *Medicago marima*, and *M. truncatula* [31]. Rejili et al. [51] reported that *Lotus creticus* is also nodulated by *N. huautlense* in the arid areas of Tunisia. *For Galegae tribe*, Mahdhi et al. [47] reported that rhizobial symbionts of *Astragalus corrugatus* are closely related to *N. galegae* under Tunisian soils.

3.6 The genus *Phyllobacterium* (*Phyllobactericeae*)

The *Phyllobacterium* genus comprises of bacteria that are well-known for their epiphytic and endophytic associations with plants [81]. Nonetheless, root-nodulating and nitrogen-fixing *Phyllobacterium* was described in Tunisia, in the nodules of genistoid legume *Lupinus micranthus* [44, 45]. Prior to this finding, endophytic *Phyllobacterium* strains were identified on the nodules of the Tunisian legumes *Genista saharae*, *Lotus creticus*, and *L. pusillus* [51, 57], but they are lacking the ability to form nodules.

3.7 The genus *Microvirga* (*Methylobacteriaceae*)

The genus *Microvirga* which comprises soil and water saprophytes was included in the alphaproteobacterial lineage of root-nodule bacteria only in 2012, although the first symbiotic strains were detected in nodules of *Lupinus texensis* [82–84]. Recently, *Microvirga* strains were only isolated from *L. micranthus* and *L. luteus* in Tunisia, belonging to the *Genisteae* tribe [44, 45].

Table 1 shows the root nodule symbionts from Northwestern African wild legumes.

4. Promising nitrogen-fixing rhizobia

The root nodule symbiosis established between legumes and rhizobia is an exquisite biological interaction responsible for fixing a significant amount of nitrogen in terrestrial ecosystems. The success of this interaction depends on the recognition of the right partner by the plant within the richest microbial ecosystems on Earth, the soil. Recent metagenomic studies of the soil biome have revealed its complexity, which includes microorganisms that affect plant fitness and growth in a beneficial, harmful, or neutral manner. In this complex scenario, understanding the molecular mechanisms by which legumes recognize and discriminate rhizobia from pathogens, but also between distinct rhizobia species and strains that differ in their symbiotic performance, is a considerable challenge.

By symbiotic efficiency and properties, strains isolated from wild legumes varied in their symbiosis effectiveness with their host plant of origin. A great diversity among and within isolates was reported by many authors. This symbiotic diversity within and between isolates growing in diverse geographical areas was also defined by Tinick and Hadobas [85] for other legume plants. All strains were capable of nodulation. Mahdhi et al. [55, 86] reported that two *Retama raetam* isolates RB3 and RM4 (*Rhizobium*) gave the highest nodule numbers per plant, 26 (±2.053) and 27 (±0.997), respectively. The effective strain LAC765 (*Ensifer*) was isolated from *Lotus creticus* with a 91.46 (±0.01%) dry biomass of the T_N control [50]. The dry matter of the aerial part is considered a criterion for assessing the efficiency of a given strain; a highly significant correlation between these two parameters has been reported. Results related to symbiotic efficiency showed that among 45 tested isolates, 20 isolates are highly efficient (relative effectiveness ≥70%), 20 isolates are partially effective (60% ≤ relative effectiveness <70%), and 5 isolates are inefficient (relative effectiveness <60%). The strain GN29 isolated from *Genista saharae*, affiliated to *Rhizobium* genus, is considered inefficient (relative effectiveness = 32.29%). Among the 20 isolates considered highly efficient, 5 isolates were isolated from *Retama retam*, five from *Lotus* sp., 4 from *Genista saharae*, 3 from *Vicia sativa*, 2 from *Argyrolobium uniflorum* and 2 from *Trigonella maritima*. From the 20 highly efficient isolates, 13 isolates belong taxonomically to *Ensifer* sp., 6 to *Rhizobium* sp., and one to *Mesorhizobium* sp.

5. Conclusion

The Mediterranean basin is a hotspot place of legume diversity and the center of diversification of many of them. Our review contributes to enlarge our knowledge on the LNB-legume symbioses. We evidenced the biodiversity among bacteria-nodulating wild legumes in Northwestern Africa and unknown associations were found. Several groups may represent new genospecies to be further characterized to assess their taxonomical status. This work thus opens further interesting perspectives and makes new models available for evolutionary studies and for understanding mechanisms involved in nitrogen-fixing symbiosis.

Author details

Mokhtar Rejili[1*], Mohamed Ali BenAbderrahim[2] and Mohamed Mars[1]

1 Laboratory of Biodiversity and Valorization of Arid Areas Bioresources (BVBAA), Faculty of Sciences of Gabes, Erriadh, Zrig, Tunisia

2 Arid and Oases Cropping Laboratory, Arid Area Institute, Gabes, Tunisia

*Address all correspondence to: rejili_mokhtar@yahoo.fr

References

[1] Lock JM. Legumes of Africa: A Check List. Kew, England, Royal Botanic Gardens; 1989

[2] Anon. Lost Crops of Africa. Washington: National Academy Press; 2006

[3] Le Houerou HN. La végétation de la Tunisie steppique (1) (Structure, écologie, sociologie, répartition, évolution, utilisation, biomasse, productivité) (avec référence aux végétations analogues d'Algérie, de Libye et du Maroc). Annales de l'Institut National de la Recherche Agronomique de la Tunisie. 1969;**42**:622

[4] Requena N, Pérez-Solis E, Azcón-Aguilar C, Jeffries P, Barea JM. Management of indigenous plant-microbe symbioses aids restoration of desertified ecosystems. Applied and Environmental Microbiology. 2001;**67**:495-498

[5] Abdel-Ghaffar AS. Aspects of microbial activities and nitrogen fixation in Egyptian desert soils. Arid Soil Research and Rehabilitation. 1989;**3**:281-294

[6] Wullstein LH. Evaluation and significance of associative dinitrogen fixation for arid soil rehabilitation. Arid Soil Research and Rehabilitation. 1989;**3**:259-265

[7] Caucas V, Abril A. *Frankia* sp. infects *Atriplex cordobensis*-cross-inoculation assay and symbiotic efficiency. Phyton. 1996;**59**:103-110

[8] Sayed WF, Wheeler CT, Zahran HH, Shoreit AAM. Effect of temperature and moisture on the survival and symbiotic effectiveness of *Frankia* spp. Biology and Fertility of Soils. 1997;**25**:349-353

[9] Freiberg C, Fellay R, Bairoch A, Broughton WJ, Rosenthal A, Perret X. Molecular basis of symbiosis between rhizobium and legumes. Nature. 1997;**387**:394-401

[10] Peoples MB, Ladha JK, Herridge DF. Enhancing legume N2 fixation through plant and soil management. Plant and Soil. 1995;**174**:83-101

[11] Ahmad MH, Rafique MU, McLaughlin W. Characterization of indigenous rhizobia from wild legumes. FEMS Microbiology Letters. 1984;**24**:197-203

[12] Jha PK, Nair S, Gopinathan MC, Babu CR. Suitability of rhizobia-inoculated wild legumes *Argyrolobium flaccidum*, *Astragalus graveolens*, *Indigofera gangetica* and *Lespedeza stenocarpa* in providing a vegetational cover in an unreclaimed lime stone quarry. Plant and Soil. 1995;**177**:139-149

[13] Sprent JI. Legume Nodulation: A Global Perspective. Oxford, UK: Wiley-Blackwell; 2009

[14] Young ND, Debellé F, Oldroyd GE, Geurts R, Cannon SB, Udvardi MK, et al. The *Medicago* genome provides insight into the evolution of rhizobial symbioses. Nature. 2011;**480**:520-524. DOI: 10.1038/nature10625

[15] Giraud E, Moulin L, Vallenet D, Barbe V, Cytryn E, Avarre JC, et al. Legumes symbioses: Absence of nod genes in photosynthetic bradyrhizobia. Science. 2007;**316**:1307-1312. DOI: 10.1126/science.1139548

[16] Wang D, Yang SM, Tang F, Zhu HY. Symbiosis specificity in the legume—Rhizobial mutualism. Cellular Microbiology. 2012;**14**:334-342. DOI: 10.1111/j.1462-5822.2011.01736

[17] Spaink HP, Wijffelman CA, Pees E, Okker RJH, Lugtenberg BJJ. Rhizobium nodulation gene *nod*D as a determinant

of host specificity. Nature. 1987;**328**: 337-340. DOI: 10.1038/328337a0

[18] Long SR. Genes and signals in the rhizobium-legume symbiosis. Plant Physiology. 2001;**125**:69-72. DOI: 10.1104/pp.125.1.69

[19] Sato S, Nakamura Y, Kaneko T, Asamizu E, Kato T, Nakao M, et al. Genome structure of the legume, *Lotus japonicus*. DNA Research. 2008;**15**: 227-239. DOI: 10.1093/dnares/dsn008

[20] Marx H, Minogue CE, Jayaraman D, Richards AL, Kwiecien NW, Sihapirani AF, et al. A proteomic atlas of the legume *Medicago truncatula* and its nitrogen-fixing endosymbiont *Sinorhizobium meliloti*. Nature Biotechnology. 2016;**34**:1198-1205. DOI: 10.1038/nbt.3681

[21] Tolin S, Arrigoni G, Moscatiello R, Masi A, Navazio L, Sablok G, et al. Quantitative analysis of the naringenin-inducible proteome in Rhizobium leguminosarum by isobaric tagging and mass spectrometry. Proteomics. 2013;**13**:1961-1972. DOI: 10.1002/pmic.201200472

[22] Čuklina J, Hahn J, Imakaev M, Omasits U, Förstner KU, Ljubimov N, et al. Genome-wide transcription start site mapping of *Bradyrhizobium japonicum* grown free-living or in symbiosis—A rich resource to identify new transcripts, proteins and to study gene regulation. BMC Genomics. 2016;**17**:302. DOI: 10.1186/s12864-016-2602-9

[23] Remigi P, Zhu J, Young JPW, Masson-Boivin C. Symbiosis within symbiosis: Evolving nitrogen-fixing legume symbionts. Trends in Microbiology. 2016;**24**:63-75. DOI: 10.1016/j.tim.2015.10.007

[24] Peix A, Ramırez-Bahena MH, Velazquez E, Bedmar EJ. Bacterial associations with legumes. Critical Reviews in Plant Sciences. 2015;**34**:17-42

[25] Pueppke SG, Broughton WJ. *Rhizobium* sp. strain NGR234 and R. fredii USDA257 share exceptionally broad, nested host ranges. Molecular Plant-Microbe Interactions. 1999;**12**:293-318

[26] Perret X, Staehelin C, Broughton WJ. Molecular basis of symbiotic promiscuity. Microbiology and Molecular Biology Reviews. 2000;**64**:180-201

[27] Rogel MA, Ormeno-Orrillo E, Martinez-Romero E. Symbiovars in rhizobia reflect bacterial adaptation to legumes. Systematic and Applied Microbiology. 2011;**34**:96-104

[28] Sprent JI, Ardley J, James EK. Tansley review: Biogeography of nodulated legumes and their nitrogen-fixing symbionts. New Phytologist. 2017;**215**:40-56. DOI: 10.1111/nph.14474

[29] Jordan DC. NOTES: Transfer of *Rhizobium japonicum* Buchanan 1980 to *Bradyrhizobium* gen. nov., a genus of slow-growing, root nodule bacteria from leguminous plants. International Journal of Systematic and Evolutionary Microbiology. 1982;**32**:136-139. DOI: 10.1099/00207713-32-1-136

[30] Djouadi S, Amrani S, Bouherama A, Nazhat-Ezzaman N, Aïd F. Nature des rhizobia associés à 15 espèces du genre *Lotus* en Algérie. Botany. 2017;**95**: 879-888. DOI: 10.1139/cjb-2017-0020

[31] Zakhia F, Jeder H, Domergue O, Willems A, Cleyet-Marel JC, Gillis M, et al. Characterization of wild legume nodulating bacteria (LNB) in the infra-arid zone of Tunisia. Systematic and Applied Microbiology. 2004;**27**:380-395. DOI: 10.1078/0723-2020-00273

[32] Amrani S, Nazhat-Ezzaman N, Bhatnagar T. Caractéristiques symbiotiques et génotypiques des Rhizobia associes a *Acacia saligna* (Labill.) Wendl. dans quelques

pépinières en Algérie. Acta Botanica Gallica. 2009;**156**:501-513

[33] Boukhatem ZF, Domergue O, Bekki A, Merabet C, Sekkour S, Bouazza F, et al. Symbiotic characterization and diversity of rhizobia associated with native and introduced acacias in arid and semi-arid regions in Algeria. FEMS Microbiology Ecology. 2012;**80**:534-547

[34] Fikri-Benbrahim K, Chraibi M, Lebrazi S, Moumni M, Ismaili M. Phenotypic and genotypic diversity and symbiotic effectiveness of rhizobia isolated from *Acacia* sp. grown in Morocco. Journal of Agricultural Science and Technology. 2017;**19**:201-216

[35] Boulila F, Depret G, Boulila A, Belhadi D, Benallaoua S, Laguerre G. *Retama* species growing in different ecological-climatic areas of northeastern Algeria have a narrow range of rhizobia that form a novel phylogenetic clade within the *Bradyrhizobium* genus. Systematic and Applied Microbiology. 2009;**32**:245-255. DOI: 10.1016/j.syapm.2009.01.005

[36] Hannane FZ, Kacem M, Kaid-Harche M. Preliminary characterization of slow growing rhizobial strains isolated from *Retama monosperma* (L.) Boiss. root nodules from northwest coast of Algeria. African Journal of Biotechnology. 2016;**15**:854-867

[37] Guerrouj K, Perez-Valera E, Chahboune R, Abdelmoumen H, Bedmar EJ, El Idrissi MM. Identification of the rhizobial symbiont of *Astragalus glombiformis* in eastern Morocco as *Mesorhizobium camelthorni*. Antonie Van Leeuwenhoek. 2013;**104**:187-198

[38] Menna P, Barcellos FG, Hungria M. Phylogeny and taxonomy of a diverse collection of *Bradyrhizobium* strains based on multilocus sequence analysis of the 16S rRNA gene, ITS region and *gln*II, *rec*A, *atp*D and *dna*K genes. International Journal of Systematic and Evolutionary Microbiology. 2009;**59**:2934-2950

[39] Chahboune R, Carro L, Peix A, Barrijal S, Velázquez E, Bedmar EJ. *Bradyrhizobium cytisi* sp. nov., isolated from effective nodules of *Cytisus villosus*. International Journal of Systematic and Evolutionary Microbiology. 2011;**61**:2922-2927

[40] Chahboune R, Carro L, Peix A, Ramirez-Bahena MH, Barrijal S, Velaszquez E, et al. *Bradyrhizobium rifense* sp. nov. isolated from effective nodules of *Cytisus villosus* grown in the Moroccan Rif. Systematic and Applied Microbiology. 2012;**35**:302-305

[41] Ahnia H, Boulila F, Boulila A, Boucheffa K, Durán D, Bourebaba Y, et al. *Cytisus villosus* from northeastern Algeria is nodulated by genetically diverse *Bradyrhizobium* strains. Antonie Van Leeuwenhoek. 2014;**105**:1121-1129. DOI: 10.1007/s10482-014-0173-9

[42] Chahboune R, El Akhal MR, Arakrak A, Bakkal M, Laglaoui A, Pueyo JJ, et al. Characterization of bradyrhizobia isolated from root nodules of *Cytisus triflorus* in the rif occidental of morocco. In: Proceedings of the 15th International Nitrogen Fixation Congress and the 12th International Conference of the African Association for Biological Nitrogen Fixation. 2008. p. 155

[43] Bourebaba Y, Durán D, Boulila F, Ahnia H, Boulila A, Temprano F, et al. Diversity of *Bradyrhizobium* strains nodulating *Lupinus micranthus* on both sides of the Western Mediterranean: Algeria and Spain. Systematic and Applied Microbiology. 2016;**39**:266-274. DOI: 10.1016/j.syapm.2016.04.006

[44] Msaddak A, Durán D, Rejili M, Mars M, et al. Diverse bacteria affiliated with the genera *Microvirga*

Phyllobacterium and *Bradyrhizobium* nodulate *Lupinus micranthus* growing in soils of Northern Tunisia. Applied and Environmental Microbiology. 2017;**83**(6). pii: e02820-16. DOI: 10.1128/AEM.02820-16

[45] Msaddak A, Rejili M, Durán D, Rey L, Palacios JM, Imperial J, et al. Definition of two new symbiovars, sv. *lupini* and sv. *mediterranense*, within the genera *Bradyrhizobium and Phyllobacterium* efficiently nodulating *Lupinus mkicranthus* in Tunisia. Systematic and Applied Microbiology. 2018;**41**:487-493

[46] Jarvis BDW, van Berkum P, Chen WX, Nour SM, Fernandez MP, Cleyet-Marel JC, et al. Transfer of *Rhizobium loti*, *Rhizobium huakuii*, *Rhizobium ciceri*, *Rhizobium mediterraneum*, and *Rhizobium tianshanense* to *Mesorhizobium* gen. nov. International Journal of Systematic Bacteriology. 1997;**47**:895-898

[47] Mahdhi M, Houidheg N, Mahmoudi N, Msaadek A, Rejili M, Mars M. Characterization of rhizobial bacteria nodulating *Astragalus corrugatus* and *Hippocrepis areolata* in Tunisian arid soils. Polish Journal of Microbiology. 2016;**65**:331-339

[48] Ourarhi M, Abdelmoumen H, Guerrouj K, Benata H, Muresu R, Squartini A, et al. *Colutea arborescens* is nodulated by diverse rhizobia in Eastern Morocco. Archives of Microbiology. 2011;**193**:115-124. DOI: 10.1007/s00203-010-0650-0

[49] Rejili M, Lorite MJ, Mahdhi M, Pinilla JS, Ferchichi A, Mars M. Genetic diversity of rhizobial populations recovered from three Lotus species cultivated in the infra-arid Tunisian soils. Progress in Natural Science. 2009;**19**:1079-1087

[50] Rejili M, Mahdhi M, Fterich A, Dhaoui S, Guefrachi I, Abdeddayem R, et al. Symbiotic nitrogen fixation of wild legumes in Tunisia: Soil fertility dynamics, field nodulation and nodules effectiveness. Agriculture, Ecosystems and Environment. 2012;**157**:60-69. DOI: 10.1016/j.pnsc.2009.02.003

[51] Rejii M, Mahdhi M, Domínguez-Núñez JA, Mars M. The phenotypic, phylogenetic and symbiotic characterization of rhizobia nodulating *Lotus* sp. in Tunisian arid soils. Annales de Microbiologie. 2013;**64**:355-362

[52] Roba M, Willems A, Le Quéré A, Maynaud G, Pervent M, et al. *Mesorhizobium delmotii* and *Mesorhizobium prunaredense* are two new species containing rhizobial strains within the symbiovar *anthyllidis*. Systematic and Applied Microbiology. 2017;**40**:135-143

[53] Chaich K, Bekki A, Bouras N, Holtz MD, Soussou S, Maure L, et al. Rhizobial diversity associated with the spontaneous legume *Genista saharae* in the northeastern Algerian Sahara. Symbiosis. 2017;**71**(2):111-120

[54] Fterich A, Mahdhi M, Caviedes MA, Pajuelo E, Rivas R, Rodriguez-Llorente ID, et al. Characterization of root-nodulating bacteria associated to *Prosopis farcta* growing in the arid regions of Tunisia. Archives of Microbiology. 2011;**193**:385-397

[55] Young JM. Renaming of *Agrobacterium larrymoorei* Bouzar and Jones 2001 as *Rhizobium larrymoorei* (Bouzar and Jones 2001) comb. nov. International Journal of Systematic and Evolutionary Microbiology. 2004;**54**:149

[56] Mahdhi M, Mars M. Genotypic diversity of rhizobia isolated from *Retama raetam* in arid regions of Tunisia. Annales de Microbiologie. 2006;**56**:305-311

[57] Mahdhi M, Nzoué A, Gueye F, Merabet C, de Lajudie P, Mars M. Phenotypic and genotypic diversity

of *Genista saharae* microsymbionts from the infra-arid region of Tunisia. Letters in Applied Microbiology. 2007;**54**:604-609

[58] Mahdhi M, de Lajudie P, Mars M. Phylogenetic and symbiotic characterization of rhizobial bacteria nodulating *Argyrolobium uniflorum* in Tunisian arid soils. Canadian Journal of Microbiology. 2008;**54**:209-217

[59] Abdelmoumen H, Filali Maltout A, Neyra M, Belabed A, El Idrissi MM. Effects of high salts concentrations on the growth of rhizobia and responses to added osmotica. Journal of Applied Microbiology. 1999;**86**:889-898

[60] Bouchiba Z, Boukhatem ZF, Ighilhariz Z, Derkaoui N, Bekki A. Diversity of nodular bacteria of *Scorpiurus muricatus* in western Algeria and their impact on plant growth. Canadian Journal of Microbiology. 2017;**63**:450-463. DOI: 10.1139/cjm-2016-0493

[61] Merabet C, Bekki A, Benrabah N, Bey M, Bouchentouf L, Ameziane H, et al. Distribution of *Medicago* species and their microsymbionts in a saline region of Algeria. Arid Land Research and Management. 2006;**20**:219-231

[62] Sebbane N, Sahnoune M, Zakhia F, Willems A, Benallaoua S, de Lajudie P. Phenotypical and genotypical characteristics of root-nodulating bacteria isolated from annual *Medicago* spp. in Soummam Valley (Algeria). Letters in Applied Microbiology. 2006;**42**:235-241

[63] Mahdhi M, Fterich A, Rejili M, Rodriguez-Llorente ID, Mars M. Legume-Nodulating bacteria (LNB) from three pasture legumes (*Vicia sativa*, *Trigonella maritima* and *Hedysarum spinosissimum*) in Tunisia. Annales de Microbiologie. 2012;**62**:61-68

[64] Fitouri SD, Trabelsi D, Saïdi S, Zribi K, Ben Jeddi F, Mhamdi R. Diversity of rhizobia nodulating sulla (*Hedysarum coronarium* L.) and selection of inoculant strains for semi-arid Tunisia. Annales de Microbiologie. 2012;**62**:77-84. DOI: 10.1007/s13213-011-0229-22012

[65] Ezzakkioui F, El Mourabit N, Chahboune R, Castellano-Hinojosa A, Bedmar EJ, Barrijal S. Phenotypic and genetic characterization of rhizobia isolated from *Hedysarum flexuosum* in northwest region of Morocco. Journal of Basic Microbiology. 2015;**55**:830-837. DOI: 10.1002/jobm.2014007

[66] Balkwill DL. Ensifer. In: Bergey's Manual of Systematics of Archaea and Bacteria. Hoboken, New Jersey-USA: John Wiley & Sons, Ltd; 2005

[67] Willems A, Fernandez-Lopez M, Munoz-Adelantado E, Goris J, et al. Description of new *Ensifer* strains from nodules and proposal to transfer *Ensifer adhaerens* Casida 1982 to *Sinorhizobium* as *Sinorhizobium adhaerens* comb. nov. request for an opinion. International Journal of Systematic and Evolutionary Microbiology. 2003;**53**:1207-1217

[68] Judicial Commission. Opinion 84 – The genus name Sinorhizobium Chen et al. 1988 is a later synonym of *Ensifer* Casida 1982 and is not conserved over the latter genus name, and the species name '*Sinorhizobium adhaerens*' is not validly published. International Journal of Systematic and Evolutionary Microbiology. 2008;**58**:1973

[69] Merabet C, Martens M, Mahdhi M, Zakhia F, et al. Multilocus sequence analysis of root nodule isolates from *Lotus arabicus (Senegal)*, *Lotus creticus*, *Argyrolobium uniflorum* and *Medicago sativa* (Tunisia) and description of *Ensifer numidicus* sp. nov. and *Ensifer garamanticus* sp. nov. International Journal of Systematic and Evolutionary Microbiology. 2010;**60**:664-674

[70] Khbaya B, Neyra M, Normand P, Zerhari K, Filali-Maltouf A. Genetic

diversity and phylogeny of rhizobia that nodulate *Acacia* spp. in Morocco assessed by analysis of rRNA genes. Applied and Environmental Microbiology. 1998;**64**:4912-4917

[71] Fterich A, Mahdhi M, Mars M. Impact of grazing on soil microbial communities along a chronosequence of *Acacia tortilis* subsp. *raddiana* in arid soils in Tunisia. European Journal of Soil Biology. 2012;**50**:56-63

[72] Sakrouhi I, Belfquih M, Sbabou L, Moulin P, Bena G, Filali-Maltouf A, et al. Recovery of symbiotic nitrogen fixing acacia rhizobia from Merzouga Desert sand dunes in south East Morocco—Identification of a probable new species of *Ensifer* adapted to stressed environments. Systematic and Applied Microbiology. 2016;**39**:22-31

[73] Ben Romdhane S, Nasr H, Samba-Mbaye R, Neyra M, Ghorbal MH. Diversity of Acacia tortilis rhizobia revealed by PCR/RFLP on crushed root nodules in Tunisia. Annals of Microbiology. 2005;**55**:249-258

[74] Jebara M, Drevon JJ, Aouani ME. Effects of hydroponic culture system and NaCl on interactions between common bean lines and native rhizobia from Tunisian soils. Agronomie. 2001;**21**:601-605

[75] Zribi K, Mhamdi R, Huguet T, Aouani ME. Distribution and genetic diversity of rhizobia nodulating natural populations of *Medicago truncatula* in Tunisian soils. Soil Biology and Biochemistry. 2004;**36**:903-908. DOI: 10.1016/j.soilbio.2004.02.003

[76] Badri M, Ilahi H, Huguet T, Aouani ME. Quantitative and molecular genetic variation in sympatric populations of *Medicago laciniata* and *M. truncatula* (Fabaceae): Relationships with eco-geographical geographical factors. Genetical Research. 2007;**89**:107-122

[77] Elboutahiri N, Thami-Alami I, Udupa SM. Phenotypic and genetic diversity in *Sinorhizobium meliloti* and *S. medicae* from drought and salt affected regions of Morocco. BMC Microbiology. 2015;**10**:15. DOI: 10.1186/1471-2180-10-15

[78] Baba Arbi S, Cheriet D, Chekireb D, Ouartsi A. Caractérisation phénotypique et génotypique des rhizobia symbiotiques des légumineuses spontanées *Melilotus indicus* et *Medicago littoralis* des palmeraies de la région de Touggourt en Algérie. In: Conférence: Journées Internationales de Biotechnologie 2014, at Hammamet, Tunisie. Vol. 2014

[79] Guerrouj K, Bouterfas M, Abdelmoumen H, Missbah El Idrissi M. Diversity of bacteria nodulating *Medicago arborea* in the northeast area of Morocco. Chiang Mai Journal of Science. 2016;**43**:440-451

[80] Mousavi SA, Österman J, Wahlberg N, Nesme X, Lavire C, Vial L, et al. Phylogeny of the *Rhizobium* clade supports the delineation of *Neorhizobium* gen. nov. Systematic and Applied Microbiology. 2014;**37**:208-215

[81] Flores-Felix JD, Carro L, Velaszquez E, Valverde A, Cerda Castillo E, Garcia-Fraile P, et al. *Phyllobacterium endophyticum* sp. nov., isolated from nodules of *Phaseolus vulgaris*. International Journal of Systematic and Evolutionary Microbiology. 2013;**63**:821-826

[82] Ardley J, O'Hara G, Reeve W, Yates R, Dilworth M, Tiwari R, et al. Root nodule bacteria isolated from South African *Lotononis bainesii*, *L. listii* and *L. solitudinis* are species of *Methylobacterium* that are unable to utilize methanol. Archives of Microbiology. 2009;**191**:311-318

[83] Andam CP, Parker MA. Novel alphaproteobacterial root nodule

symbiont associated with *Lupinus texensis*. Applied and Environmental Microbiology. 2007;**73**:5687-5691. DOI: 10.1128/AEM.01413-07

[84] Reeve W, Chain P, O'Hara G, Ardley J, Nandesena K, Bräu L, et al. Complete genome sequence of the *Medicago* microsymbiont *Ensifer* (*Sinorhizobium*) *medicae* strain WSM419. Standards in Genomic Sciences. 2010;**2**:77

[85] Tinick MJ, Hadobas PA. Nodulation of Trifolium repens with modified Bradyrhizobium and the nodulation of *Parasponia* with *Rhizobium leguminosarum biovar trifolii* 1990a. Plant and Soil. 1990;**125**:49-61

[86] Mahdhi M, Nzoué A, de Lajudie P, Mars M. Characterization of root-nodulating bacteria on *Retama raetam* in arid Tunisian soils. Progress in Natural Science. 2008;**18**:43-49

7

Organic Nitrogen in Agricultural Systems

Eulene Francisco da Silva, Marlenildo Ferreira Melo, Kássio Ewerton Santos Sombra, Tatiane Severo Silva, Diana Ferreira de Freitas, Maria Eugênia da Costa, Eula Paula da Silva Santos, Larissa Fernandes da Silva, Ademar Pereira Serra and Paula Romyne de Morais Cavalcante Neitzke

Abstract

This work summarizes information about organic nitrogen (N) in the agricultural system. The organic N forms in soils have been studied by identifying and quantifying the released organic compounds when soils are acid treated at high temperature, in which the following organic N fractions are obtained: hydrolyzable total N, subdivided into hydrolyzable NH_4^+-N, amino sugars-N, amino acids-N, and unidentified-N and acid insoluble N, a fraction that remains associated with soil minerals after acid hydrolysis. Nitrogen mineralization and immobilization are biochemical processes in nature. This chapter summarizes how these processes occur in the agricultural system. Then, soluble organic nitrogen (SON), volatilization and denitrification processes, and biological nitrogen fixation (BNF) as a key component of the nitrogen cycle and how it makes N available to plants are also discussed. Finally, we discuss the use of organic fertilizers as N source to satisfy the worldwide demand for organic foods produced without synthetic inputs.

Keywords: biological N fixation, immobilization, mineralization, organic fertilization

1. Introduction

Nitrogen (N) is the fourth most abundant element in cellular biomass and comprises most of the Earth's atmosphere. In the surface layer of most soils, over 90% of N occurs in organic forms. Soil organic N can be divided into two categories: (1) N from organic residues and (2) N from soil organic matter or humus [1]. All these materials are important in maintaining or improving soil fertility and plant nutrition through direct and indirect effects on microbial activity and nutrient availability [2]. Analysis of organic fractions has been highlighted due to the increasing application of organic fertilizers and their direct and indirect effects on crop growth and yield and soil attributes. Thus, we will discuss about organic N forms, N mineralization and immobilization, volatilization and denitrification, soluble organic N, biological N fixation, and organic fertilization with emphasis on N.

2. Organic nitrogen

Nitrogen is an essential element for plants, being constituent of important biomolecules such as adenosine triphosphate (ATP), reduced nicotinamide adenine dinucleotide (NADH), nicotinamide adenine dinucleotide phosphate (NADPH), chlorophylls, amino acids and proteins (glyco- and lipoproteins), nitrogenous bases and nucleic acids, and various enzymes [3, 4]. Soil organic N consisting of proteins, chitins, amino acids, and nucleic acids represents about 90–98% of total soil N [1, 5]. Mineralized N forms are transient in the soil so that the existing amount depends on numerous processes such as mineralization, immobilization, nitrification, denitrification, leaching, and plant uptake. Therefore, the study of mineral N may not represent the N availability during the crop growing. On the other hand, the study of organic N fractions and their transformations over time can help in predicting the N availability for crops, in estimating the N supply to the soil, and in evaluating the potential release of mineral N by organic fertilizers.

Many compounds account the soil organic N, being approximately 40% protein material (proteins, peptides, and amino acids), 5–6% amino sugars, 35% heterocyclic nitrogen compounds (including purines and pyrimidines), and 19% NH_3, with ¼ fixed as NH_4^+. Thus, protein materials and heterocyclic compounds predominate in the total soil N, and organic N fractionation may inform about the mineralization susceptibility of compounds [6]. The organic N forms in soil have been studied by identifying and quantifying the released organic compounds when soils are acid treated at high temperature. The organic N fractions obtained by acid hydrolysis are hydrolyzable total N, subdivided into hydrolyzable NH_4^+-N, amino sugars-N, amino acids-N, and unidentified-N and acid insoluble N, a fraction that remains associated with soil minerals after acid hydrolysis [7].

The fractionation allows separating the labile N forms from the soil, such as amide-N and amino-N (acid hydrolyzable), which can be rapidly synthesized in the mineralization process, releasing inorganic N (NH_4^+ and NO_3^-) to the soil solution. However, most of the organic N can compose more stable fractions in the soil, such as non-hydrolyzable-N and unidentified-N. Variation in the non-hydrolyzable-N may be related to soil management, because the higher the hydrolysis intensity of organic N fractions in the soil, the higher the presence of finer particles that form clay-metal-humus complexes that constitute the non-hydrolyzed N. In Brazil, studies are reported in soils from Amazônia [8], São Paulo [9–12], and Espirito Santo [13].

In Latosols and Argisols from Amazônia, determination of the organic N forms indicated that the immobilization was mainly from microbial origin and the ^{15}N immobilized in the soil was found as acid-soluble N and undistilled-N [8]. In São Paulo, in sugarcane-cultivated soil, amino acid-N fractions predominated, and, after 12 weeks incubation, the total hydrolyzable-N did not vary, but the hydrolyzable NH^{4+}-N decreased [9]. In soil samples under different cover plants [10], the amino acid-N fraction predominated, with the following distribution: 14–38% hydrolyzable NH_4^+-N, 36–52% adenosine triphosphate as NH_4^+-N + amino sugars, 10–32% amino sugar-N, 26–46% amino acid-N, and 3–28% unidentified-N.

Moreover, in São Paulo, in a soil under maize cultivation, it was observed that topdressing N fertilization decreased the N content of the most labile fractions (hydrolyzable NH_4^+-N and amino sugars-N) in the surface layer of the soil, and the amino acid-N and amino sugar-N fractions were considered the organic N reservoirs that control the soil N availability [11]. In contrast, fertilization with cattle manure [12] increased the most easily mineralized (up to 100 days) organic N fractions and subsequently increased the more stable organic N fractions, mainly in clay soil. In Espirito Santo, in soil under eucalyptus, [13] observed that the amino-N was predominant (39%), followed by unidentified-N (27%), amide-N (18%), and hexosamine-N (15%).

Several theories have been developed to explain the resistance of some N compounds to microbial attack. It is mentioned that N compounds are probably protein constituents (amino acids, peptides, and proteins) that are stabilized by reactions with lignins, tannins, quinones, and reducing sugars. Moreover, N compounds would adsorb to the clay fraction of soil and thereby would be protected against the action of protease enzymes. Also, the formation of organic N complexes and polyvalent cations (iron and aluminum) is another biologically stable form of protection [14]. Accumulation and/or decrease of organic C and N is more dynamic in sandy soils than in clayey ones, probably due to the highest oxygenation capacity and lower residue input of sandy soils due to its low productive potential, which gives it less resilience.

3. Nitrogen mineralization and immobilization in the agricultural systems

Nitrogen mineralization and immobilization are biochemical processes widely discussed in the literature. We will focus on how these processes occur in the agricultural system. N mineralization occurs through hydrolysis and biodegradation of soil organic matter when N content in the substrate exceeds the metabolic N requirement by microbial cells. The process is mediated by heterotrophic soil microorganisms [15] that use nitrogenous organic substances as a source of C, N, and energy, releasing NH_4^+ ions as a residue (ammonification). In its turn, immobilization is defined as the transformation of inorganic N (NH_4^+, NH_3, NO_3^-, NO_2^-) to microbial forms. Microbiota assimilates inorganic forms of N by incorporating them into the amino acids, which will participate in protein synthesis during soil biomass formation [14].

N mineralization and immobilization occur simultaneously and oppositely in the soil. The net balance between these processes is controlled by several factors: (a) environmental, such as soil temperature, aeration, and moisture; (b) soil physical, such as texture, structure, and size of aggregates [16]; (c) soil chemical, such as pH; (d) agricultural management system adopted [17]; and (e) quality parameters of the decomposing waste (such as C/N, C/P, and C/S ratios), content of easily decomposable and recalcitrant fractions, type of associated decomposers, size and activity of microbial biomass, and inorganic N availability [18]. Carbon/nitrogen (C/N) ratio less than 25 in organic waste favors N mineralization and fast decomposition, while greater than 30 strongly favors N immobilization and fast decomposition [19]. The crop developmental stage also influences waste C/N ratio. For instance, wastes from millet plants cut at the flowering or milky grain stages present high C/N ratio which delays mineralization. On the other hand, wastes from millet cut at the flag leaf stage, even though phytomass is lower, present less C/N ratio which favors N mineralization for the next crop [20].

In residue plant, considering ^{13}C-CPMAS NMR spectral regions [21], observed that the carbonyl C and N-alkyl and methoxyl C regions had the most significant positive correlation with N mineralization, while the di-O-alkyl C and O-alkyl C were strongly associated with N immobilization. This study demonstrates that the biochemical quality of organic C defined by 13C-CPMAS NMR is capable of predicting N dynamic pattern better than C/N ratio. Abbasi et al. [22] observed positively correlated with the initial residue N contents and negatively correlated with lignin content C/N ratio, lignin/N ratio, polyphenol/N ratio, and (lignin +

polyphenol)/N ratio indicating a significant role of residue chemical composition and quality in regulating N transformations and cycling in soil.

In the N compartments, N from the most labile fractions is released in the early mineralization process, and its mineralization estimate can be used to adjust the nitrogen fertilization recommendations. In fact, it was observed that the mineralization potential and the respective mineralization rate can be used to predict the N availability for plants in the agricultural system. Camargo et al. [23] found that the potentially mineralizable nitrogen values in 10 soils from Rio Grande do Sul ranged from 108.6 to 210.8 mg kg^{-1}.

In respect to the management system adopted, time is essential for N mineralization, mainly in the no-tillage (NT) system. Siqueira et al. [24] found that in soil under NT system for 12 and 22 years, the averages for N mineralization were 0.19 and 0.26 g m^{-2} day^{-1}, respectively. For organic compounds such as sludge, the N mineralization rate is generally below 50%, 5–38% [25], 14–43% [26], 7–16% [27], and 24–31% [28]. Among the species used in straw production, Fabaceae plants stand out for fixing atmospheric N_2 and presenting low C/N ratio tissues, in addition to the high soluble compound content and low lignin and polyphenol contents. This fact favors the fast decomposition and mineralization, with significant N input to the soil–plant system, but with reduced soil cover, which is essential for NT system [29]. On the other hand, Poaceae plants present relatively high dry matter content and high C/N ratio (> 30), which increase the persistence of soil cover although increase N immobilization [30, 31].

4. Nitrogen volatilization and denitrification

Volatilization is the main cause of N loss where ammonia gas (NH_3) is produced according to the simplified equation: $NH_4^+ + OH^- \leftrightarrow NH_{3(g)} + H_2O$. NH_3 loss increases with increasing soil pH. Ammonium ion (NH_4^+) can be adsorbed by soil colloids (clays in humus); thus the largest losses are found in sandy soils and poor in soil organic matter (SOM). Denitrification is another factor that favors N loss, which is mainly controlled by organic matter content, pH, and soil temperature. This process is performed by anaerobic bacteria such as *Pseudomonas*, *Bacillus*, *Micrococcus*, and *Achromobacter*, which are heterotrophic and get energy from carbon, through oxidation of organic compounds. Some autotrophic species also participate in the process such as *Thiobacillus denitrificans* and *T. thioparus* [32].

NH_3 losses by volatilization in agriculture occur due to many factors: ambient temperature, soil moisture at fertilization time, urease enzyme activity, soil pH, cation exchange capacity, soil cover, rainfall after fertilization, and SOM content [33, 34]. Tasca et al. [34] reported 4.6-fold less NH_3 volatilization when topdressing urea was performed at 18°C temperature, compared to 35° C, which demonstrates that N losses increase with increasing temperature. Low volatilization rates are also reported under higher soil moisture values, around 20%, because fertilizer hydrolysis facilitates the NH_4^+ diffusion, making it less susceptible to volatilization, even considering the increased soil biological activity in that moisture. In contrast, higher N losses occur under around 10% humidity values, because the NH_4^+ incorporation is inefficient, resulting in higher N-NH_3 emissions [34]. Moreover, NH_3 losses by volatilization are higher during the driest periods of the year. Soil moisture at fertilization time directly interferes with urea hydrolysis and conse-quently with NH_3 volatilization losses. Thus, soil wetting soon after urea application is more important than the soil moisture at the application time [35]. According to Ros et al. [36], water applied after urea fertilization or the occurrence of rainfall may decrease NH_3 volatilization if it is sufficient to dilute the hydroxyl (OH^-)

concentration around the urea granules produced during the hydrolysis, besides providing the incorporation of urea in the soil.

Plant cover also influences N-NH_3 volatilization. Pinheiro [37] found the removal of sugarcane straw from the soil decreased NH_3 volatilization rates. The analysis of topsoil and straw indicated higher urea and NH_4^+ retention in the largest amounts of straw on the soil, besides effective urea hydrolysis occurring directly in the straw. These results demonstrated a direct contribution of the straw mulches on NH_3 volatilization. However, despite NH_3 volatilization decreases with straw removal, the choice of straw amount to be removed cannot be based only on NH_3 volatilization of N fertilizer. Analyzing fertilizer mixtures in laboratory, Vitti et al. [38] found that mixing urea (330 mg) with ammonium sulfate (300 mg) significantly reduced N-NH_3 losses (97.47 mg) relative to urea (121.52 mg), without affecting the physicochemical quality attributes of the mixture for technical and agronomic efficiency purpose. In Brazil, urea is the most used mineral N fertilizer, but it has volatilization losses due to the enzymatic hydrolysis that consumes H^+ and increases soil pH. For that reason, even in acidic soils, urea is subject to N losses by volatilization [39]. In agricultural systems, the largest N losses by volatilization occur 3–5 days after fertilizer application [40]. Santos [41] observed that from total N-NH_3 loss by volatilization, 92.5% occurred until the fifth day after fertilization, negatively affecting the corn grain yield.

Fertilizer type may also influence N-NH_3 volatilization. The application of polymer and organic compound-coated urea promoted the lowest ammonia losses by volatilization [42, 43]. In soil under pasture (*Brachiaria decumbens*), Lana et al. [44] observed NH_3 losses 2 days after urea application (2765 mg) and that the use of an inhibitor (NBPT) reduced the volatilization peak by up 4 days. The use of urea plus Uremax NBPT 500® decreased volatilization by approximately 75% after 11 days. Also, adding acid fertilizers may reduce NH_3 losses by 29% [45]. According to Gurgel et al. [46], mineral fertilizers mixed with urea and humic acid (5 and 10%) and urea and zeolite (10%) reduced N-NH_3 losses up to 38%. Results were even more effective in sandy soils.

The use of liquid and solid organic biofertilizers such as poultry and swine residues are also alternative means to reduce N losses, since N is present in biofertilizers as organic form, thereby requiring more time to be mineralized by microorganisms for plant uptake. Niraula et al. [47] reported that cattle manure applied in corn had 11% lower cumulative NH_3 emission than urea, without affecting grain yield, despite having higher CO_2 and CH_4 emissions. Thus, after comparing the ammonia volatilization levels reported in 92 studies, Bouwman et al. [48] concluded that the average NH_3 emissions from the synthetic urea fertilizer and manure slurry were 21.0 and 21.2% from applied N fertilizer, respectively. Moreover, acidification has been a resource used to minimize urea volatilization with liquid waste. Park et al. [49] observed the application of acidified slurry reduced NH_3 emissions by 78.1%, N_2O emissions by 78.9%, and NO_3^- leaching by 17.81% compared to control (non-pH-controlled pig slurry), over the course of the experiment.

Quantifying ammonia volatilization from various organic N sources (castor bean cake, bokashi, legume fertilizers, cattle manure), Rocha et al. [50] observed (i) the N loss rate by NH_3 volatilization varies from 3 to 25% in winter/spring and 2 to 38% in summer/autumn among the studied organic fertilizers; (ii) when incorporating organic fertilizers into the soil, volatilization was significantly lower than when they are maintained on the soil surface, with a volatilization reduction by 80% for castor cake, 78% for bokashi, and 67% for legume fertilizer, while for cattle manure there was no difference; and (iii) when on surface, potential NH_3 volatilization from the total N applied in winter/spring and summer/autumn seasons, respectively, was 25.5 and 38.1% for castor cake, 16.6 and 13.7% for bokashi, 8.2 and 8.8% for legume fertilizer, and 3.4 and 2.4% for cattle manure.

In Planosol under irrigated rice, the addition of cover plants on the soil and water management by intermittent irrigation were practices that mitigated N_2O emissions. Zschornack et al. [51] observed an increase in N_2O emissions by more than 200% in a drained area than continuous water blade area. Thus, soil drainage during rice cultivation increases N_2O emissions by stimulating nitrification and denitrification processes. In addition, N_2O emissions depend on the input waste quality and increase significantly when legumes are inserted into cover plants. Moreover, analyzing biochar in rice, He et al. [52] suggested that the combination of biochar and HQ (urease inhibitor-hydroquinone) or the combined application of urease and nitrification inhibitors to soil enriched with biochar at least 1 year previously could be an effective practice for reducing NH_3 emissions and increasing rice yields.

Finally, microorganism respiration may also contribute to retaining N into the soil. By dissimilatory nitrate reduction to ammonium (DNRA), a respiratory process antagonistic to denitrification, nitrate is used by microorganisms, mainly *Bradyrhizobium* and *Mesorhizobium* bacteria, as electron acceptors. This process results in N retention and production of the less mobile ammonium cation (NH_4^+), thereby reducing the contribution to the total N_2O pool [53]. In addition to N fixation, the potential N retention by microorganisms through DNRA becomes a relevant feature in the reduction of N losses by denitrification [54]. This suggests DNRA may act as a mechanism for conserving N in agricultural systems.

5. Soluble nitrogen

Soluble organic nitrogen is a labile source of N for microorganisms and is an important soluble N reservoir in agricultural soils. Plant species (associated or not with mycorrhizae) can directly uptake simple organic N present in the SON pool [55]. The SON pool is composed of high (protein oligomers), medium (small peptides) [56], and low molecular weight compounds (monomers such as amino acids) [57]. As plants uptake organic and inorganic N, the relative proportion of these different N sources in soils is a determinant of N management.

SON is suggested as a transitional phase during N transformation between soil organic matter and inorganic N (NH_4^+-N) and considered an intermediate step in microbial mineralization of organic N [58]. The SON pool can regulate the N transformation rate in the soil, i.e., the ammonification and nitrification rates, affecting the substrate associated with different plant species. Thus, soil organic N fractions and SON pools are important indicators of soil fertility and plant nutrition requirements [59], inferring the potential supply of N mainly in low N mineralization soils [60].

Besides an important component of soil total soluble N, SON plays a key role in N cycling and therefore in determining soil N availability in agricultural sys-tems [61]. The amount of SON represents a relatively high proportion of the total soluble nitrogen (TSN) pool. It has been reported that SON constitutes 17–90% and 32–50% of TSN in pasture and agricultural soils, respectively [46, 47, 62, 63]. Like in mineral N, SON dynamics are affected by mineralization, immobilization, leaching, and plant uptake, but its pool size is more constant than mineral N [64]. Although remains unclearly understood, SON is an important pool in N transformations and plant uptake.

Biotic and abiotic processes are involved in the SON generation in soil [58]. By biotic processes, SON can be produced directly from microbial turnover and indirectly through the microbial excretion of extracellular enzymes [61]. However, as plants and microorganisms can compete for soil organic N, it is also possible that SON reservoirs vary spatially due to the variation in activity and density of microbial population between different types of agricultural management. Zhang et al. [65] reported that SON fractions were significantly and positively correlated with the no-tillage system practices and that this agricultural system is beneficial and effective for increasing soil N turnover.

Proteins are the most abundant nitrogen compounds in SON. Depolymerization of these organic macromolecules in monomeric SON (amino acids) can be considered rate-limiting for the total N cycling in soils [66]. Soil amino acids can contribute, in relative and absolute terms, to the SON pool in agricultural soils, which was observed in soil under fertilized sugarcane [55]. Also, plants can use proteins as N source without the help of other organisms [67]. Although the relative contribution of amino acids to N supply for crops remains unclear, all studied plants have shown the ability to uptake and metabolize amino acids as well as soils containing amino acids [68].

Organic agriculture practices can increase the content of SON, protein, and free amino acids in the soil as a result of frequent and long-term inputs of organic matter. In addition, agricultural production quantity may also influence the SON pool abundance. However, the effect of organic cultivation on specific free amino acids and protein pools remains unclear [66].

Soil organic matter, pH, total C, total N, and C/N ratio are the main factors affecting soil SON abundance. SON dynamics can be significantly affected by mineralization and immobilization during microbial growth and decomposition of organic matter. Besides that, agricultural practices such as irrigation manage-ment, fertilization, plowing, harrowing, harvesting, and the plant growth stage can also play an important role in SON dynamics [59, 63]. Furthermore, high temperatures may increase the SON content by stimulating decomposition of organic matter [69]. Knowing the temporal dynamics of organic N pools in the soil may help to understand how these pools are affected by soil properties, climate and crop management, and whether SON can contribute to N supply of crops.

6. Biological nitrogen fixation

Nitrogen in the gaseous form (N_2) represents 78% of the atmospheric gases but is inert and unavailable to plants. Only nitrogen-fixing microorganisms, including bacteria, cyanobacteria, and fungi, are able to break the triple bond between the atoms ($N \equiv N$) of the atmospheric nitrogen, thus transforming it into ammonia (NH_3) through the nitrogenase enzyme ($N_2 + 8H^+ + 6e^- \rightarrow 2NH_3 + H_2$) [70]. Biological nitrogen fixation is a key component of the nitrogen cycle and responsible for most of the nitrogen available to plants.

BNF is performed by symbiotic, endophytic, or free-living microorganisms [71, 72]. Symbiotic bacteria associate with plants forming root nodules (rhizobia), where they fix nitrogen while benefiting from plant photoassimilates. It has been observed that this symbiosis occurs not only in plants from the Leguminosae family [71] but also in cereals such as rice, maize, and wheat from the Poaceae family [73]. BNF also occurs in nonsymbiotic associations. Endophytic bacteria colonize plant tissues and fix N while benefiting from plant photoassimilates, although the amount of N fixed is lower than in symbiosis [73, 74]. Also, free-living microorganisms inhabiting rhizosphere, soil region around plant roots, fix nitrogen while feeding on root exudates (amino acids, peptides, proteins, enzymes, vitamins, and hormones), which stimulate growth of diazotrophic bacteria from genera *Acetobacter*, *Azoarcus*, *Azospirillum*, *Azotobacter*, *Beijerinckia*, *Burkholderia*, *Enterobacter*, *Herbaspirillum*, *Klebsiella*, *Paenibacillus*, and *Pseudomonas* [71].

Nitrogen-fixing microorganisms occur naturally in soil [71] and in water [72] or colonize seeds [74]. However, in the agricultural environment, conventional practices such as plowing, harrowing, chemical fertilization, and pesticide application reduce the soil microorganism populations, which make these areas depending on the application of nitrogen fertilizers [75, 76]. Chemical fertilizers require a great amount of energy to be produced, energy that is derived from fossil fuels. Moreover, they are potential soil and water contaminants and expensive and scarce for many developing country farmers [77]. Therefore, strategies have been studied to increase BNF by plants and thus reduce dependence on chemical fertilization.

Conservation practices such as minimum tillage, no tillage, and cover crops stimulate BNF as they increase the population and activity of soil microorganisms (bacteria, actinomycetes, and mycorrhizae) [78, 79]. In addition to capturing soil N, reducing N loss by leaching, and becoming an N source for succeeding crops, mixing cover crops (legumes and grasses) provide additional N through BNF [79, 80].

Another alternative for increasing BNF is to inoculate nitrogen-fixing microorganisms in crops. Inoculated into the seeds, roots, or leaves, these microorganisms may increase the formation of root nodules, stimulate root growth, improve nutrient uptake, stimulate antioxidant defense system, increase tolerance to biotic (pest and pathogen) and abiotic (drought and salinity) stresses, and thereby increase crop productivity. Inoculation of nodulating as well as endophytic fungi or bacteria stimulates growth in both legumes and grasses and represents a viable and sustainable alternative (**Table 1**). Among the most used microorganisms are *Rhizobium* and *Bradyrhizobium* genera bacteria inoculated in legumes and *Azospirillum* and *Enterobacter* genera in grasses (**Table 1**).

Studies also focus on the application of nitrogen-fixing microorganisms through irrigation water, on the genetic improvement for BNF by legume crops [96], on becoming plants able to self-fertilize by stimulating root fungal associations in grasses, and on providing cereals with the nitrogen-fixing enzyme (nitrogenase) [77]. Estimations indicate these practices can reduce fertilizer application costs by billions of dollars annually.

Crop	Scientific name	Inoculated microorganism	Reference
Rice	*Oryza sativa*	*Bacillus amyloliquefaciens*, *Enterobacter cloacae*, *Klebsiella variicola*	[81, 82]
Sugarcane	*Saccharum officinarum*	*Gluconacetobacter diazotrophicus*, *Herbaspirillum seropedicae*, *H. rubrisubalbicans*, *Burkholderia tropica* e *Azospirillum amazonense*	[83]
Cowpea	*Vigna unguiculata*	*Actinomadura*, *Bradyrhizobium elkanii*, *B. pachyrhizi*, *B. yuanmingense*, *Paenibacillus graminis*, *Rhizophagus irregularis*	[84–88]
Common bean	*Phaseolus vulgaris*	*Rhizobium leguminosarum* bv. phaseoli, *R. tropici*	[89]
Maize	*Zea mays*	*Azospirillum brasilense*, *Herbaspirillum seropedicae*	[90, 91]
Soybean	*Glycine max*	*Bradyrhizobium japonicum*, *Bacillus megaterium*, *Methylobacterium oryzae*,	[92, 93]
Wheat	*Triticum aestivum*	*Azospirillum brasilense*, *A. insolitus*, *Enterobacter* sp., *Microbacterium arborescens*, *Serratia marcescens*, *Zoogloea ramigera*	[94, 95]

Table 1.
Legume and cereal crops and nitrogen-fixing microorganisms used for inoculation.

7. Nitrogen and organic fertilization

The worldwide demand for organic foods, produced without the use of synthetic inputs, has driven the use of conservation practices, especially fertilization using organic wastes. The application of organic wastes to the soil improves soil fertility by increasing the organic matter (OM) and nutrient contents, such as N and phosphorus (P), and soil microbiota population, as well as improving the cation exchange capacity (CEC) [97].

Organic fertilization improves yield and quality of vegetables such as lettuce (*Lactuca sativa* L.) [98], tomato (*Solanum lycopersicum* Mill.) [99], and carrot (*Daucus* carota L.) [100]; fruits such as papaya (*Carica papaya* L.) [101], citrus (*Citrus* spp.) [102], and raspberry (*Rubus* idaeus L.) [103]; and annual crops such as maize (*Zea mays* L.) [104] and cowpea (*Vigna unguiculata* (L.) Walp) [105]. Most organic fertilizers used as N source are derived from (a) agricultural wastes (cattle, swine and poultry manure), slaughterhouses (bone and blood meal), composting, and vermicomposting; (b) agro-industrial wastes (oilseed pies, sugarcane bagasse, and vinasse) and biochar; and (c) household wastes and sewage sludge composting (**Table 2**).

N input by organic fertilizers occurs predominantly through mineralization of organic N, although some mineral N fractions may be released [107, 119]. The organic N mineralization rate is regulated by N fractions and C/N ratio of the decomposing waste, as well as by environmental temperature and humidity [120, 121]. Under favorable conditions, high N content organic fertilizers mineralize quickly similarly to synthetic fertilizers, while those with low N content and high C/N ratio mineralize slowly [122]. Thus, knowing the mineralization rate allows choosing the best organic fertilizer to be used in agriculture (**Table 2**).

Manures are the main used organic fertilizers worldwide, especially as N source, though the amount and quality of N in manure may vary according to animal species, age, and feed. Forage-based diets increase the residue production, although reduce the quality that is provided by a concentrate-based diet [97, 119]. Cattle, equine, sheep, goat, and swine manures present similar N content, ranging from 0.77 to 3.90%. In its turn, poultry litter may have 2.80–4.60% N content, due to concentrate-based feed supplied to poultries, being a fast mineralizing fertilizer [106, 107]. Thus, manure fertilization has been efficient for many crops, such as sweet pepper (*Capsicum annuum* L.) [123] and radish (*Raphanus sativus* L.) [124].

Residues from the castor bean (*Ricinus communis* L.; *Euphorbiaceae*) chain stand out due to the high N content which is found in the pie (7.54% N), in the oil extraction residue (12.82% N), and in the pulp from direct oil transesterification for biodiesel production [106, 125–127]. Castor pie mineralization rate is more intense than in other composts and thus quickly releases N and other readily available nutrients to plants. As reported by [126], evaluating microbial respiration, who obtained mineralization rates 6 times faster than those obtained in cattle manure and 14 times faster than in sugarcane bagasse, other pies, such as peanut (*Arachis* spp.) and cotton (*Gossypium* spp.), may also have high N (4.0–7.0%) content and similar mineralization characteristics [128, 129].

The product obtained from the composting of organic wastes is rich in stable organic matter. Wastes are transformed through biological decomposition, and the process is affected by environmental conditions and N content. As nitrogen compounds are food for microbiota, N deficiency in waste may retard the maturation process, and the excess may increase the N volatilization as ammonia (NH_3), consequently affecting N stabilization processes in composting [130]. Also, humus from vermicomposting (usually by using *Eisenia foetida* species) is highly stable and presents high contents of N and humic acids, which indicate a better relationship between the mineralization and humification processes of OM, with decreasing C/N ratio [115, 131].

Source	N content (%)	C/N ratio	Reference
Cattle manure	0.8–3.2	16.0–21.0	[97, 106]
Equine manure	1.4–3.9	21.9–25.0	[97, 107]
Sheep manure	1.2–1.8	9.0–29.0	[108, 109]
Swine manure	1.9–2.8	10.0–12.0	[97, 107]
Poultry litter	2.8–4.6	4.2–22.0	[97, 106, 107]
Blood meal[1]	11.8–12.9	—	[110, 111]
Bone meal	4.1–4.2	4.0–7.0	[97, 112]
Meat and bone meal	5.5–6.6	6.0	[106, 112, 113]
Castor pulp[1]	12.8	—	
Castor pie	5.2–7.5	6.0–9.0	[97, 106, 112]
Cotton pie[1]	4.5	—	[106]
Filter pie	1.5–1.8	21.0–24.0	[97, 112]
Sugarcane bagasse	0.9–1.5	85.0	[106, 111, 114]
Vinasse	0.3–1.2	4.0–17.0	[97, 112]
Compost	0.7–2.6	11.3–64.0	[107, 115]
Humus	1.3–2.6	11.0–34.0	[115, 116]
Millicompost	2.0–2.2	15.0–19.0	[98, 117]
Biochar	0.1–5.0	7.0–400.0	[118]
Sewage sludge	0.8–3.5	9.0–50.0	[97, 112]
Household waste	0.9–2.6	7.0–27.0	[97, 107, 112]

[1]*C/N ratio not found.*

Table 2.
Nitrogen content and carbon/nitrogen ratio (C/N) in organic fertilizers.

In addition to the earthworms, arthropods that constitute the edaphic macrofauna [87, 132, 133] are also of great interest. Millipedes (Myriapoda: Diplopoda) fragment and feed on organic wastes and excrete low C/N ratio feces (2.2% N) producing the millicompost [134–136]. Studies suggest that millicompost is similar to vermicompost and commercial substrates in relation to N supply and other macro- and micronutrients for seedling production, such as in lettuce (*Lactuca sativa*) [98] and pitaya (*Hylocereus* spp.) (Cactaceae) [137].

In relation to slow-release organic fertilizers, biochar is an alternative. A by-product from carbonization (pyrolysis) of biomass under low-oxygen atmosphere, biochar is fine-grained carbonaceous material with decomposition resistance [118]. N content in biochar depends on the source material (biomass) as well as on the pyrolysis temperature. Biochars from wood have high C/N ratio and low N content (0.1%), while those from manures have low C/N ratio and high N content (5.0%). For instance, biochar from eucalyptus wood (*Eucalyptus urophylla* S. T. Blake and *Corymbia citriodora* (Hook.) K.D. Hill and L.A.S. Johnson) contains 0.66 and 0.48% N, respectively, while from coffee husks (*Coffea* spp.) contains 2.74% N [138]. Besides slowly releasing nutrients, the use of biochars increases N uptake via ion exchange and NH_3 removal by adsorption, stimulates immobilization (reducing NO_3^- losses), and reduces N_2O emissions [139–142]. Moreover, biochar improves mycorrhizal associations and nitrogen biological fixation [118].

Urban wastes have also been used in agriculture. Sewage sludge showed to be an excellent N source (0.80 and 3.47% N) besides slowly mineralizing. N mineralization rates from 20 to 38% were found after 105 days [143], which depends on source material characteristics and treatment processes as well as on heavy metal content that accelerate or limit mineralization [107, 144]. Slaughterhouse residues, such as bone and blood meal, present high N rates, but they are not yet used in agriculture because studies on its adoption and behavior as organic fertilizer are scarce [97, 110, 112].

8. Concluding remarks

In the surface layer of most soils, the soil organic N can be divided into two categories: N from organic residues and N from soil organic matter or humus. N mineralization and immobilization processes occur simultaneously and oppositely in the soil. The net balance between these processes is controlled by several factors such as environmental conditions, soil physicochemical factors, agricultural management adopted, quality of the decomposing residues, and content of easily decomposable and recalcitrant fractions. As organic agriculture increases soluble organic nitrogen content, this fraction has been extensively studied. Also, being biological nitrogen fixation a key component of the nitrogen cycle and responsible for most of the nitrogen available to plants, it was also discussed in this chapter.

Finally, we discussed nitrogen and organic fertilization, since the worldwide demand for organic foods produced without the use of synthetic inputs has driven the use of conservation practices, especially fertilization using organic wastes. Most organic fertilizers used as N source is derived from agricultural and agro-industrial wastes, slaughterhouse wastes, composting and vermicomposting, biochars, household wastes, and sewage sludge composting.

Author details

Eulene Francisco da Silva[1*], Marlenildo Ferreira Melo[1],
Kássio Ewerton Santos Sombra[1], Tatiane Severo Silva[1], Diana Ferreira de Freitas[2],
Maria Eugênia da Costa[1], Eula Paula da Silva Santos[1], Larissa Fernandes da Silva[1],
Ademar Pereira Serra[3] and Paula Romyne de Morais Cavalcante Neitzke[1]

1 Federal Rural University of the Semi-arid Region (UFERSA), Mossoró, Brazil

2 Federal Rural University Pernambuco (UFRPE/UAST), Serra Talhada, Brazil

3 Brazilian Agricultural Research Corporation (EMBRAPA), Campo Grande, Brazil

*Address all correspondence to: eulenesilva@ufersa.edu.br

References

[1] Kelley KR, Stevenson FJ. Forms and nature of organic N in soil. Fertilizer Research. 1995;**42**:1-11. DOI: 10.1007/BF00750495

[2] Nguyen TH, Shindo H. Effects of different levels of compost application on amounts and distribution of organic nitrogen forms in soil particle size fractions subjected mainly to double cropping. Agricultural Sciences. 2011;**2**:213-219. DOI: 10.4236/as.2011.23030

[3] Harper JE. Nitrogen metabolism. In: Boote KJ, Bennett JM, Sinclair TR, et al., editors. Physiology and Determination of Crop Yield. Madison: ASA/CSSA/SSSA. Chapt. 11A; 1994. pp. 285-302

[4] Bredemeier C, Mundstock CM. Regulation of nitrogen absortion and assimilation in plants. Ciência Rural. 2000;**30**:365-372

[5] Olk CO. Organic forms of soil nitrogen. In: Schepers JS, Raun WR, editors. Nitrogen in Agricultural Systems. Madison, USA: American Society of Agronomy; 2008. pp. 57-100

[6] Schulten HR, Schnitzer M. The chemistry of soil organic nitrogen: A review. Biology and Fertility of Soils. 1998;**26**:1-15

[7] Stevenson FJ. Nitrogen-organic forms. In: Page AL, editor. Methods of Soil Analysis: Chemical and Microbiological Properties. Part 2. Madison, USA: Soil Science Society of America; 1982. pp. 625-641

[8] Alfaia S. Caracterização e distribuição das formas do nitrogênio orgânico em três solos da Amazônia Central. Acta Amazonica. 2006;**36**:135-140

[9] Otto R, Mulvaney RL, Khan SA, Trivelin PCO. Quantifying soil nitrogen mineralization to improve fertilizer nitrogen management of sugarcane. Biology and Fertility of Soils. 2013;**1**:1-12. DOI: 10.1007/s00374-013-0787-5

[10] Kuhnen F. Mineralização de Nitrogênio de solos e de resíduo orgânico em laboratório e em campo [thesis]. Jaboticabal-SP: Universidade Estadual Paulista "Júlio d Mesquita Filho"; 2013. 64p

[11] Bergamasco MAM. Formas de N-orgânico em Latossolo em função de Nitrogênio e de plantas de cobertura em pré-safra do milho [thesis]. Jaboticabal-SP: Universidade Estadual Paulista "Júlio d Mesquita Filho"; 2015. 44p

[12] Adame CR. Formas de nitrogênio orgânico em solos tratados com esterco bovino. [thesis]. Jaboticabal-SP: Universidade Estadual Paulista "Júlio d Mesquita Filho"; 2018. 45p

[13] Pegoraro RF, Silva IR, Novais RF, Barros NF, Cantarutti RB, Fonseca S. Abundância natural de ^{15}N e formas de nitrogênio em Argissolo cultivado com Eucalipto e Acácia. Ciência Florestal. 2016;**26**:295-305

[14] Camargo FAO, Silva LS, Gianello C, Tedesco MJ. Nitrogênio orgânico do solo. In: Santos GA et al., editors. Fundamentos da matéria orgânica do solo. Porto Alegre-RS, Brazil: Metropole; 2008. pp. 87-100

[15] Rigby H, Clarke BO, Pritchard DL, Meehan B, Beshah F, Smith SR, et al. A critical review of nitrogen mineralization in biosolids-amended soil, the associated fertilizer value for crop production and potential for emissions to the environment. The Science of the Total Environment. 2016;**541**:1310-1338. DOI: 10.1016/j.scitotenv.2015.08.089

[16] Bimüller C, Kreyling O, Kölbl A, von Lützow M, Kögel-Knabner I. Carbon

and nitrogen mineralization in hierarchically structured aggregates of different size. Soil and Tillage Research. 2016;**160**:23-33

[17] Kristensen HL, Debosz K, Mccarty GW. Short-term effects of tillage on mineralization of nitrogen and carbon in soil. Soil Biology and Biochemistry. 2003;**35**:979-986. DOI: 10.1016/S0038-0717(03)00159-7

[18] Moreira FMS, Siqueira JO. Microbiologia e bioquímica do solo. 2nd ed. Atual. e ampl. Editora UFLA: Lavras; 2006. 729p

[19] Cantarella H. Nitrogênio. In: Novais RF, Alvarez VVH, Brarros NF, Fontes RLF, Cantaruti RB, Neves JCL, editors. Fertilidade do solo. 1st ed. Viçosa, MG: Sociedade Brasileira de Ciência do Solo; 2007. pp. 375-470

[20] Folini JSS, Catuchi TA, Barbosa AM, Calonego JC, Tititan CS. Acúmulo de nutrientes e relação C/N em diferentes estádios fenológicos do milheto submetido à adubação nitrogenada. Revista Agro@mbiente Online. 2016;**10**:1-9

[21] Bonanomi G, Sarker TC, Zotti M, Cesarano G, Allevato E, Mazzoleni S. Predicting nitrogen mineralization from organic amendments: Beyond C/N ratio by ^{13}C-CPMAS NMR approach. Plant and Soil. 2019;**101**:1-18. DOI: 10.1007/s11104-019-04099-6

[22] Abbasi MK, Tahir MM, Sabir N, Khurshid M. Impact of the addition of different plant residues on nitrogen mineralization–immobilization turnover and carbon content of a soil incubated under laboratory conditions. Solid Earth. 2015;**6**:197-205. DOI: 10.5194/se-6-197-2015

[23] Camargo FAO, Gianello C, Vidor C. Potencial de mineralização do nitrogênio em solos do Rio Grande do Sul. Revista Brasileira de Ciência do Solo. 1997;**21**:575-579

[24] Siqueira Neto M, Piccolo MC, Venzke Filho SP, Feigl BJ, Cerri CC. Mineralização e desnitrificação do nitrogênio no solo sob sistema plantio direto. Bragantia. 2010;**69**:923-936

[25] Alcântara MAK, Aquino Neto V, Camargo OA, Cantarella H. Mineralização do nitrogênio em solos tratados com lodos de curtume. Pesquisa Agropecuária Brasileira. 2007;**42**:547-555. DOI: 10.1590/S0100-204X2007000400013

[26] Boeira RC, Maximiliano VCB. Mineralização de compostos nitrogenados de lodos de esgoto na quinta aplicação em Latossolo. Revista Brasileira de Ciência do Solo. 2009;**33**:711-722. DOI: 10.1590/S0100-06832009000100022

[27] Andrade CA, Silva LFM, Pires AMM, Coscione AR. Mineralização do carbono e do nitrogênio no solo após sucessivas aplicações de lodo de esgoto. Pesquisa Agropecuária Brasileira. 2013;**48**:536-544. DOI: 10.1590/S0100-204X2013000500010

[28] Pires AMM, Andrade CA, Souza NAP, Carmo JB, Coscione AR, Carvalho CS. Pesquisa Agropecuária Brasileira. 2015;**50**:333-342. DOI: 10.1590/S0100-204X2015000400009

[29] Ferreira EPB, Stone LF, Partelli FL, Didonet AD. Produtividade do feijoeiro comum influenciada por plantas de cobertura e sistemas de manejo do solo. Revista Brasileira de Engenharia Agrícola e Ambiental. 2011;**15**:695-701

[30] Perin A, Santos RHS, Urquiaga S, Guerra JGM, Cecon PR. Produção de fitomassa, acúmulo de nutrientes e fixação biológica de nitrogênio por adubos verdes em cultivo isolado e consorciado. Pesquisa Agropecuária

Brasileira. 2004;**39**:35-40. DOI: 10.1590/S0100-204X2004000100005

[31] Favarato LF, Souza JL, Galvão JCC, Souza CM, Guarconi RC, Balbino JMS. Crescimento e produtividade do milho-verde sobre diferentes coberturas de solo no sistema plantio direto orgânico. Bragantia. 2016;**75**:497-506. DOI: 10.1590/1678-4499.549

[32] Machado CJ. Aplicação de fertilizantes com diferentes tecnologias:volatilização de NH_3 [thesis]. Uberlândia-MG: Universidade Federal de Uberlândia- MG; 2015. 62p

[33] Rodrigues JO, Partelli FL, Pires FR, Oliosi G, Espindula MC, Monte JA. Volatilização de amônia de ureias protegidas na cultura do cafeeiro conilon. Coffee Science. 2016;**11**:530-537

[34] Tasca FA, Ernani PR, Rogeri DA, Gatiboni LC, Cassol PC. Volatilização de amônia do solo após a aplicação de ureia convencional ou com inibidor de urease. Revista Brasileira de Ciência do Solo. 2011;**35**:493-502

[35] Lima JES, Nascente AS, Silveira PM, Leandro WM. Volatilização da amônia da ureia estabilizada com NBPT na adubação em cobertura da *Urochloa ruziziensis*. Colloquium Agrariae. 2018;**14**:92-100

[36] Ros COD, Aita C, Giacomini SJ. Volatilização de amônia com aplicação de uréia na superfície do solo, no sistema plantio direto. Ciência Rural. 2005;**35**:799-805. DOI: 10.1590/S0103-84782005000400008

[37] Pinheiro PL. Interação entre remoção de palha e adubação nitrogenada sobre a volatilização de NH_3 e emissão de N_2O na cultura da cana-de-açúcar[thesis]. Santa Maria-RS: Universidade Federal de Santa Maria; 2018. 82p

[38] Vitti GC, Tavares Jr JE, Luz PHC, Favarin JL, Costa MCG. Influência da mistura de sulfato de amônio com uréia sobre a volatilização de nitrogênio amoniacal. Revista Brasileira de Ciência do Solo. 2002;**15**:663-671

[39] Barberena IM, Espindula MC, Araújo LFB, Marcolan AL. Use of urease inhibitors to reduce ammonia volatilization in Amazonian soils. Pesquisa Agropecuária Brasileira. 2019;**54**:1-9. DOI: 10.1590/s1678-3921.pab2019.v54.00253

[40] Cancellier EL, Silva DRG, Faquin V, Gonçalves B, Almeida B, Cancellier LL, et al. Ammonia volatilization from enhanced-efficiency urea on no-till maize in brazilian cerrado with improved soil fertility. Ciência e Agrotecnologia. 2016;**40**:133-144. DOI: 10.1590/1413-70542016402031115

[41] Santos WM. Desempenho agronômico e volatilização da amônia de fertilizantes pastilhados e convencionais na cultura de milho[thesis]. São Cristóvão-SE: Universidade Federal de Sergipe; 2017. 69p

[42] Lorensini F, Ceretta CA, Girotto E, Cerini JB, Lourenzi CR, Trindade LCMM, et al. Lixiviação e volatilização de nitrogênio em um Argissolo cultivado com videira submetida à adubação nitrogenada. Ciência Rural. 2012;**42**:1173-1179

[43] Siqueira ETJ. Avaliação de fontes de nitrogênio na produção de cana-de-açúcar, aporte de matéria orgânica no solo e perdas por volatilização de amônia [thesis]. Chapadinha-MA: Universidade Federal do Maranhão; 2018. 62p

[44] Lana RMQ , Pereira VJ, Leite CN, Teixeira GM, Gomes JS, Camargo R. NBPT (urease inhibitor) in the dynamics of ammonia volatilization. Revista Brasileira de Ciência Agrária. 2018;**13**:1-8

[45] Oliveira JÁ, Stafanato JB, Goulart RS, Zonta E, Lima E,

Mazur N, et al. Volatilização de amônia proveniente de ureia compactada com enxofre e bentonita, em ambiente controlado. Revista Brasileira de Ciência do Solo. 2014;**38**:1558-1564. DOI: 10.1590 S0100-06832014000500021

[46] Gurgel GCS, Ferrari AC, Fontana A, Polidoro JC, Coelho LAM, Zonta E. Volatilização de amônia proveniente de fertilizantes minerais mistos contendo ureia. Pesquisa Agropecuária Brasileira. 2016;**51**:1686-1694

[47] Niraula S, Rahman S, Chatterjee A, Cortus EL, Mehata M, Spiehs MJ. Beef manure and urea applied to corn show variable effects on nitrous oxide, methane, carbon dioxide, and ammonia. Agronomy Journal. 2018;**11**:1448-1467

[48] Bouwman AF, Boumans LJM, Batjes NH. Estimation of global NH3 volatilization loss from synthetic fertilizers and animal manure applied to arable lands and grasslands. Global Biogeochemical Cycles. 2002;**16**:8-13

[49] Park SH, Lee BR, Jung KH, Kim TH. Acidification of pig slurry effects on ammonia and nitrous oxide emissions, nitrate leaching, and perennial ryegrass regrowth as estimated by 15N-urea flux. Asian-Australasian Journal of Animal Sciences. 2018;**31**:457-466

[50] Rocha AA, Araújo ES, Santos SS, Goulart JM, Espindola JAA, Guerra JGM, et al. Ammonia volatilization from soil-applied organic fertilizers. Revista Brasileira de Ciência do Solo. 2019;**43**:1-10. DOI: 10.1590/18069657rbcs20180151

[51] Zschornack T, Rosa CM, Camargo ES, Reis CES, Schoenfeld R, Bayer C. Impacto de plantas de cobertura e da drenagem do solo nas emissões de CH_4 e N_2 O sob cultivo de arroz irrigado. Pesquisa Agropecuária Brasileira. 2016;**51**:1163-1171. DOI: 10.1590/S0100-204X2016000900016

[52] He T, Deyan L, Yuan J, Ni K, Zaman M, LUO J, et al. A two years study on the combined effects of biochar and inhibitors on ammonia volatilization in an intensively managed rice field. Agriculture, Ecosystems and Environment. 2018;**264**:44-53

[53] Otto R, Zavaschi E, Souza Netto GJM, Machado BA, Mira AB. Ammonia volatilization from nitrogen fertilizers applied to sugarcane straw. Revista Ciência Agronômica. 2017;**48**:413-418.b. DOI: 10.5935/1806-6690.20170048

[54] Suleiman MF, Wagner-Riddle C, Brown SE, Warland J. Greenhouse gas mitigation potential of annual and perennial dairy food production systems. Agriculture, Ecosystems and Environment. 2017;**245**:52-62. DOI: 10.1016/j.agee.2017.05.001

[55] Holst J, Brackin R, Robinson N, Lakshmanan P, Schmidt S. Soluble inorganic and organic nitrogen in two Australian soils under sugarcane cultivation. Agriculture, Ecosystems and Environment. 2012;**155**:16-26. DOI: 10.1016/j.agee.2012.03.015

[56] Moran-Zuloaga D, Dippold M, Glaser B, Kuzyakov Y. Organic nitrogen uptake by plants: Reevaluation by position-specific labeling of amino acids. Biogeochemistry. 2015;**125**:359-374. DOI: 10.1007/s10533-015-0130-3

[57] Brackin R, Näsholm T, Robinson N, Guillou S, Vinall K, Lakshmanan P, et al. Nitrogen fluxes at the root-soil interface show a mismatch of nitrogen fertilizer supply and sugarcane root uptake capacity. Scientific Reports. 2015;**5**:15727. DOI: 10.1038/srep15727

[58] Chen CR, Xu ZH. Analysis and behavior of soluble organic nitrogen in forest soils. Journal of Soils and

Sediments. 2008;**8**:363-378. DOI: 10.1007/s11368-008-0044-y

[59] Wu H, Du S, Zhang Y, An J, Zou H, Zhang Y, et al. Effects of irrigation and nitrogen fertilization on greenhouse soil organic nitrogen fractions and soil-soluble nitrogen pools. Agricultural Water Management. 2019;**216**:415-424. DOI: 10.1016/j.agwat.2019.02.020

[60] Jones DL, Healey JR, Willett VB, Farrar JF, Hodge A. Dissolved organic nitrogen uptake by plants—An important N uptake pathway? Soil Biology and Biochemistry. 2005;**37**:413-423. DOI: 10.1016/j.soilbio.2004.08.008

[61] Yang K, Zhu J, Yan Q, Zhang J. Soil enzyme activities as potential indicators of soluble organic nitrogen pools in forest ecosystems of Northeast China. Annals of Forest Science. 2012;**69**:795-803. DOI: 10.1007/s13595-012-0198-z

[62] Bhogal A, Murphy DV, Fortune S, Shepherd MA, Hatch DJ, Jarvis SC, et al. Distribution of nitrogen pools in the soil profile of undisturbed and reseeded grasslands. Biology and Fertility of Soils. 2000;**30**:356-362. DOI: 10.1007/s003740050016

[63] Zhang Y, Xu W, Duan P, Cong Y, An T, Yu N, et al. Evaluation and simulation of nitrogen mineralization of paddy soils in Mollisols area of Northeast China under waterlogged incubation. PLoS One. 2017;**12**:e0171022. DOI: 10.1371/journal. pone.0171022

[64] Murphy DV, Macdonald AJ, Stockdale EA, Goulding KWT, Fortune S, Gaunt JL, et al. Soluble organic nitrogen in agricultural soils. Biology and Fertility of Soils. 2000;**30**:374-387. DOI: 10.1007/s003740050018

[65] Zhang H, Zhang Y, Yan C, Liu E, Chen B. Soil nitrogen and its fractions between long-term conventional and no-tillage systems with straw retention in dryland farming in northern China. Geoderma. 2016;**269**:138-144. DOI: 10.1016/j.geoderma.2016.02.001

[66] Wang XL, Ye J, Perez PG, Tang DM, Huang DF. (2013). The impact of organic farming on the soluble organic nitrogen pool in horticultural soil under open field and greenhouse conditions: A case study. Soil Science & Plant Nutrition. 2013;**59**:237-248. DOI: 10.1080/00380768.2013.770722

[67] Paungfoo-Lonhienne C, Lonhienne TG, Rentsch D, Robinson N, Christie M, Webb RI, et al. Plants can use protein as a nitrogen source without assistance from other organisms. Proceedings of the National Academy of Sciences. 2008;**105**:4524-4529. DOI: 10.1073/pnas.0712078105

[68] Paungfoo-Lonhienne C, Visser J, Lonhienne TG, Schmidt S. Past, present and future of organic nutrients. Plant and Soil. 2012;**359**:1-18. DOI: 10.1007/s11104-012-1357-6

[69] Conant RT, Ryan MG, Ågren GI, Birge HE, Davidson EA, Eliasson PE, et al. Temperature and soil organic matter decomposition rates–synthesis of current knowledge and a way forward. Global Change Biology. 2011;**17**:3392-3404. DOI: 10.1111/j.1365-2486.2011.02496.x

[70] Newton WE. 2015. Chapter 2 Recent advances in understanding nitrogenases and how they work. Available from: https://onlinelibrary.wiley.com/doi/pdf/10.1002/9781119053095.ch2 [Accessed: 03 July 2019]

[71] Chanway CP, Anand R, Yang H. Nitrogen fixation outside and inside plant tissues. In: Advances in Biology and Ecology of Nitrogen Fixation. London: InTech. Epub ahead of print; 2014. DOI: 10.5772/57532

[72] Hemida AA, Ohyam T. Nitrogen fixing cyanobacteria: Future prospect. In: Advances in Biology and Ecology of Nitrogen Fixation. London: InTech. Epub ahead of print; 2014. DOI: 10.5772/56995

[73] Rosenblueth M, Ormeño-Orrillo E, López-López A, Rogel MA, Reyes-Hernández BJ, Martínez-Romero JC, et al. Nitrogen fixation in cereals. Frontiers in Microbiology. 2018;**9**:1794. DOI: 10.3389/fmicb.2018.01794

[74] Verma SK, Kingsley K, Irizarry I, Bergen M, Kharwar RN, White JF. Seed-vectored endophytic bacteria modulate development of rice seedlings. Journal of Applied Microbiology. 2017;**122**:1680-1691. DOI: 10.1111/jam.13463

[75] Zheng M, Zhou Z, Luo Y, Zhao P, Mo J. Global pattern and controls of biological nitrogen fixation under nutrient enrichment: A meta-analysis. In: Global Change Biology. 2019. DOI: 10.1111/gcb.14705

[76] Gelfand I, Philip Robertson G. A reassessment of the contribution of soybean biological nitrogen fixation to reactive N in the environment. Biogeochemistry. 2015;**123**:175-184. DOI: 10.1007/s10533-014-0061-4

[77] Stokstad E. The nitrogen fix. Science. 2016;**353**:1225-1227

[78] Mbuthia LW, Acosta-Martínez V, DeBruyn J, et al. Long term tillage, cover crop, and fertilization effects on microbial community structure, activity: Implications for soil quality. Soil Biology and Biochemistry. 2015;**89**:24-34. DOI: 10.1016/j.soilbio.2015.06.016

[79] Blesh J. Functional traits in cover crop mixtures: Biological nitrogen fixation and multifunctionality. Journal of Applied Ecology. 2018;**55**:38-48. DOI: 10.1111/1365-2664.13011

[80] Li X, Sørensen P, Li F, Li F, Petersen SO, Olesen JE. Quantifying biological nitrogen fixation of different catch crops, and residual effects of roots and tops on nitrogen uptake in barley using in-situ 15N labelling. Plant and Soil. 2015;**395**:273-287. DOI: 10.1007/s11104-015-2548-8

[81] Shahzad R, Khan AL, Bilal S, Waqas M, Kang SM, Lee IJ. Inoculation of abscisic acid-producing endophytic bacteria enhances salinity stress tolerance in Oryza sativa. Environmental and Experimental Botany. 2017;**136**:68-77. DOI: 10.1016/j.envexpbot.2017.01.010

[82] Defez R, Andreozzi A, Bianco C. The overproduction of indole-3-acetic acid (IAA) in endophytes upregulates nitrogen fixation in both bacterial cultures and inoculated rice plants. Microbial Ecology. 2017;**74**:441-452. DOI: 10.1007/s00248-017-0948-4

[83] Gava GJC, Scarpare FV, Cantarella H, Kölln OT, Ruiz-Corrêa ST, Arlanch AB, et al. Nitrogen source contribution in sugarcane-inoculated plants with diazotrophic bacterias under urea-N fertigation management. Sugar Tech. 2019;**21**:462-470. DOI: 10.1007/s12355-018-0614-2

[84] Leite J, Passos SR, Simões- Araújo JL, Rumjanek NG, Xavier GR, Zilli JÉ. Genomic identification and characterization of the elite strains Bradyrhizobium yuanmingense BR 3267 and Bradyrhizobium pachyrhizi BR 3262 recommended for cowpea inoculation in Brazil. Brazilian Journal of Microbiology. 2018;**49**:703-713. DOI: 10.1016/j.bjm.2017.01.007

[85] Oliveira RS, Carvalho P, Marques G, Ferreira L, Pereira S, Nunes M, et al. Improved grain yield of cowpea (*Vigna*

unguiculata) under water deficit after inoculation with *Bradyrhizobium elkanii* and *Rhizophagus irregularis*. Crop & Pasture Science. 2017;**68**:1052. DOI: 10.1071/CP17087

[86] Santos AA, Silveira JAG, Bonifacio A, Rodrigues AC, Figueiredo MDVB. Antioxidant response of cowpea co-inoculated with plant growth-promoting bacteria under salt stress. Brazilian Journal of Microbiology. 2018;**49**:513-521a. DOI: 10.1016/j.bjm.2017.12.003

[87] Santos JBD, Ramos AC, Azevedo Júnior R, Oliveira Filho LCI, Baretta D, Cardoso EJBN. Soil macrofauna in organic and conventional coffee plantations in Brazil. Biota Neotropica. 2018b;**18**:1-13. DOI: 10.1590/1676-0611-BN-2018-0515

[88] Souza EM, Bassani VL, Sperotto RA, Granada CE. Inoculation of new rhizobial isolates improve nutrient uptake and growth of bean (*Phaseolus vulgaris*) and arugula (*Eruca sativa*). Journal of the Science of Food and Agriculture. 2016;**96**:3446-3453. DOI: 10.1002/jsfa.7527

[89] Abou-Shanab RAI, Wongphatcharachai M, Sheaffer CC, Sadowsky MJ. Response of dry bean (*Phaseolus vulgaris* L.) to inoculation with indigenous and commercial Rhizobium strains under organic farming systems in Minnesota. Symbiosis. 2019;**78**:125-134. DOI: 10.1007/s13199-019-00609-3

[90] Curá JA, Franz DR, Filosofía JE, Balestrasse K, Burgueño L. Inoculation with Azospirillum sp. and Herbaspirillum sp. bacteria increases the tolerance of maize to drought stress. Microorganisms. 2017;**5**:41. DOI: 10.3390/microorganisms5030041

[91] Bertoncelli P, Martin TN, Stecca J, Deak E, Pinto MAB, Schonell A. O manejo de inverno e inoculação de sementes influenciam na produtividade e qualidade da silagem de milho sob sistema plantio direto. Revista Ceres. 2017;**64**:523-531. DOI: 10.1590/0034-737X201764050010

[92] Sanz-sáez Á, Heath KD, Burke PV, Ainsworth EA. Inoculation with an enhanced N_2-fixing *Bradyrhizobium japonicum* strain (USDA110) does not alter soybean (Glycine max Merr.) response to elevated [CO_2]. Plant, Cell & Environment. 2015;**38**:2589-2602. DOI: 10.1111/pce.12577

[93] Subramanian P, Kim K, Krishnamoorthy R, Sundaram S, Sa T. Endophytic bacteria improve nodule function and plant nitrogen in soybean on co-inoculation with *Bradyrhizobium japonicum* MN110. Journal of Plant Growth Regulation. 2015;**76**:327-332. DOI: 10.1007/s10725-014-9993-x

[94] Silveira APD, Sala VMR, Cardoso EJBN, Labanca EG, Cipriano MAP. Nitrogen metabolism and growth of wheat plant under diazotrophic endophytic bacteria inoculation. Applied Soil Ecology. 2016;**107**:313-319. DOI: 10.1016/j. apsoil.2016.07.005

[95] Kumar A, Maurya BR, Raghuwanshi R, Meena VS, Islam MT. Co-inoculation with enterobacter and rhizobacteria on yield and nutrient uptake by wheat (*Triticum aestivum* L.) in the alluvial soil under indo-gangetic plain of India. Journal of Plant Growth Regulation. 2017;**36**:608-617. DOI: 10.1007/s00344-016-9663-5

[96] Alcantara RMCM, Xavier GR, Rumjanek NG, Rocha MM, Carvalho JS. Eficiência simbiótica de progenitores de cultivares brasileiras de feijão-caupi. Revista Ciência Agronômica. 2014;**45**:1-9

[97] Trani PE, Terra MM, Tecchio MA, Teixeira LAJ, Hanasiro J. Adubação

orgânica de hortaliças e frutíferas. IAC: Campinas; 2013. 16p

[98] Antunes LFS, Scoriza FN, França EM, Silva DG, Correia MEF, Leal MAA, et al. Desempenho agronômico da alface crespa a partir de mudas produzidas com gongocomposto. Revista Brasileira de Agropecuária Sustentável. 2018;**8**:57-65

[99] Arjune YP, Ansari AA, Jaikishun S, Homenauth O. Effect of vermicompost and other fertilizers on soil microbial population and growth parameters of f1 Mongal tomato (*Solanum lycopersicum* mill.). Pakistan Journal of Botany. 2019;**51**:1883-1889. DOI: 10.30848/PJB2019-5(1)

[100] Rahman MA, Islam MT, Al Mamun MA, Rahman MS, Ashraf MS. Yield and quality performance of carrot under different organic and inorganic nutrient sources with mulching options. Asian Journal of Agricultural and Horticultural Research. 2018;**1**:1-8

[101] Mirza A, Jakhar R, Singh J. Response of organic practices, mulching and plant growth regulators on growth, yield and quality of papaya (*Carica papaya* L) cv. Taiwan Red Lady. Indian Journal of Agricultural Research. 2019;**96-99**(2019):**53**

[102] Escanhoela ASB, Pitombo LM, Brandani CB, Navarrete AA, Bento CB, Carmo JB. Organic management increases soil nitrogen but not carbon content in a tropical citrus orchard with pronounced N_2O emissions. Journal of Environmental Management. 2019;**234**:326-335

[103] Stojanov D, Milošević T, Mašković P, Milošević N, Glišić I, Paunović G. Influence of organic, organo-mineral and mineral fertilisers on cane traits, productivity and berry quality of red raspberry (*Rubus idaeus* L.). Scientia Horticulturae. 2019;**252**:370-378. DOI: 10.1016/j.scienta.2019.04.009

[104] Menezes JFS, Berti MPS, Vieira Junior VD, Ribeiro RL, Berti CLF. Extração e exportação de nitrogênio, fósforo e potássio pelo milho adubado com dejetos de suínos. Revista de Agricultura Neotropical. 2018;**5**:55-59

[105] Magalhães ACM, Blum J, Lopes FB, Tornquist CG. Production components of the cowpea under different doses of organic fertiliser. Journal of Experimental Agriculture International. 2018;**26**:1-9

[106] Severino LS, Lima RDLS, Beltrão NDM. Composição química de onze materiais orgânicos utilizados em substratos para produção de mudas. Campina Grande: Embrapa Algodão; 2006. 5p

[107] Carneiro WJDO, Silva CA, Muniz JÁ, Savian TV. Mineralização de nitrogênio em Latossolos adubados com resíduos orgânicos. Revista Brasileira de Ciência do Solo. 2013;**37**:715-725. DOI: http://dx.doi.org/10.1590/S0100-068320130003

[108] Figueiredo CC, Ramos MLG, Mcmanus CM, Menezes AM. Mineralização de esterco de ovinos e sua influência na produção de alface. Horticultura Brasileira. 2012;**30**:175-179. DOI: 10.1590/S0102-05362012000100029

[109] Peixoto Filho JU, Freire MBS, Freire FJ, Miranda MF, Pessoa LG, Kamimura KM. Produtividade de alface com doses de esterco de frango, bovino e ovino em cultivos sucessivos. Revista Brasileira de Engenharia Agrícola e Ambiental-Agriambi. 2013;**17**:419-424

[110] Sorrenti GB, Fachinello JC, Castilhos DD, Bianchi VJ, Marangoni B. Influência da adubação orgânica no crescimento de tangerineira cv Clemenules e nos atributos

químicos e microbiológicos do solo. Revista Brasileira de Fruticultura. 2008;**30**:1129-1135. DOI: 10.1590/ S0100-29452008000400047

[111] Zamberlam J, Froncheti A. Agroecologia-Caminho de Preservação do Agricultor e do Meio Ambiente. Editora Vozes Ltda: Petrópolis; 2012. 196p

[112] Chacón EAV, Mendonça ES, Silva RR, Lima PC, Silva IR, Cantarutti RB. Decomposição de fontes orgânicas e mineralização de formas de nitrogênio e fósforo. Ceres. 2011;**58**:373-383. DOI: 10.1590/ S0034-737X2011000300019

[113] Pires AA, Monnerat PH, Marciano CR, Rocha Pinho LG, Zampirolli PD, Rosa RCC, et al. Efeito da adubação alternativa do maracujazeiro-amarelo nas características químicas e físicas do solo. Revista Brasileira de Ciência do Solo. 2008;**32**:1997-2005

[114] Yamaguchi CS, Ramos NP, Carvalho CS, Pires AMM, Andrade CA. Decomposição da palha de cana-de-açúcar e balanço de carbono em função da massa inicialmente aportada sobre o solo e da aplicação de vinhaça. Bragantia. 2017;**76**:135-144

[115] Cotta JAO, Carvalho NLC, Brum TDS, Rezende MOO. Compostagem versus vermicompostagem: comparação das técnicas utilizando resíduos vegetais, esterco bovino e serragem. Engenharia Sanitária e Ambiental. 2015;**20**:65-78. DOI: 10.1590 S1413-415220150200001 11864

[116] Lisboa CC, Lima FRD, Reis RHCL, Silva CA, Marques JJ. Taxa de mineralização do nitrogênio de resíduos orgânicos. Cultura Agronômica. 2018;**27**:341-355

[117] Antunes LFS, Scoriza FN, Silva DG, Fernandes MEC. Production and efficiency of organic compost generated by millipede activity. Ciência Rural. 2016;**46**:815-819. DOI: 10.1590/0103-8478cr20150714

[118] Lehmann J, Joseph S, editors. Biochar for Environmental Management: Science, Technology and Implementation. London: Routledge; 2015. p. 976

[119] Schröder JJ, De Visser W, Assinck FBT, Velthof GL. Effects of short-term nitrogen supply from livestock manures and cover crops on silage maize production and nitrate leaching. Soil Use and Management. 2013;**29**:151-160. DOI: 10.1111 / sum.12027

[120] Amlinger F, Götz B, Dreher P, Geszti J, Weissteiner C. Nitrogen in biowaste and yard waste compost: Dynamics of mobilisation and availability–A review. European Journal of Soil Biology. 2003;**39**:107-116. DOI: 10.1016/S1164-5563(03)00026-8

[121] Alves RN, Menezes RS, Salcedo IH, Pereira WE. Relação entre qualidade e liberação de N por plantas do semiárido usadas como adubo verde. Revista Brasileira de Engenharia Agrícola e Ambiental. 2011;**15**:1107-1114

[122] Zandvakili OR, Barker AV, Hashemi M, Etemadi F, Autio WR, Weis S. Growth and nutrient and nitrate accumulation of lettuce under different regimes of nitrogen fertilization. Journal of Plant Nutrition. 2019;**42**:1575-1593

[123] Rodrigues RMP, França KS, Didolanvi OD, Oliveira RL, Sousa MLL, Carvalho RS. Rendimento do pimentão em função de diferentes doses de esterco caprino. Cadernos de Agroecologia. 2018;**13**:1-7

[124] Lima DC, Lopes HLS, Sampaio ASO, Souto LS, Pereira ACS, Silva AM, et al. Crescimento inicial da cultura do rabanete (*Raphanus sativus* L.) submetida a níveis e fontes de fertilizantes orgânicos.

Revista Brasileira de Gestão Ambiental. 2019, 2019;**13**:19-24. DOI: 10.18378/rbga. v13i1.6152

[125] Severino LS. O que sabemos sobre a torta da mamona. Campina Grande: Embrapa Algodão; 2005. 31p

[126] Severino LS, Costa FX, Beltrão NEM, Lucena AMA, Guimarães MM. Mineralização da torta de mamona, esterco bovino e bagaço de cana estimada pela respiração microbiana. Revista de Biologia de Ciências da Terra. 2005;**5**:1-6

[127] Alves FQG, Soares EPS, Sobral RRS, Melo ADD, Duarte ABM, Rocha MR, et al. Diferentes doses de torta de mamona no desempenho de bulbos de rabanete consorciado com alface. Horticultura. 2012;**30**:5464-5471

[128] Costa FX, Severino LS, Beltrão NM, Freire RMM, Lucena AMA, Guimarães MMB. Avaliação de teores químicos na torta de mamona. Revista de Biologia e Ciências da Terra. 2004;**4**:1-7

[129] Lima RL, Severino LS, Sampaio LR, Sofiatti V, Gomes JA, Beltrão NE. Blends of castor meal and castor husks for optimized use as organic fertilizer. Industrial Crops and Products. 2011;**33**:364-368

[130] Franco GG, Silva SL, Emiliano ED, Silva MVS, Costa FS. Produção agroecológica de compostagem de folhas, frutos e madeira triturada. Cadernos de Agroecologia. 2018;**13**:1-5

[131] Dores-Silva PR, Landgraf MD, Rezende MOO. Processo de estabilização de resíduos orgânicos: vermicompostagem versus compostagem. Química Nova. 2013;**36**:640-645

[132] Souza MH, Vieira BCR, Oliveira PG, Amaral AA. Macrofauna do solo. Enciclopédia Biosfera. 2015;**11**:115-131

[133] Suárez LR, Pinto SPC, Salazar JCS. Soil macrofauna and edaphic properties in coffee production systems in Southern Colombia. Floresta e Ambiente. 2019;**26**:1-8. DOI: 10.1590/2179-8087.033418

[134] Garcia FRM, Campos JV. Biologia e controle de artrópodes de importância fitossanitária (Diplopoda, Symphyla, Isopoda), pouco conhecidos no Brasil. Biológico. 2001;**63**:7-13

[135] Thakur PC, Shailendra PA, Sinha K. Comparative study of characteristics of biocompost produced by millipedes and earthworms. Advances in Applied Science Research. 2011;**2**:94-98

[136] Ramanathan B, Alagesan P. Evaluation of millicompost versus vermicompost. Current Science. 2012;**103**:140-143

[137] Cruvinel FF, Antunes LFS, Vasconcellos MAS, Rangel Júnior IM, Martelleto LAP. Produção de mudas orgânicas de pitaia em diferentes substratos. Cadernos de Agroecologia. 2018;**13**:1-5

[138] Veiga TRLA, Lima JT, Dessimoni ALA, Pego MFF, Soares JR, Trugilho PF. Different plant biomass characterizations for biochar production. Cerne. 2017;**23**:529-536. DOI: 10.1590/01047760201723042373

[139] Cayuela ML, Sánchez- Monedero MA, Roig A, Hanley K, Enders A, Lehmann J. Biochar and denitrification in soils: When, how much and why does biochar reduce N_2O emissions? Scientific Reports. 2013; **3-1732**:1-7. DOI: 10.1038/srep01732

[140] Clough T, Condron L, Kammann C, Müller C. A review of biochar and soil nitrogen dynamics. Agronomy. 2013;**3**:275-293

[141] El-Naggar A, Lee SS, Rinklebe J, Farooq M, Song H, Sarmah AK, et al. Biochar application to low fertility soils: A review of current status, and future prospects. Geoderma. 2019;**337**:536-554. DOI: 10.1016/j.geoderma.2018.09.034

[142] Fungo B, Lehmann J, Kalbitz K, Thionģo M, Tenywa M, Okeyo I, et al. Ammonia and nitrous oxide emissions from a field Ultisol amended with tithonia green manure, urea, and biochar. Biology and Fertility of Soils. 2019;**55**:135-148. DOI: 10.1007/s00374-018-01338-3

[143] Boeira RC, Ligo MAV, Dynia JF. Mineralização de nitrogênio em solo tropical tratado com lodos de esgoto. Pesquisa Agropecuária Brasileira. 2002;**37**:1639-1647. DOI: 10.1590/S0100-204X2002001100016

[144] Cabrera ML, Kissel DE, Vigil MF. Nitrogen mineralization from organic residues: Research opportunities. Journal of Environmental Quality. 2005;**34**:75-79. DOI: 10.2134/jeq2005.0075

8

Comprehensive Account of Inoculation and Coinoculation in Soybean

Muhammad Jamil Khan, Rafia Younas, Abida Saleem, Mumtaz Khan, Qudratullah Khan and Rehan Ahmed

Abstract

This chapter elaborates dependency of leguminous plants on rhizobia to carry out dynamic process of nitrogen fixation. Soybean, an extensively grown leguminous crop with 30% share in world's vegetable oil, is taken into account to understand its symbiotic relationship with plant growth-promoting rhizobacteria (PGPRs). This chapter narrates colonization of PGPRs on soybean roots and single and mixed inoculation and coinoculation of certain strains of specialized bacteria with rhizobia. PGPRs' coinoculation seemed more effective than mono-inoculation and is discussed in Ref. to nodulation rate. Moreover, dynamic linear models for quantification of leguminous biological nitrogen fixation (BNF) are reviewed. This chapter further uncoils the relevance of foliar application to the release of phytohormones by PGPRs, resulting in situ biosynthesis of active metabolites in phyllosphere. Inoculation of phytohormones is compared to their exogenous application for nodule organogenesis. Finally, the influence of coinoculation on enhanced micronutrient bioavailability is relayed. The chapter is concluded with technical and economic aspects of coinoculation in soybean.

Keywords: legumes, nodulation, BNF, phytohormones, mixed inoculation

1. Introduction

Better plant growth is ensured by the balanced availability of essential nutrients in soil. Each nutrient has its own function and is required in different amount depending on the plant demand. Nitrogen (N), one of the most essential macronutrients, is routinely applied through chemical fertilizer as most field crops require large amounts of it. Nitrogen, the fifth most abundant element in the universe, was first discovered in 1772 by a Scottish physician, Daniel Rutherford. Due to its essentiality for survival of life on earth, it was called as "azote," meaning "without life," by Antoine Lavoisier about 200 years. Nitrogen is essential for the sustenance of life on this planet as it serves as building block for the synthesis of proteins. The inevitable role of N is well acknowledged in several biochemical processes such as cell division, growth promotion, and photosynthesis, as part of vitamins and carbohydrates and energy reactions in the plant body [1, 2]. Deficiency of N in plants is recognized by the symptom of delaying maturity of plant which leads to the late blooming. Deficiency symptoms also include chlorosis of leaves (light green or yellowing of leaves) and retarded plant growth. Due to high mobility of N, these deficiency symptoms first appear in older leaves of the plant [3].

The gaseous form of N is termed as dinitrogen (N_2) which accounts for 78% of the total gaseous content of the atmosphere. This form of N is unavailable for plants until it is fixed and converted into ammonium and nitrates, the forms in which plants can uptake N [4]. Soils contain both organic and inorganic N; however, organic form constitutes a major part of total soil N content. Plants, on the other hand, can use only specific inorganic forms of N like nitrate and ammonium. Like phosphorus (P) and carbon (C), N undergoes biogeochemical conversion from gaseous state to mineralized form in soil followed by its return to the atmosphere in the gaseous phase. The net concentration of N_2 per year was estimated to be 3×10^9 tons on global basis [5]. Nitrogen cycle is considered to be a biogeochemical cycle, where the N changes into different chemical forms and shifts to different ecological spheres of the earth. The fundamental components of N cycle are decomposers and N-fixing bacteria. Nitrogen cycle initiates with microbial fixation of N in the soil, where mineralization of N takes place by conversion of atmospheric or organic N into ammonium, a process known as ammonification. Further, ammonium is converted into nitrate by soil microbes and nitrifying bacteria, e.g., *Nitrobacter* and *Nitrosomonas* species. Denitrification is the ultimate step carried out by the denitrifying bacteria such as *Pseudomonas* and *Clostridium*, which decompose nitrate and convert it into N_2, thus returning N_2 back to the atmosphere.

2. Nitrogen fixation

The fixation of N involves conversion of N_2 into various nitrogenous compounds such as ammonium and nitrate, so that they may become more reactive and plant available.

2.1 Industrial N fixation

Industrial N fixation involves the Haber-Bosch process which is an energy-inefficient method for making nitrogen fertilizers:

$$N_2 + 3H_2 \xrightarrow{200°C,\quad 200\ atm} 2NH_3 \tag{1}$$

2.2 Natural N fixation

N fixation can be biological and nonbiological in natural environment.

2.2.1 Nonbiological N fixation (lightning)

In nonbiological fixation, a relatively small amount of N is fixed by a spontaneous reaction that occurs during lightning. It is estimated that about 10% of the world's supply of fixed N comes from lightning [6]. Lightning can be described as occurrence of a sudden electrostatic discharge during a thunderstorm. During lightning, atmospheric nitrogen reacts with oxygen to form nitric oxide (NO). In the presence of excessive O_2, nitric oxide oxidizes to nitrogen dioxide (NO_2). In the presence of water, NO_2 may react to form nitrous (HNO_2) and nitric acid (HNO_3) or may react with rainwater and oxygen to produce nitric acid. These acids find their way to reach the soil with rainwater, interaction with alkaline substrates occurs, and hydrogen is released forming nitrate (NO_{3-}) and nitrite ions (NO_{2-}). The nitrate ions can be readily consumed by microbes and plants. However, soil microbes are not directly involved in this kind of N fixation. The chemical reactions involved in such N fixation are presented below:

$$N_2 + O_2 \xrightarrow{lighting} 2NO(Nitric\ oxide) \quad (2)$$

$$2\,NO + O_2 \xrightarrow{oxidation} 2\,NO_2(Nitrogen\ dioxide) \quad (3)$$

$$2\,NO_2 + H_2O \rightarrow HNO_2(Nitrous\ acid) + HNO_3(Nitric\ acid) \quad (4)$$

OR

$$4\,NO_2 + 2H_2O + O_2 \rightarrow 4HNO_3(Nitric\ acid) \quad (5)$$

$$HNO_3 \rightarrow H^+ + NO_{3-}(nitrate\ ions) \quad (6)$$

$$HNO_2 \rightarrow H^+ + NO_{2-}(nitrite\ ions) \quad (7)$$

2.2.2 Biological N fixation

Biological fixation of N_2 is carried out by N-fixing bacteria in soil. This fixation accounts for approximately 60% of fixed N in soil. Fixation of N_2 by microbes is termed as biological N fixation (BNF). Soil microbes are diazotrophs (bacteria and archaea) that contain enzyme nitrogenase, capable of converting N_2 into ammonium and nitrates, a process termed as nitrification. Common diazotrophs are rhizobia, blue-green algae (cyanobacteria), *Azotobacter*, *Frankia*, and green sulfur bacteria. Diazotrophs usually have a symbiotic relationship with leguminous family of plants. The major legumes are flowering plants like soybean, peanuts, clover, and lupines, tea plants like rooibos, and grasses such as alfalfa. The roots of legumes contain small protrusions called as nodules. These nodules are anchored by diazotrophs, providing anaerobic conditions for diazotrophs, further necessary for nitrogen fixation. Plants in turn use this fixed N for different functions. Upon death of the plants, this fixed N is released to the soil and acts as a nitrogen source for soil and non-leguminous plants. Nitrogen fixation is an energy-intensive process. One molecule of nitrogen gas breaks into its atoms and combines with hydrogen to form 2 molecules of ammonia at the expense of 16 molecules of ATP and a complex set of enzymes. Its reduction reaction can be written as:

$$N_2 + 3H_2 \xrightarrow{Energy} 2NH_3 \quad (8)$$

2.3 Classification of biological nitrogen fixation (BNF)

BNF can be classified into nonsymbiotic (free-living) and symbiotic (in association).

2.3.1 Nonsymbiotic biological nitrogen fixation

Microorganisms that fix atmospheric nitrogen independently are known as free-living diazotrophs. This type of fixation is carried out by free-living microorganisms. Examples of free-living organisms, which fix N, are cyanobacteria (blue-green algae, e.g., *Anabaena*, *Calothrix*, *Gloeothece*, and *Nostoc*), aerobic (*Azotobacter*, *Azospirillum*, Beijerinckia, *Derxia*), facultative (*Bacillus*, *Klebsiella*), and anaerobic (non-photosynthetic such as *Clostridium* and *Methanococcus* and photosynthetic such as *Chromatium* and *Rhodospirillum*).

2.3.2 Symbiotic biological nitrogen fixation

Symbiotic nitrogen fixation, carried out by specialized soil bacteria as discussed above, is the good source of N for plants. In return, plants provide required nutrients and energy for bacterial growth. Upon the death of nitrogen-fixing bacteria, nitrogen is released to the environment, and some non-leguminous plants may benefit from that nitrogen. In leguminous plants, nitrogen-fixing bacteria colonize on plant roots forming nodules. Within these nodules, nitrogen fixation is carried out by the bacteria, and the end product, NH_3, produced is absorbed by the plant [7].

2.4 Legumes

Legumes belong to Fabaceae or Leguminosae family and are primarily grown for human consumption, as forage and silage for livestock, and act as a green manure for enhancing soil fertility. Some common legumes include alfalfa, soybeans, chick peas, pigeon peas, clovers, cow peas, kidney, lentils, mung beans, peanuts, peas, and vetches. These are native to tropical rain forests and dry forests in America and Africa [8]. Legumes consist of 750 genera and 19,000 species of herbs, shrubs, trees, and climbers.

Legume seeds (pulses or grain legumes) are the major part of human diet. Nutritionally, legume seeds are rich in protein contents as compared to cereal grains. The combined use of legumes and cereals may provide necessary dietary proteins. Legumes are also used as pasture and animal fodder in which soybeans are most commonly used. Legumes, as green manure, improve soil quality by adding nitrogen and organic matter. Legumes are used in crop rotation for the sustainable crop production. About 2500 species of Leguminosae produce root nodules.

2.5 Soybean

The soybean (*Glycine max* L.), commonly called soja bean or soya bean, is a legume species native to East Asia. It is enormously grown for edible seeds and oil extraction. The major countries involved in cultivation of soybean are the United States, Brazil, and Argentina. Soybean is the most economical source of vegetable protein around the world. It is also involved in the production of several chemical products. Many botanists believed that soybeans were first cultivated in central China earlier in 7000 BC and in the United States in 1804 [9]. Soybean appears to be an erect branching plant with length more than 2 m. It is a self-pollinated plant with adoption to various cultivable lands. This plant conveniently cultivate in fertile, well-drained, and sandy loam with relatively warm conditions. The vital source of N in legumes is nodulation prevailed by N-fixing bacteria. Soybean can fulfill 50–70%of its N demand from the air by establishing root nodules through adequate popu-lation of N-fixing bacteria.

2.6 Nodule formation

Nitrogen fixation in legumes starts with the formation of small, knob-like protuberances called nodules. The bacteria get all the necessary nutrients and energy from the plants. The roots of legumes release chemicals known as flavonoids to attract the bacteria [10]. In response to flavonoids, the soil bacteria produce nod factors. Nod factors are signaling molecules which are sensed by the roots. As a result, a series of biochemical modifications lead to cell division in the root to create the nodule. Lectins, a sugar-binding protein in root hairs of legumes, are activated by nod factors. This helps in the recognition and attachment of rhizobial cells to the root hairs whose tips in turn become curved. The growing root hair curls around the bacteria in several attempts until one or more bacteria are enclosed. The enclosed bacteria colonize and eventually enter the developing nodule through infection thread. Infection thread is a structure extended through the root hair into the epidermis cell and then comes out of the root cortex. The bacteria are then surrounded by plant-derived membrane. Rhizobial multiplication starts in cortical cells which results in the formation of nodule on the surface. In side nodules, the bacterial cells continue multiplication and colonization until host cells are completely filled. After that bacterial cell becomes dormant bacteroids and starts floating in leghemoglobin. Leghemoglobin is a reddish pigment in cytoplasm of host cells which efficiently scavenges O_2 so that maintenance of the steady state of oxygen and stimulation of ATP production is possible. Plants provide shelter and organic compounds to the rhizobia, and in turn rhizobia provide fixed nitrogen to the plant. Among leguminous crops, soybean takes great consideration due to higher contribution of BNF. Normally, nodulation occurs after 4 weeks of plantation. The small nodules become visible after 1 week of the infection. The color of nodule appears white or gray when nitrogen fixation is insufficient, whereas color changes to pink or reddish as N_2 fixation progresses. This color change is attributed to the occurrence of leghemoglobin which is similar to blood hemoglobin that regulates the flow of oxygen to the rhizobia.

Perennial legumes such as alfalfa, clover, etc. develop nodule about half an inch capable of fixing N throughout the growing season. Annual legumes like beans, soybeans, and peanuts have short-lived nodule, round in shape with size of pea. These nodules are continuously replaced during the growing season. Annual legumes provide nourishment to developing seed instead of nodules; therefore, nodules cannot fix N anymore. The number of nodules varies per plant species, e.g., on average beans comprised of <100 nodules per plant, soybean can have several 100 nodules per plant, and peanut may have >1000 nodules per plant. Nodules on annual legumes, such as beans, peanuts, and soybeans, are short-lived and round in shape and can reach the size of a large pea and will be replaced constantly during the growing season. At the time of pod fill, nodules on annual legumes generally lose their ability to fix nitrogen because the plant feeds the developing seed rather than the nodule. Beans have less than 100 nodules per plant, soybeans will have several hundred per plant, and peanuts may have 1000 or more nodules on a well-developed plant.

Nodulation is regulated by both external and internal processes. Soil temperature, soil N mineral content, acidity of soils, and water scarcity can be categorized as external factors, whereas autoregulation and ethylene are the most influential internal factors. Autoregulation of nodule (AON) specifies the number of nodules per plant. Leaf tissue via chemical signal can sense the onset of nodulation and inhibit it in the developing root. Such chemicals are leucine-rich repeat (LRR) receptor kinases that are crucial for autoregulation of nodule formation. The mechanism for nodule formation is coded by *enod40* gene also called nodulin 40. Its expression leads to relocalization of nuclear proteins.

Microbes inhabiting soil can be termed as plant growth-promoting rhizobia (PGPR) due to their multifuntionality in symbiotic relationship with plant. PGPRs play role in plant nutrition by mineralizing nutrients in rhizosphere. PGPRs as indicated by name actively participate in phosphate solubilization and production of siderophore, phytohormones, and several enzymes. The biochemical characteristics of PGPR, for instance, lipopolysaccharides (endotoxins), homoserine lactones (signaling molecules), acetoin (preventing over-acidification in cytoplasm), and flagella (locomotive and sensory organs) help plants to develop systematic resistance against pests and pathogens. The PGPRs enhance tolerance against extremity of environmental conditions such as drought, nutrient deficiency, and prevalence of organic (pesticides) and inorganic (heavy metals) toxicity. PGPR, therefore, are considered as biofertilizers for sustainable agricultural practices.

3. Inoculation and coinoculation of PGPRs

Soybean develops symbiotic relationship with a range of PGPRs to fix nitrogen (N) and improve plant growth [11–13]. Establishment of symbiotic relationship between roots of the host plant and symbiont is a two-step process. In first step, host tissue is infected with rhizobacteria and in second nodule formation occurs. Plant roots contribute in the symbiotic relationship by releasing flavonoids, while rhizobacterium produces nodulation factors. Rhizobacterium is entrapped in plant hairs' curls, and infection threads are formed at the root hair curls, permitting bacterial invasion of the root tissue. The process of nodulation is initiated just below the infected point. Rhizobacterium may be restricted to infection threads, but mostly, they are released into nodule cells where nitrogen fixation occurs.

3.1 Inoculation

Inoculation and coinoculation of PGPRs have become a popular research area in recent crop production. The interest in rhizosphere microbiology was developed due to the beneficial effects of some free-living strains of bacteria on plant growth and disease control and maintaining good soil health. Initial studies were focused on bacterial genera including *Pseudomonas*, *Rhizobium* spp. *Azotobacter*, *Bacillus*, and *Azospirillum* to enhance plant growth by fixing atmospheric nitrogen [14, 15]. However, later research was shifted to elucidate the role of PGPRs in promoting plant growth by mineralizing organic phosphorous, solubilizing inorganic soil phosphorous, modulating plant hormones, and rendering plant tolerance to adverse environmental conditions [11, 16]. This has triggered diversified application of PGPRs' inoculation and coinoculation in various field crops. The term "inoculation" may refer to "natural or deliberate application of certain beneficial strains of bacteria to plant seeds or soil to enhance plant growth." Inocula, the strain of bacteria used in inoculation and coinoculation, may be native or alien, with inherent or engineered ability to colonize plant roots and promote plant growth. Plant growth-promoting genera may include different strains of *Bacillus*, *Pseudomonas*, *Agrobacterium*, *Rhizobium*, *Mycobacterium*, etc. Bacterial inoculation has increased yield in many crops using *Azotobacter* and *Bacillus* strains. Steadily, the research focus has been shifted to *Azospirillum* from *Azotobacter* due to better crop yields reported with the later. Similarly, *Bacillus subtilis* and *Pseudomonas* spp. have been proven to be effective in controlling plant pathogens of soil origins. One major positive effect of inoculation is the solubilization of inorganic phosphate in the soil to make it plant available. Root exudates greatly influence the colonization of rhizobacteria. However, one major challenge in successful inoculation is the colonization of PGPRs in the rhizosphere where indigenous microbes may limit survival of the introduced bacteria. This has been addressed through introduction of antibiotic-resistant *rhizobacteria*. Besides these, soils are complex heterogeneous environments with great variations in particle size distribution, pH, organic matter content, temperature, water, and availability of nutrients that may greatly influence inoculation success.

3.2 Coinoculation

To overcome some limitations of inoculation and increase PGPRs' efficiency, coinoculation is now commonplace in experimental and field trials. The objective is to increase the consistency and frequency of nodulation rate in various plant species. By definition, coinoculation is the combined application of PGPRs and other bacteria, bestowed with some specialized functions, to increase the nodulation rate, plant growth, and plant tolerance to adverse environmental conditions. For example, coinoculation of PGPRs with nitrogen-fixing bacteria has caused earlier nodulation and greater intensity, better uptake of nutrients and water, and improved plant growth [11]. In another study, coinoculation of soybean plants with strains of *Pseudomonas* and *Bacillus*, in combination with *Sinorhizobium meliloti*, has improved plant phosphorous uptake [16]. Moreover, *Azospirillum* has been used to increase the rhizobia-legume symbiotic relationship in soybean to improve its nutritive value [12]. For coinoculation, in vitro strain selection or genetically engineered strains are commonly used [17]. It is generally assumed that one rhizobacterium may be less effective in diverse environmental conditions. Thus, mixtures of various rhizobacterial species are promising in enhancing plant growth. But the coexistence of different bacterial strains under normal and adverse field conditions may be a challenge. Nowadays, coinoculation of PGPRs with mycorrhizal fungi is being practiced to promote growth in various plant species [13, 18–19]. Moreover, some studies have been focused on combining free-living bacteria, PGPRs, and mycorrhizal fungi [20].

3.3 Efficiency of coinoculation for enhancing nodulation rate

Inoculation and coinoculation of plants with single or multiple PGPRs may bring changes in the number of root hairs, nodule formation, root exudation, and release of phytohormones in addition to several physiological and metabolic changes. Generally, the potential of a specific PGPR strain to enhance nodulation rate can be best judged in a single experiment; however, consistent performance needs multiple field trials. The initial study on the role of PGPRs in enhancing nodulation rate was conducted on *Rhizobium trifolii*. The efficiency of coinoculation may also be dependent on the hormones and enzymes produced by PGPRs. For example, *Azospirillum* produces indoleacetic acid and pectinase which affect the development of symbi-otic relationship and ultimately the nodulation efficiency [21]. Corporate research is focused on developing commercial inocula; however, several challenges need to be overcome before the product can make sense to the users. These include but are not limited to explaining exact mode of action of PGPRs under individual circumstances, persistency of performance over different ecological environments, and the optimization of the fermentation systems.

4. Phytohormones released by PGPRs

Rhizosphere is the soil adjacent to the growing roots of a plant. A strong interaction exists between the roots and soil. The microbial activity in the rhizosphere makes the interaction even stronger. The interaction between the plants and microbes can be symbiotic, nonsymbiotic, neutral, and parasitic. There are a number of microbes that are found in the rhizosphere; these include bacteria, fungi, actinomycetes, protozoa, and algae. Among these the most common is the bacterial population. Plant growth-promoting rhizobacteria are the bacterial biomass that colonizes the plant roots in the rhizosphere [22]. PGPRs have been reported to play many important functions in plants; these include nitrogen fixation and uptake, tolerance under stress conditions, and production of certain phytohormones, i.e., plant growth regulators, siderophores (iron-binding protein compounds), volatile substances, and also certain enzymes, i.e., glucanase and chitinase to protect plants against disease [23, 24].

Phytohormones are produced in low concentration but have greater influence on the biochemical, physiological, and morphological functions of plants. They function as chemical messengers to transfer cellular activities in higher plants [25]. During the abiotic stress condition, these phytohormones play vital roles through communicating different transducing signals, which may control the external and internal stimuli [26]. Also some of the phytohormones are identified as stress hormones like abscisic acid (ABA). These phytohormones have a significant role in various plant processes. ABA besides facilitation during biotic and abiotic stress also is critical for maintaining seed dormancy, growth regulation, inhibiting germination, controlling the stomatal closure, and fruit abscission [27]. The plant growth regulators produced include auxin, gibberellic acid, cytokinins, and ethylene. Ahmad and Hasnain (2010) [28] have reported that *Bacillus* spp. producing auxin showed positive effect on the growth of potatoes. Earlier research work has revealed that PGPRs inoculation improved plant tolerance to stress condition due to enhanced production of growth regulators [29, 30].

4.1 Effect of synthetic PGRs on release of phytohormones

Plant growth regulators (PGRs) are synthetically available and are used in commercial agriculture extensively. Through various investigations, it has been found that application of growth regulators at pre-sowing stage to the seeds may enhance the nutrient reserves, tissue hydration, growth, and yield of crops [31]. Khan et al. (2018) [32] found synergistic effects of PGPRs and PGRs on different qualitative parameters of crops, i.e., chlorophyll, sugar, and protein contents. They concluded that application of PGRs to the plants inoculated with PGPRs helped plants under stress conditions. Also the amount of PGRs applied exogenously to the plants may be stored as reversible conjugates, and they also release phytohormones as required by the plants at different growth stages. Also these PGRs are found effective in transferring accumulates from source to the sink [33, 34].

Also some of the researchers have reported that the release of phytohormones may be enhanced several times by the applications of some suitable precursor of the plant hormones. These precursors are utilized by the rhizobacteria and converted into active phytohormones, and they are continuously used by the plants [35]. Among these precursors, L-methionine is an important precursor of ethylene (C_2H_4), a gaseous plant hormone that positively affects at almost all stages of growth and developmental processes [36]. Application of L-methionine to the rhizosphere enhanced the ethylene production and has shown significant increase in the growth and yield traits of soybean [36].

4.2 Effect of coinoculation on release of phytohormones

The bacterial population in the rhizosphere sometimes modifies the formation of nodules when they are coinoculated. The mechanism behind this process is that the coinoculation may directly enhance the growth and development of plant by the increase in microbial biomass, extending the root system by release of phytohormones, solubilization of phosphate in the rhizosphere, etc. Moreover, development of roots provides additional sites for nodule formation [37, 38]. Indole-acetic acid (IAA) is an important metabolite of auxin group produced by the *Azospirillum brasilense* bacteria in the presence of tryptophan. Also the *A. brasilense* may produce the IAA in the absence of tryptophan under aerobic condition in the presence of NH_4 [39, 40].

Some PGPRs produce allelochemicals which are phytotoxic in nature. Production of these allelochemicals may adversely affect the soil health [41], by having negative effect on the enzymatic activity and plant functions, and may also hamper the nutrient availability to plants. The number of allelochemicals has been isolated from the bacterial strain present in the rhizosphere. It has been reported that a single strain of bacteria may produce a wide range of allelochemicals, e.g., *Streptomyces hygroscopicus* may produce nigericin and geldanamycin; these may be isolated and utilized as herbicides [42].

5. Influence of coinoculation on bioavailability of micronutrients

An exponential increase in the world's population will demand a higher production of food crops. By 2050, it is projected that the world's food demand will reach up to 3 billion tons. This high demand for food has been resulted in the excessive use of chemical fertilizer (nitrogen, phosphorus, potassium) in combination with advancements in technology to enhance the plant growth and production. Nitrogen is a vital nutrient in plant growth and productivity. Unfortunately, when a recommended dose of fertilizers is applied to crops for an average yield, less than 50% of applied nitrogen fertilizer is consumed by plants [43]. This low use efficiency of N causes the high fertilizer consumption and nitrate contamination of groundwater and soil which finally resulted in environmental degradation and health problems. Inoculation with microbes has been considered as an environmentally friendly alternative to minimize the use of synthetic nitrogen fertilizer without compromising the crop growth and yield [44, 45]. By biological nitrogen fixation, atmospheric nitrogen is converted to plant-utilizable forms, which is performed by microorganisms which convert the nitrogen to ammonia [46]. These microorganism generally is categorized into two groups: (i) nitrogen-fixing bacteria which generally includes the Rhizobiaceae family members and forms symbiotic associations with legume plants [47] and other non-leguminous plants and (ii) nonsymbiotic nitrogen-fixing bacteria (free-living, associative, and endophytic) such as *Cyanobacteria*, *Azotobacter*, *Gluconacetobacter diazotrophicus*, *Azocarus*, *Azospirillum*, etc. [48]. Rhizobia (including *Sinorhizobium*, *Bradyrhizobium*, *Rhizobium*, *Mesorhizobium*) are considered as symbiotic partners of legume plants and known by their role in the formation of N-fixing nodules in plant rhizosphere [49], while the nonsymbiotic N-fixing bacteria deliver only a small amount of fixed N which is required by the associated plant [50]. N-fixing PGPB strains and their effects on leguminous plants have been tabulated in **Table 1**.

Plant growth-promoting bacteria (PGPB) comprise a group of microorganisms that colonize the internal plant tissue and root surface and provide many benefits to host plants [51, 52]. These microorganisms can improve plant growth by contributing several mechanisms and processes including synthesis of hormones such as cytokinins, auxins [53], ethylene [54], gibberellins [55], and a variety of other molecules [56], biological control of pathogens [57, 58], and solubilization of phosphate [59]. Combinations of these mechanisms finally benefit the plant by

Bacterial strains	Plant	Effect	References
Azotobacter chroococcum	*Avena sativa* L.	Improved growth and yield	Devi et al. [70]
Azospirillum brasilense	Maize	Improved N use efficiency and improved yield	Morais et al. [71]
Bradyrhizobium spp. + *Azospirillum brasilense*	Soybean	Promoted growth and yield with N application	Hungria et al. [79]
Rhizobium	Chickpea	Promoted growth in combination with N application	Namvar et al. [80]
Bradyrhizobium, Azospirillum	Soybean	Significantly improved nodule biomass	Chibeba et al. [81]
Rhizobium sp. *BARIRGm901*	Soybean	Increased nodule weight and crop yield, improved the activity of nitrogenase enzyme and nitrogen assimilation	Alam et al. [82]
Diazotrophic bacteria	Rice	Increased grain yield	Araujo et al. [83]
Ochrobactrum ciceri Ca-34, Mesorhizobium ciceri TAL-1148	Chickpea	Improved nodule biomass and crop yield	Imran et al. [84]

Table 1.
N-fixing PGPB strains and their respective effect on leguminous plants.

improving growth [60, 61] and biological nitrogen fixation and increase the activity of nitrate reductase when growing as plant endophytes [62]. These bacteria also produce the siderophores and synthesize enzymes, antibiotics, or fungicidal compounds that protect the plants against phytopathogenic microorganisms [63, 64]. There are several factors such as agricultural practices, plant genotype, bacteria species, and strain that may affect the success of inoculation and plant response to these PGPB [65, 66]. Chickpea and *Rhizobium leguminosrum* subsp. *Cicero* associations, for instance, produce up to 176 kg/ha annually depending on environmental factors, cultivars, and bacterial strain [67]. *Azorhizobium caulinodans* is root- and stem-nodulating nitrogen-fixing bacterium which has been isolated from the stem nodules of *Sesbania rostrata (Bremek and Oberm.)* [68]. By endophytic colonization of non-legume roots, i.e., wheat, it can stimulate root growth and increase nitrogen content and yield [69]. Devi et al. [70] reported that the growth and yield of oats (*Avena sativa* L.) were increased due to the seed inoculation with *Azotobacter chroococcum* combined with the nitrogen fertilizer as compared to control and nitrogen fertilizer alone. Highest yield (239.02 quintal per hectare) was observed in *Azotobacter* seed inoculated +80 kg N as compared to control (111 q/ha) and nitrogen 80 kg/ha (205 q/ha) alone. In another study, Morais et al. (2016) [71] reported the effects of *Azospirillum brasilense* (inoculated in seed furrow) on maize growth and yield. Average maize grain productivity was observed to be 12.76 and 13.06 ton/ha when nitrogen is applied at the rate of 100 and 200 kg/ha, respectively. However, with the addition of seed furrow inoculation at the rate of 200 ml/ha, average grain productivity of maize was increased up to 13.21 and 14.0 ton/ha under the nitrogen application at the rate of 100 and 200 kg/ha, respectively. This PGPR improves the growth and yield by increasing the N and P content in plant, higher phosphate solubilization, ammonia, indoleacetic acid (IAA), and siderophore production [72]. Inoculation of seeds with *Rhizobium* increases the protein, chlorophyll content, nitrogen uptake, and growth parameter in legume crops [73, 74].

Now scientists have developed new microbial associations to avoid such negative interrelations and increase the effectiveness of biofertilizers. Consortia of PGPR with mycorrhizal algae [75] or fungi [51] can show a better performance as a result of cumulative or synergistic interactions between beneficial mechanisms of different microorganisms. Mycorrhiza is a symbiotic interaction between plants and soil fungi called as arbuscular mycorrhizal fungi (AMF). Both associates get benefits for this relationship by improving nutritional status, which reduces the needs of fertilizers for crops [76, 77]. Vesicular-arbuscular mycorrhizal fungi improved the availability of nitrogen and phosphorus to support the plant to survive in different environmental severe conditions [78].

6. Quantification of nodulation process by dynamic linear models

The symbiotic relationship between N_2-fixing bacteria and leguminous plants is a core factor in enhancing soybean crop yield around the world. The atmospheric nitrogen captured by these bacteria is enzymatically reduced to ammonia. This ammonia is assimilated by plant tissues in the form of nitrogenous compounds. Around 20–22 million tons of N is fixed by symbiotic rhizobia [85], while 17 million tons is removed or assimilated by aerial biomass of legumes [86]. The fixed N can serve as an inevitable resource of N depending on net N fixation in soil as compared to its removal or assimilation in aerial parts of legumes which is estimated to be 45–75% [87]. Nonetheless, the cropping systems with legumes have high crop yield as compared to non-legumes [88]. The fixation of N_2 can be maximized by sustainable and organic farming practices. However, legume specie, soil type and climatic conditions can also impact fixation rate of N_2 [89].The production of soybean as cash crop is evident in Brazil, Argentina, Russia, Ukraine, and the United States [90]. In Asia, North China and Japan chiefly cultivate soybean along with wheat [91].

Quantification of leguminous biological nitrogen fixation (BNF) can be beneficial for sustaining N demand and supply which can increase productivity and ability to combat environmental stresses. The techniques available for quantifying legume BNF are costly and protracted. Moreover, the data provided by such techniques are pertinent to limited time and space. Simulation of legume BNF is attainable by empirical and dynamic modeling. Empirical modeling is based on observation and experiment, while dynamic modeling is capable of representing a pattern or behavior over a time period. In case of legume BNF simulation, dynamic modeling can be desirable as it can correlate various environmental factors and legume growth status with N fixation. Broadly, legume BNF is discussed in relation to demand, uptake, and assimilation of N in biomass of root, nodule, and aerial parts of leguminous plants. Moreover, concentration of N accumulated in soil, along with soil's environmental parameters such as water content, N mineral concentration, internal substrate, C substrate and supply, and temperature are essential to quantify N fixation. Last but not the least growth rate of leguminous plant is a dynamic indicator in estimation of fixed N [92–95].

6.1 Estimation of N fixation by considering economic yield or aboveground biomass

During growing period, N fixation can be estimated by considering economic yield or dry matter of aerial biomass [96–98]. For this purpose, the equation can be:

$$N_{fix} = \alpha.DM.f_{leg}.N_{con}.\%Ndfa.(1 + R_{root}) \tag{9}$$

where DM represents dry matter of aerial biomass or yield, f_{leg} is proportion of legume crop in intercropping system, N_{con} is concentration of N assimilated in legume plant, and %Ndfa indicates proportion of N in crop which is derived from fixation of N_2, whereas R_{root} is a ratio of N fixed in belowground parts to the N fixed in aerial parts of legumes. α is a parameter which can have different definition depending on the researcher. For example, α can be used to represent correlation between decline in %Ndfa and high soil N content. In order to estimate total N input, α can be calculated as:

$$\alpha = 1 - \beta.N_{net.inorg} \quad (10)$$

where ß evaluates the responsiveness of legume for N fixation to already present mineral N (nitrate and ammonia) in the soil [98]. This method can directly estimate N fixation. Its parameter values can be taken both as estimated values from literature or measured values from on-site analysis. This method can work in the absence of previous data from past years. In these equations, environmental and weather conditions are not considered; therefore, this method can only be suitable for soils with similar properties and with exposure to moderate weather conditions. Moreover, the parameter values can be accustomed according to soil condition.

6.2 Linear empirical model

The empirical model can be used to explicit correlation between amount of N fixed in legumes and the total harvested part of legumes. In the case of intercropping system, fixed N in legumes can be correlated to the present fraction of legumes in the field. The equation is devised to calculate N fixation in kg N ha^{-1}, such as:

$$N_{fix} = c + d.Leg \quad (11)$$

where Leg denotes excess in harvested biomass (kg ha^{-1}) while c and d comprise the selected parameters.

The empirical model is based on statistical correlation with speculation of strong linear relationship between N fixation and variables. The applicability of this model is on wide variety of soils. This model requires adequate amount of data to constitute a correlation study and to determine the values for the selected parameter. The linear empirical model, however, does not account environmental conditions
[99, 100].

6.3 Crop models as example of dynamic models

Leguminous N fixation in soybean was first simulated by Duffy et al. (1975) [101]. He estimated rate of N fixation by measuring root growth rate after specific days of planting. Crop models being dynamic in nature involve the potential impacts of soil environmental conditions for estimating N fixation. However, soil salinity, pH and availability of other nutrients are exempted in such models. Examples of crop models are Sinclair [102, 103], EPIC [104–106], Hurley Pasture model [107–110], Schwinning model [111, 112], CROPGRO [113–115, 93, 116], SOILN[117], APSIM [95, 118], Sousanna model [94] and STICS [119–121]. These crop models are applicable in varying environmental conditions; therefore, each model can have different versions for calculating N fixation. Thus, Liu et al. (2011) [122] devised a general equation for these crop models:

$$N_{fix} = N_{fixpot} f_T f_W f_N f_C f_{gro} \tag{12}$$

where N_{fixpot} indicates the potential rate of N to be fixed by legumes (g N fixed day^{-1}), f represents the influence function of environmental conditions, f_T is impact of soil temperature, f_W can be taken as impact of water deficiency or flooding in soil, f_N can estimate impact of availability of mineral N (nitrate and ammonia) in soil or N availability in root substrate, f_C represents effect of C concentration in root and aerial parts of legume plant, and f_{gro} is the effect of plant's growth stage on potential rate of N fixation. In the case of Environmental Policy Integrated Climate (EPIC) model and Simulateur mulTIdisciplinaire pour les Cultures Standard (STICS), the equation is generalized as:

$$N_{fix} = N_{fixpoint} f_T \ min \ (f_W, f_N) \ f_{gro} \tag{13}$$

where min indicates the minimum value that can be assumed between f_W and f_N. If applying STICS model, the limitation by anoxia is represented by extra function, i.e., f_a.

6.3.1 *Potential N fixation*

In dynamic models, the potential rate of N fixation is estimated on the basis of demand or uptake of N by legume plant or on the ability of root nodules to fix atmospheric N_2. In EPIC, the potential rate of N fixation is equal to the demand of N by legume plant [107]. The higher the demand of the N in legume plant, the higher will be the potential of N fixation. In contrast, according to Agricultural Production Systems siMulator (APSIM), the internal concentration of N in plant tissues governs the N demand of legume plant, which in turn defines the potential rate of N fixation in legumes. However, APSIM is applicable when plant has sufficient N concentration which can fulfill N demand of new tissues by uptake N from the soil [122]. N uptake is relatively passive and much preferable than N fixation; therefore, N fixation is only estimated when plant's demand for N is not fulfilled by N uptake [123]. Potential rate of N fixation, therefore, can be defined as difference between N demand and uptake [95, 118]. On the other hand, some researchers claim that the potential of N fixation is dependent on size and biomass of root and nodules, i.e., above- and underground biomass [124, 125]. However, estimation of N fixation using aboveground biomass is more convenient to handle than underground biomass [126].

6.3.2 *Soil temperature*

Soybean being plant of tropical and sub-tropical regions requires warm conditions for growing. The favorable temperature for soybean root zone ranges from 25–30°C [127]. Crop models such as Hurley Pasture model, CROPGRO, SOILN, and STICS estimated the effect of soil temperature on rate of N fixation by specifying certain temperature range. The generalized forms of equations are:

$$f_T = \begin{cases} 0 & (T < T_{min} \ or \ T > T_{max}) \\ \dfrac{T - T_{min}}{T_{optL} - T_{min}} & (T_{min} \le T \le T_{optL}) \\ 1 & (T_{optL} \le T \le T_{optH}) \\ \dfrac{T_{max} - T}{T_{max} - T_{optH}} & (T_{optH} < T < T_{max}) \end{cases} \tag{14}$$

where T represents soil temperature in °C, T_{min} is the minimum temperature below which N fixation can stop, T_{max} is the maximum temperature above which N fixation can stop to occur, and T_{optL} and T_{optH} indicate low and high values of optimal temperature range. In optimal condition, the optimum response to soil temperature becomes equal to the unit. Depending on location and legume species, the temperature range can vary in different models [109].

6.3.3 Soil water content

The excessive and deficient amount of soil water in the *Rhizobium* can negatively impact N fixation by the nodule. In STICS model, water deficit point is defined as segment of soil layers with water content above permanent wilting point [119]. Sinclair model, on the other hand, correlated transpirable water with nitrogenase activity of nodules [103, 104]. The nitrogenase activity is assessed by the reduction of acetylene which is used to explicit the proposed mechanism of BNF. The reduction in transpiration rate < 10% determines the transpirable water in soil, which in turn is stipulated by comparing the field capacity of soil and soil water content [102]:

$$f = -1 + \frac{2}{1 + e^{\left(-m * f_{TSW} + n\right)}} \tag{15}$$

where f_{TSW} represents the fraction of transpirable water in soil, whereas m and n are constants defining responsiveness of legumes for N fixation in low soil water content. APSIM, EPIC, and SOILN formulated linear function, which is expressed as:

$$f_{\omega} = \begin{cases} 0 & \left(W_f \leq W_a\right) \\ \varphi_1 + \varphi_2 . W_f & \left(W_a < W_f < W_b\right) \\ 1 & \left(W_f \geq W_b\right) \end{cases} \tag{16}$$

where W_f is the ratio of relative availability of water content in soil at a given field capacity, W_a is the minimum value of water content below which N fixation cannot occur, ϕ_1 and ϕ_2 are the coefficients, and W_b is the threshold value of W_f above which N fixation is not impeded by water content of soil.

However, researchers with special focus on water stress conditions revealed that the top layer of soil around 30 cm is susceptible to dryness or wetness during dry spell or irrigation period. This can influence the access of water to root nodules [128]. Therefore, the presence of water within the roots is a more reliable factor in quantifying N fixation in limited water supply. Contrarily, in Hurley Pasture model, the chemical activity in the roots is assumed to control N fixation, wherein the chemical activity indirectly relies on probable water content in the root and tem-perature of soil [107]. So the effect of water is correlated with the thermal condition of soil such as:

$$f_W = e^{20 * \left[\frac{18 * \varphi rt}{8314 * (T_s + 273.15)}\right]} \tag{17}$$

where Φ_{rt}, probable water content in the root (J Kg^{-1}) and T_s is termed as thermal value of water content in soil (°C).

Excessive water can cause anoxic conditions in soil. In such condition, N fixation is assumed to be at zero in Sinclair model [103]. In anaerobic conditions, pore spaces become occupied with water; therefore, N fixation cannot occur.

6.3.4 Mineral N/internal substrate

The availability of N in the form of nitrates and ammonia is said to be mineral N in soil. In SOILN model, mineral N is incorporated for estimating N fixation in nodules such as:

$$f_N = \begin{cases} 1 - 0.0784 In\ N_s & (N_s \geq 1) \\ 1 & (N_s < 1) \end{cases} \tag{18}$$

where Ns is mineral N content of soil (mg N m^{-3}). The N uptake can be influenced by mineral N in soil; therefore, Schwinning model estimates potential of N fixation as:

$$f_N = \varepsilon \times \left(1.0 - f_{N_{up}}\right) = \varepsilon \times \left(1.0 - f_{max} \frac{1}{1 + K_N/N_s}\right) \tag{19}$$

where ε is the efficiency of legume BNF, f_{max} is the maximum amount of N derived from the uptake of mineral N from soil, K_N indicates the concentration of nitrate in soil (g N m^{-2}) with N uptake reaching at half of its maximum rate, and Ns is the actual concentration of nitrate in soil (g N m^{-2}). In the given soil conditions, if nitrate concentration (N_{sNitra}) lies between 10 and 30 g (Nm^{-3}) within 30 cm topsoil layer, the EPIC model can be represented as:

$$f_N = \begin{cases} 1 & (N_{sNitra} \leq 10) \\ 1.5 - 0.05 N_{sNitra} & (10 < N_{sNitra} < 30) \end{cases} \tag{20}$$

In STICS model, high nitrate concentration in soil is assumed to inhibit nodulation progress which ultimately reduces potential rate of N fixation. If the concentration of nitrate in soil is higher than critical value, N_{fixpot} is set at baseline value; otherwise, N_{fixpot} is set at normal value [119]. In Hurley Pasture and Soussanna models, the plant substrate N concentration is included, such as:

$$f_N = \frac{1}{1 + N_{inter}/K_r} \tag{21}$$

where N_{inter} (g N g^{-1}r.wt) is assumed to be the N concentration in the root substrate (in Hurley Pasture model), or N concentration in plant substrate (in Soussanna model), and Kr is the coefficient for stating inhibition of N fixation at high nitrate concentration level in soil.

6.3.5 C in plant substrate or C supply

In plants, C is the source of energy for N fixation. Carbohydrate supports nodule biomass accumulation. The effect of C in estimating potential rate of N fixation is incorporated in Hurley Pasture and CROPGRO models such as:

$$f_C = \frac{1}{1 + K_C/C_r} \tag{22}$$

where Cr indicates concentration of C and Kc stands for Michaelis–Menten constant.

6.3.6 *Plant growth stage*

The impact of seasonal change on N fixation is incorporated in EPIC and STICS [106] such as:

$$f_{gro} = \begin{cases} 0 & (g < g_{min} \; or \; g > g_{max}) \\ \dfrac{g - g_{min}}{g_{optL} - g_{min}} & (g_{min} \leq g \leq g_{optL}) \\ 1 & (g_{optL} \leq g \leq g_{optH}) \\ \dfrac{g_{max} - g}{g_{max} - g_{optH}} & (g_{optH} < g < g_{max}) \end{cases} \tag{23}$$

where G_{min} is indicating the time period before which N fixation does not occur. This happens because of insufficient nodulation (expressed as % of total time period required for growing); g_{optL} is the initial time of growth and g_{optH} is the final time of growth. The time period between g_{optL} and g_{optH} represents N fixation by legumes, which is independent of growth stage. g_{max} is the growth time where N fixation stops due to deterioration of nodule.

The influence of symbiosis on metabolic fluxes and plant growth is quantified by a flux balance analysis. A genome-scale compartmentalized model for the clover (*Medicago truncatula*) as model plant has been devised by Pfau et al. (2018) [129]. The model predicted that nitrate uptake is significantly inhibited by the presence of ammonium in soil. When both nitrate and ammonium are available in soil, the uptake of ammonium is much favorable due to its integration into amino acids with fewer reductants and energy than nitrates.

The simulation of BNF by the abovementioned models included various biotic and abiotic factors to simulate and predict N fixation. Nodule biomass is more reliable to estimate N_{fixpot} than root and aerial biomass. C supply is considered to be the prominent factor in estimating N_{fixpot}. High concentration of nitrate in soil as mineral N can act as inhibitor for N fixation by nodules. Although empirical and dynamic models incorporated several factors such as soil temperature, water content, C, and other mineral contents, all the models lack information regarding the influence of soil pH and O_2 permeability. Therefore, adequate experimental work is required to cumulate the effect of such factors on biological fixation of N in legumes.

7. Technical and economic aspects

The impact of inoculation and coinoculation with elite strains such as *Azospirillum* species *(A. brasilense)* and *Bradyrhizobium* species (*B. japonicum*, *B. elkanii, and B. diazoefficiens*) has been extensively studied [130, 131]. Inoculation of *Azospirillum* spp. directly influences grain yield by improving N availability and its uptake. Moreover, this strain is helpful in the synthesis of phytohormones and developing pest resistance [132]. Crop yield is considered to be a primal factor for estimating profitability; therefore, increments in revenue are based on increments in grain yield [133, 134]. Coinoculation, regardless of cultivar, is reported to increase profitability by 14.4% as compared to non-inoculated treatments [134]. The economic evaluation of soybean plant is based on variables such as number of pods per plant, 100 grain weight, and yield. The data is usually quantified in kg ha^{-1} at wet basis [134].

The production of soybean crop can be estimated by total operating cost (TOC) method [135]. TOC is the sum of cost of fertilizers, heavy machinery, labor, pesticides, interests, etc. The major expenses are contributed by mechanization and fertilizers besides the cost of desiccation, control of weeds, pests, and pathogens. The inoculation of *Azospirillum brasilense* has increased the TOC, whereas the lowest TOC was reported with inoculation of *Bradyrhizobium* strain. However, the highest soybean yield was obtained with coinoculation of *A. brasilense*, leading to higher financial returns. Inoculation with *Bacillus* and *Pseudomonas* led to significant improvement in protein and nitrogen contents in grains in addition to high yield [136]. Similar results were reported when *Rhizobium* and *Pseudomonas fluorescens* improved yield and protein content when inoculated in beans [137].

In some studies, foliar inoculation of PGPR is found to be more effective than inoculation or coinoculation. For instance, foliar inoculation of *Azospirillum* in later stages of plant growth is correlated to high N content in developing grains [12]. This is because of the release of IAA by *Azospirillum*, which instigated nodulation in secondary nodules, thereby facilitating N fixation and its uptake in growing soybean plants. Likewise, foliar inoculation of *Azospirillum brasilense* at advanced growing stage of soy plant proved to be much more effective than its inoculation and coinoculation with *Bacillus japonicum* at sowing stage [12, 138]. However, foliar application at sowing stage is unable to produce any noticeable improvement in grain yield [12]. Moreover, coinoculation of *A. brasilense* and *B. japonicum* is reported to increase leghemoglobin by 39%, leading to high proportion of active nodules which in turn increased N fixation [139].

Organic and inorganic fertilizers such as NPK fertilizer and farmyard manure used along with PGPR, i.e., *Azotobacter* and *Trichoderma*, are reported to produce the highest biomass yield [140]. However, the inoculation of *Bradyrhizobium japonicum* on seeds increased grain yield of soybean by 8.4% (222 kg/ha), while its coinoculation with *A. brasilense* in furrow yielded 16.1% (427 kg/ha) without applying any external N source [141]. Similarly, Hungria et al. (2015) [79] coinoculated seeds of soybean with *Azospirillum* and *Bradyrhizobium* which resulted in high crop yield (388 kg/ha) without using any N fertilizer. The onset of earlier nodulation in soybean crop has been observed by the coinoculation of *Bradyrhizobium* and *Azospirillum* [81]. Moreover, these researchers claimed that the presence of *Azospirillum* after 18 days after emergence (DAE) facilitated plants to environmental stresses. Phosphorus as an essential nutrient for root growth is also necessary for rhizobia to convert N_2 into mineralized N [142]. Depending on the genotype of soybean, other nutrients like P can be influential in nodulation [143]. These researchers carried out coinoculation of rhizobia with arbuscular mycorrhizal fungi (AMF) in deep and shallow root genotypes of soybean. Regardless of soil N content, P was found to be a limiting factor in increasing nodulation, with low P colonization of AMF increased, whereas with high P, nodulation progressed in deep root soybean. Microbial inoculants are quite economical, making inoculation as a sustainable approach in soybean production [144]. Hence, the introduction of PGPR at appropriate stage of plant cycle can be a beneficial and reliable procedure for low-cost investment and sustainable agriculture.

8. Future prospects and conclusions

The reliance on N fixation is inevitable in spite of application of inorganic N fertilizer in huge amount (18 million tons/year) [86]. Legume plants being highly nodulated have high potential for N fixation which can be further facilitated by sustainable agricultural practices for high crop production. Inoculation and coinoculation with different strains not only positively impact crop yield but also improve nutrient value of grains. PGPRs are natural source of plant growth hormones especially IAA, prompting nodule growth whether applied at sowing or in later stages of plant growth. In some cases, foliar inoculation was more effective for nitrogen and protein assimilation in soybeans than inoculation and coinoculation at sowing phase. Among PGPRs, certain strains of *Azospirillum* have a great potential of replacing inorganic sources of N, making inoculation a more economical approach toward sustainable agriculture.

The viability of PGPR inoculants is susceptible to rhizospheric conditions of soil, for which the compatibility studies are a compulsion [145]. When applied in the field, certain bacterial species (endophytes and rhizosphere-restricted bacteria) become VBNC, i.e., viable but not cultivable [63]. This might occur due to stress encounter by bacteria while colonizing host cell. The reason for VBNC is still unknown, but it is common to most rhizobial species. The research at molecular and genetic levels might solve this mystery. Soils with high mineral N content (ammonium and nitrate ions) are more prone to N reduction, as PGPR can readily consume it. Therefore, the viability of an applied farming approach can indicate the accessibility of organic N content in soil [98]. Moreover, the soils with common physico-chemical features and exposure to similar climatic conditions may differ in net reduction of N content. This may be due to probable surface or drainage runoff of organic N during agricultural practices [146]. However, the estimation of soil N mass balance (input and output) requires long-term study which in turn will be helpful in the election of suitable cropping system. The use of economical viable PGPR inoculants along with efficient cropping systems can increase the probability of stable N retention in soils. In the case of developing countries, the lack of knowledge and relevant technological restrains demand an immediate implication of research (i.e., PGPR inoculation at sowing or spraying on leaves) in field conditions, thus providing cost reduction benefits to farmers and empowering local communities.

Author details

Muhammad Jamil Khan[1*], Rafia Younas[2], Abida Saleem[1], Mumtaz Khan[2], Qudratullah Khan[1] and Rehan Ahmed[2]

1 Department of Soil Science, Institute of Soil and Environmental Sciences, Gomal University, Dera Ismail Khan, Khyber Pakhtunkhwa, Pakistan

2 Department of Environmental Sciences, Institute of Soil and Environmental Sciences, Gomal University, Dera Ismail Khan, Khyber Pakhtunkhwa, Pakistan

*Address all correspondence to: mkhan@gu.edu.pk

References

[1] Bloom AJ. The increasing importance of distinguishing among plant nitrogen sources. Current Opinion in Plant Biology. 2015;**25**:10-16. DOI: 10.1016/j.pbi.2015.03.002

[2] Hemerly A. Genetic controls of biomass increase in sugarcane by association with beneficial nitrogen-fixing bacteria. In: Plant and Animal Genome XXIV-Plant and Animal Genome Conference. San Diago, USA; 2016

[3] Bianco MS, CecílioFilho AB, deCarvalho LB. Nutritional status of the cauliflower cultivar Verona grown with omission of out added macronutrients. PLoS One. 2015;**10**:e0123500. DOI: 10.1371/journal.pone.0123500

[4] Fields S. Global nitrogen: Cycling out of control. Environmental Health Perspectives. 2004;***112***:A556-A563. DOI: 10.1289/ehp.112-a556

[5] Postgate JR. The Fundamentals of Nitrogen Fixation.FEBS Letters. United Kingdom: Cambridge University Press; 1982. p. 162

[6] Sprent P. Editor. Nitrogen Fixing Organisms.Pure and Applied Aspects. 1st Ed. Netherland: Springer; 1990. ISBN 978-0-412-34690-3

[7] Mus F, Crook MB, Garcia K, Costas AG, Geddes BA, Kouri ED, et al. Symbiotic nitrogen fixation and the challenges to its extension to non-legumes. Applied and Environmental Microbiology. 2016;***82***:3698-3710. DOI: 10.1128/AEM.01055-16

[8] Burnham RJ, Johnson KR. South American palaeobotany and the origins of neotropical rain forests. Philosophical Transactions of the Royal Society, B: Biological Sciences. 2004; **359**:1595-1610. DOI: 10.1098/rstb.2004.1531

[9] Hymowitz T. On the domestication of the soybean. Economic Botany. 1970; **24**:408-421

[10] Liu CW, Murray JD. The role of flavonoids in nodulation host-range specificity: An update. Plants (Basel). 2016;**5**:33. DOI: 10.3390/plants5030033

[11] Rosas SB, Andrés JA, Rovera M, Correa NS. Phosphate-solubilizing pseudomonas putida can influence the rhizobia–legume symbiosis. Soil Biology and Biochemistry. 2006;**38**:3502-3505. DOI: 10.1016/j.soilbio.2006.05.008

[12] Puente ML, Zawoznik M, de Sabando ML, Perez G, Gualpa JL, Carletti SM, et al. Improvement of soybean grain nutritional quality under foliar inoculation with *Azospirillum brasilense* strain Az39. Symbiosis. 2018: 1-7. DOI: 10.1007/s13199-018-0568-x

[13] Ballesteros-Almanza L, Altamirano-Hernandez J, Peña-Cabriales JJ, Santoyo G, Sanchez-Yañez JM, et al. Effect of coinoculation with Mycorrhiza and rhizobia on the nodule trehalose content of different bean genotypes. The Open Microbiology Journal. 2010;**4**: 83-92. DOI: 10.2174/18742858010040 10083

[14] Institute on the Plant R, Its E, Carson EW, Southern Regional Education B. The Plant Root and its Environment; Proceedings. Charlottesville: University Press of Virginia; 1974

[15] Steenhoudt O, Vanderleyden J. Azospirillum, a free-living nitrogen-fixing bacterium closely associated with grasses: Genetic, biochemical and ecological aspects. FEMS Microbiology Reviews. 2000;**24**:487-506

[16] Guiñazú LB, Andrés JA, Del Papa MF, Pistorio M, Rosas SB. Response of alfalfa (*Medicago sativa L.*)

to single and mixed inoculation with phosphate-solubilizing bacteria and *Sinorhizobium meliloti*. Biology and Fertility of Soils. 2010;**46**:185-190

[17] Gómez-Sagasti MT, Marino D. PGPRs and nitrogen-fixing legumes: A perfect team for efficient Cd phytoremediation? Frontiers in Plant Science. 2015;**6**:81. DOI: 10.3389/fpls.2015.00081

[18] Marulanda-Aguirre A, Azcón R, Ruiz-Lozano JM, Aroca R. Differential effects of a *Bacillus megaterium* strain on *Lactuca sativa* plant growth depending on the origin of the arbuscular mycorrhizal fungus coinoculated: Physiologic and biochemical traits. Journal of Plant Growth Regulation. 2007;**27**:10. DOI: 10.1007/s00344-007- 9024-5

[19] Pathak D, Lone R, Koul KK. Arbuscular Mycorrhizal fungi (AMF) and plant growth-promoting Rhizobacteria (PGPR) Association in Potato (*Solanum tuberosum* L.): A brief Review. In: Kumar V, Kumar M, Sharma S, Prasad R, editors. Probiotics and Plant Health. Singapore: Springer Singapore; 2017. pp. 401-420. DOI: 10.1007/978-981-10-3473-2_18

[20] Perotto S, Bonfante P. Bacterial associations with mycorrhizal fungi: Close and distant friends in the rhizosphere. Trends in Microbiology. 1997;**5**:496-501. DOI: 10.1016/S0966-842X(97)01154-2

[21] Somers E, Ptacek D, Gysegom P, Srinivasan M, Vanderleyden J. *Azospirillum brasilense* produces the Auxin-like Phenylacetic acid by using the key enzyme for Indole-3-acetic acid biosynthesis. Applied and Environmental Microbiology. 2005;**71**: 1803-1810. DOI: 10.1128/AEM.71.4.1803-1810.2005

[22] Ahmad F, Ahmad I, Khan MS. Screening of free-living rhizospheric bacteria for their multiple plant growth promoting activities. Microbiological Research. 2008;**163**:173-181. DOI: 10.1016/j.micres.2006.04.001

[23] Choudhary DK, Sharma KP, Gaur RK. Biotechnological perspectives of microbes in agro-ecosystems. Biotechnology Letters. 2011;**33**: 1905-1910. DOI: 10.1007/s10529-011-0662-0

[24] García-Fraile P, Menéndez E, Rivas R. Role of bacterial biofertilizers in agriculture and forestry. AIMS Journal. 2015;**2**:183-205. DOI: 10.3934/bioeng.2015.3.183

[25] Vob U, Bishopp A, Farcot E, Bennett MJ. Modelling hormonal response and development. Trends in Plant Science. 2014;**19**:311-319. DOI: 10.1016/j.tplants.2014.02.004

[26] Kazan K. Diverse roles of jasmonates and ethylene in abiotic stress tolerance. Trends in Plant Science. 2015; **20**:219-229. DOI: 10.1016/j.tplants.2015.02.001

[27] Li XJ, Yang MF, Chen H, Qu LQ, Chen F, Shen SH. Abscisic acid pretreatment enhances salt tolerance of rice seedlings: Proteomic evidence. Biochemicaet Biophysica Acta. 2010; **1804**:929-940. DOI: 10.1016/j.bbapap.2010.01.004

[28] Ahmed A, Hasnain S. Auxin producing *Bacillus* sp.: Auxin quantification and effect on the growth *Solanum tuberosum*. Pure and Applied Chemistry. 2010;**82**:313-319. DOI: 10.1351/PAC-CON-09-02-06

[29] Marulanda A, Porcel R, Barea JM, Azcón R. Drought tolerance and antioxidant activities in lavender plants colonized by native drought-tolerant or drought-sensitive Glomus species. Microbial Ecology. 2007;**54**:543-552. DOI: 10.1007/s00248-007-9237-y

[30] Asgher M, Khan MI, Anjum NA, Khan NA. Minimizing toxicity of cadmium in plants—Role of plant growth regulators. Protoplasma. 2015; **252**:399-413. DOI: 10.1007/s00709-014- 0710-4

[31] Vamil R, Ul-Haq A, Agnihotri RK, Sharma R. Effect of certain plant growth regulators on the seedling survival, biomass production and proline content of *Bambusa arundinacea*. Science Research Reporter. 2011;**1**:44-48

[32] Khan N, Bano A, Zandi P. Effects of exogenously applied plant growth regulators in combination with PGPR on the physiology and root growth of chickpea (*Cicer arietinum*) and their role in drought tolerance. Journal of Plant Interactions. 2018;**13**:239-247. DOI: 10.1080/17429145.2018.1471527

[33] Solaimalai A, Sivakumar C, Anbumani S, et al. Agricultural Reviews. 2001;**22**:33-40. Print ISSN : 0253–1496

[34] Senthil A, Djanaguiraman M, Chandrababu R. Effect of root dipping of seedlings with plant growth regulators and chemicals on yield and yield components of rice (*Oryza sativa* L.) transplanted by broadcast method. The Madras Agricultural Journal. 2003;**90**:383-384

[35] Arshad M, Frankenberger WT Jr. Production and stability of C_2H_4 in soil. Biology and Fertility of Soils. 1990;**10**: 29-34. DOI: 10.1007/BF00336121

[36] Arshad M, Frankenberger WT Jr. Ethylene: Agricultural Sources and Applications. New York: Kluwer Academic Publishers; 2002

[37] Gull M, Hafeez FY, Saleem M, Malik KA. Phosphorus uptake and growth promotion of chickpea by co – Inoculation of mineral phosphate solubilizing bacteria and a mixed rhizobial culture. Australian Journal of Experimental Agriculture. 2004;**44**: 623-628. DOI: 10.1071/EA02218

[38] Mirza BS, Mirza MS, Bano A, Malik K. Coinoculation of chickpea with rhizobium isolates from roots and nodules and phytohormone-producing Enterobacter strains. Australian Journal of Experimental Agriculture. 2007;**47**: 1008-1015. DOI: 10.1071/EA06151

[39] Carreno-Lopez R, Campos-Reales N, Elmerich C, Baca BE. Physiological evidence for differently regulated tryptophan-dependent pathways for indole-3-acetic acid synthesis in *Azospirillum brasilense*. Molecular & General Genetics. 2000;**264**:521-530. DOI: 10.1007/s004380000340

[40] Lynch JM. Origin, nature and biological activity of aliphatic substances and growth hormones found in soil. In: Vaughan D, Malcom RE, editors. Soil Organic Matter and Biological Activity. Dordretch: Springer; 1985. pp. 151-174. DOI: 10.1007/978-94-009-5105-1_5

[41] Karen S, Udo B, Frank L, Dominique R. Can simultaneous inhibition of seedling growth and stimulation of rhizosphere bacterial populations provide evidence for phytotoxin transfer from plant residues in the bulk soil to the rhizosphere of sensitive species. Journal of Chemical Ecology. 2011;**27**:807-829. DOI: 10.1023/A:1010362221390

[42] Heisey RM, Putnam AR. Herbicidal effects of Gedanamycin and Nigercin, antibiotics from *Streptomyces hygroscopicus*. Journal of Natural Products. 1986;**49**:859-865

[43] Halvorson AD, Peterson GA, Reule CA. Tillage system and crop rotation effects on dryland crop yields and soil carbon in the central Great Plains. Agronomy Journal. 2002;***94***: 1429-1436

[44] Hungria M, Campo RJ, Souza EM, Pedrosa FO. Inoculation with selected strains of *Azospirillum brasilense* and A.

lipoferum improves yields of maize and wheat in Brazil. Plant and Soil. 2010;**331**: 413-425

[45] Hagh ED, Khoii FR, Valizadeh M, Khorshidi MB. The role of *Azospirillum lipoferum* bacteria in sustainable production of maize. Journal of Food, Agriculture and Environment. 2010;**8**: 702-704

[46] Biswas B, Gresshoff PM. The role of symbiotic nitrogen fixation in sustainable production of biofuels. International Journal of Molecular Sciences. 2014;**15**:7380-7397

[47] Ahemad M, Khan MS. Effects of pesticides on plant growth promoting traits of Mesorhizobium strain MRC4. Journal of the Saudi Society of Agricultural Sciences. 2012;**11**:63-71

[48] Bhattacharyya PN, Jha DK. Plant growth-promoting rhizobacteria (PGPR): Emergence in agriculture. World Journal of Microbiology and Biotechnology. 2012;*28*:1327-1350

[49] Antoun H, Prévost D. Ecology of plant growth promoting rhizobacteria. In: PGPR: Biocontrol and Biofertilization. Dordrecht: Springer; 2005. pp. 1-38

[50] Glick BR. Plant growth-promoting bacteria: Mechanisms and applications. Scientifica. *2012*

[51] Pérez-Montaño F, Alías-Villegas C, Bellogín RA, Del Cerro P, Espuny MR, Jiménez-Guerrero I, et al. Plant growth promotion in cereal and leguminous agricultural important plants: From microorganism capacities to crop production. Microbiological Research. 2014;**169**:325-336

[52] Kloepper JW, Lifshitz R, Zablotowicz RM. Free-living bacterial inocula for enhancing crop productivity. Trends in Biotechnology. 1989;**7**:39-44

[53] Tien TM, Gaskins MH, Hubbell DH. Plant growth substances produced by *Azospirillum brasilense* and their effect on the growth of pearl millet (*Pennisetum americanum* L.). Applied and Environmental Microbiology. 1979; **37**:1016-1024

[54] Strzelczyk E, Kampert M, Li CY. Cytokinin-like substances and ethylene production by *Azospirillum* in media with different carbon sources. Microbiological Research. 1994;**149**: 55-60

[55] Bottini R, Fulchieri M, Pearce D, Pharis RP. Identification of gibberellins A1, A3, and iso-A3 in cultures of *Azospirillum lipoferum*. Plant Physiology. 1989;**90**:45-47

[56] Perrig D, Boiero ML, Masciarelli OA, Penna C, Ruiz OA, Cassán FD, et al. Plant-growth-promoting compounds produced by two agronomically important strains of *Azospirillum brasilense*, and implications for inoculant formulation. Applied Microbiology and Biotechnology. 2007;**75**: 1143-1150

[57] Correa OS, Romero AM, Soria MA, de Estrada M. *Azospirillum brasilense* plant genotype interactions modify tomato response to bacterial diseases, and root and foliar microbial communities. In: Cassan FD, Garcia de Salmone I, editors. *Azospirillum* Spp: Cell Physiology, Plant Interactions and Agronomic Research in Argentina. Argentina, AsociaciónAgronómica de Microbiología; 2008. pp. 85-94

[58] Ahemad M, Khan MS. Ameliorative effects of Mesorhizobium sp. MRC4 on chickpea yield and yield components under different doses of herbicide stress. Pesticide Biochemistry and Physiology. 2010;**98**:183-190

[59] Kumar A, Maurya BR, Raghuwanshi R. Isolation and characterization of PGPR and their effect on growth, yield and nutrient content in wheat (*Triticum aestivum* L.). Biocatalysis and Agricultural Biotechnology. 2014;**3**:121-128

[60] Marques AP, Pires C, Moreira H, Rangel AO, Castro PM. Assessment of the plant growth promotion abilities of six bacterial isolates using Zea mays as indicator plant. Soil Biology and Biochemistry. 2010;**42**:1229-1235

[61] Dobbelaere S, Vanderleyden J, Okon Y. Plant growth-promoting effects of diazotrophs in the rhizosphere. Critical Reviews in Plant Sciences. 2003; **22**:107-149

[62] Huergo LF, Monteiro RA, Bonatto AC, Rigo LU, Steffens MBR, Cruz LM, et al. Regulation ofnitrogen fixation in *Azospirillum brasilense*. In: Cassán FD, GarciaSalamone I, editors. *Azospirillum* Spp.: Cell Physiology, Plant Interactions and Agronomic Research in Argentina. Argentina: AsociaciónArgentina de Microbiología; 2008. pp. 17-36

[63] Compant S, Clement C, Sessitsch A. Plant growth promoting bacteria in the rhizo- and endosphere of plants: Their role, colonization, mechanisms involved and prospects for utilization. Soil Biology and Biochemistry. 2010;**42**: 669-678

[64] Zahir ZA, Arshad M, Frankenberger WT. Plant growth promoting rhizobacteria: Applications and perspectives in agriculture. Advances in Agronomy. 2004;**81**:98-169

[65] Tahir M, Mirza MS, Hameed S, Dimitrov MR, Smidt H. Cultivation-based and molecular assessment of bacterial diversity in the rhizosheath of wheat under different crop rotations. PLoS One. 2015;**10**:e0130030

[66] Roesti D, Gaur R, Johri BN, Imfeld G, Sharma S, Kawaljeet K, et al. Plant growth stage, fertilizer management and bio-inoculation of arbuscular mycorrhizal fungi and plant growth promoting rhizobacteria affect the rhizobacterial community structure in rain-fed wheat fields. Soil Biology and Biochemistry. 2006;**38**:1111-1120

[67] Ogutcu H, Kasimoglu C, Elkoca E. Effects of rhizobium strains isolated from wild chickpeas on the growth and symbiotic performance of chickpeas (*Cicer arietinum L.*) under salt stress. Turkish Journal of Agriculture and Forestry. 2010;**34**:361-371

[68] Dreyfus B, Garcia JL, Gillis M. Characterization of *Azorhizobium caulinodans* gen. Nov., sp. nov., a stem-nodulating nitrogen-fixing bacterium isolated from *Sesbania rostrata*. International Journal of Systematic and Evolutionary Microbiology. 1988;**38**:89-98

[69] Qiang L, Hua-wei L, Wei-ling W. Colonization of *Azorhizobium caulinodans* in wheat and nutrient-related miRNA expression. Journal of Plant Nutrition and Fertilization. 2014; **20**:930-937

[70] Devi U, Singh KP, Kumar S, Sewhag M. Effect of nitrogen levels, organic manures and Azotobacter inoculation on yield and economics of multi-cut oats. Forage Research. 2014; **40**:36-43

[71] Morais TPD, Brito CHD, Brandão AM, Rezende WS. Inoculation of maize with *Azospirillum brasilense* in the seed furrow. Revista Ciência Agronômica. 2016;**47**:290-298

[72] Stajkovic O, Delic D, Josic D, Kuzmanovic D, Rasulic N, Knezevic-Vukcevic J. Improvement of common bean growth by coinoculation with rhizobium and plant growth-promoting bacteria. Romanian Biotechnology Letters. 2011;**16**:5919-5926

[73] Namvar A, Sharifi SR, Sedghi M, Asghari M, Zakaria R, Khandan T, et al. Study the effects of organic and inorganic nitrogen fertilizer on yield, yield components and nodulation state of chickpea (*Cicer arietinum* L.).

Communications in Soil Science and Plant Analysis. 2011;**42**(9):1097-1109

[74] Erman M, Demir S, Ocak E, Tufenkci S, Oguz F, Akkopru A. Effects of rhizobium, arbuscular mycorrhiza and whey applications on some properties in chickpea (*Cicer arietinum L.*) under irrigated and rainfed conditions 1-yield, yield components, nodulation and AMF colonization. Field Crops Research. 2011;**122**:14-24

[75] Nain L, Rana A, Joshi M, Shrikrishna JD, Kumar D, Shivay YS, et al. Evaluation of synergistic effects of bacterial and cyanobacterial strains as biofertilizers for wheat. Plant and Soil. 2010;**331**:217-230

[76] Almagrabi OA, Abdelmoneim TS. Using of arbuscular mycorrhizal fungi to reduce the deficiency effect of phosphorous fertilization on maize plants (*Zea mays* L.). Life Science Journal. 2012;**9**:1648-1654

[77] Abdelmoneim TS, Moussa TA, Almaghrabi OA, Alzahrani HS, Abdelbagi I. Increasing plant tolerance to drought stress by inoculation with arbuscular mycorrhizal fungi. Life Science Journal. 2014;**11**:10-17

[78] Bargali K. Screening of leguminous plants for VAM association and their role in restoration of degraded lands. Journal of American Science. 2011;**7**:7-11

[79] Hungria M, Nogueira MA, Araujo RS. Soybean seed coinoculation with Bradyrhizobium spp. and *Azospirillum brasilense*: A new biotechnological tool to improve yield and sustainability. American Journal of Plant Sciences. 2015;**6**:811-816

[80] Namvar A, Sharif RS, Khandani T, Moghadam MJ. Seed inoculation and inorganic nitrogen fertilization effects on some physiological and agronomical traits of chickpea (*Cicer arietinum L.*) in irrigated condition. Journal of Central European Agriculture. 2013;**14**(3):28-40

[81] Chibeba AM, de Fátima Guimarães M, Brito OR, Nogueira MA, Araujo RS, Hungria M. Coinoculation of soybean with Bradyrhizobium and Azospirillum promotes early nodulation. American Journal of Plant Sciences. 2015;**6**:1641-1649

[82] Alam F, Bhuiyan MAH, Alam SS, Waghmode TR, Kim PJ, Lee YB. Effect of rhizobium sp. BARIRGm901 inoculation on nodulation, nitrogen fixation and yield of soybean (*Glycine max*) genotypes in gray terrace soil. Bioscience, Biotechnology, and Biochemistry. 2015;**79**:1660-1668

[83] Arouja AES, Baldani VLD, Galisa PS, Pereira JA, Baldani JI. Response of traditional upland rice varieties to inoculation with selected diazotrophic bacteria isolated from rice cropped at the northeast region of Brazil. Applied Soil Ecology. 2013;**64**: 49-55

[84] Imran A, Mirza MS, Shah TM, Malik KA, Hafeez FY. Differential response of kabuli and desi chickpea genotypes towards inoculation with PGPR in different soils. Frontiers in Microbiology. 2015;**6**:859

[85] Herridge DF, Peoples MB, Marschner Review BRM. Global inputs of biological nitrogen fixation in agricultural systems. Plant and Soil. 2008;**311**:1-18

[86] Peoples MB, Brockwell J, Herridge DF, Rockester IJ, Alves BJR, Urquiaga S, et al. The contributions of nitrogen-fixing crop legumes to the productivity of agricultural systems. Symbiosis. 2009;**48**:1-17

[87] Salvagiotti F, Cassman KG, Spetch JE, Walter DT, Weiss A, Dobermann A. Nitrogen uptake, fixation and response to fertilizer N in

soybeans: A review. Field Crops Research. 2008;**108**:1-13

[88] Kirkegaard J, Christen O, Krupnisky J, Layzell D. Break crop benefits in temperate wheat production. Field Crops Research. 2008;**107**:185-195

[89] Peoples MB, Herridge DF. Quantification of biological nitrogen fixation in agricultural ecosystems. In: Pedrosa FO, Hungria M, Yates MG, Newton WE, editors. Nitrogen Fixation: From Molecules to Crop Productivity. The Netherlands: Kluwer Academic Publishers; 2000. pp. 519-524

[90] Purdy R, Langemeier M. International Benchmark for Soybean Production. farmdoc daily, 2018;8(120). Retrieved from https://farmdocdaily.illinois.edu/2018/06/international-benchmarks-soybean-production.html

[91] Spencer JE, Owen L. 2018. Asia. Encyclopedia Britannica. Retrieved from: https://www.britannica.com/place/Asia/Agriculture

[92] Cabelguenne M, Debaeke P, Bouniols A. EPIC phase, a version of the EPIC model simulating the effects of water and nitrogen stress on biomass and yield, taking account of developmental stages: Validation on maize, sunflower, sorghum, soybean and winter wheat. Agricultural Systems. 2002;**60**:175-196

[93] Boote KJ, Mínguez MI, Sau F. Adapting the CROPGRO legume model to simulate growth of faba bean. Agronomy Journal. 2002;**94**:743-756

[94] Soussana JF, Minchin FR, Macduff JH, Raistrick N, Abberton MT, Michaelson-Yeates TPT. A simple model of feedback regulation for nitrate uptake and N_2 fixation in contrasting phenotypes of white clover. Annals of Botany. 2002;**90**:139-147

[95] Robertson MJ, Carberry PS, Huth NI, Turpin JE, Probert ME, Poulton PL, et al. Simulation of growth and development of diverse legume species in APSIM. Australian Journal of Agricultural Research. 2002;**53**:429-446

[96] Ledgard SF, Brier GJ, Littler RA. Legume production and nitrogen fixation in hill pasture communities. New Zealand Journal of Agricultural Research. 1987;**30**:413-421

[97] Høgh-Jensen H, Loges R, Jørgensen FV, Vinther FP, Jensen ES. An empirical model for quantification of symbiotic nitrogen fixation in grass-clover mixtures. Agricultural Systems. 2004;**82**:181-194

[98] Korsaeth A, Eltun R. Nitrogen mass balances in conventional, integrated and ecological cropping systems and the relationship between balance calculations and nitrogen runoff in an 8-year field experiment in Norway. Agriculture, Ecosystems and Environment. 2000;**79**:199-214

[99] Watson CA, Goss MJ. Estimation of N_2-fixation by grass-white clover mixtures in cut or grazed swards. Soil Use and Management. 1997;**13**:165-167

[100] Kristensen ES, Høgh-Jensen H, Kristensen IS. A simple model for estimation of atmospherically-derived nitrogen in grass-clover systems. Biological Agriculture and Horticulture. 1995;**12**:263-276

[101] Duffy J, Chung C, Boast C. Franklin M. a simulation model of biophysiochemical transformations of nitrogen in tile-drained corn belt soil. Journal of Environmental Quality. 1975; **4**:477-486

[102] Sinclair TR. Water and nitrogen limitations in soybean grain production I. model development. Field Crops Research. 1986;**15**:125-141

[103] Sinclair TR, Muchow RC, Ludlow MM, Leach GJ, Lawn RJ,

Foale MA. Field and model analysis of the effect of water deficits on carbon and nitrogen accumulation by soybean, cowpea and black gram. Field Crops Research. 1987;**17**:121-140

[104] Sharpley AN, Williams JR. EPIC-erosion/productivity impact calculator 1.Model documentation, USDA Tech. 1990. Bull. No. 1768, pp. 1-235

[105] Bouniols A, Cabelguenne M, Jones CA, Chalamet A, Charpenteau JL, Marty JR. Simulation of soybean nitrogen nutrition for a silty clay soil in southern France. Field Crops Research. 1991;**26**:19-34

[106] Cabelguenne M, Debaeke P, Bouniols A. EPIC phase, a version of the EPIC model simulating the effects of water and nitrogen stress on biomass and yield, taking account of developmental stages: Validation on maize, sunflower, sorghum, soybean and winter wheat. Agricultural Systems. 1999;**60**:175-196

[107] Thornley JHM. Plant Submodel. In: Thornley JHM, editor. Grassland Dynamics: An Ecosystem Simulation Model. Wallingford, Oxon, UK: CAB International; 1998. pp. 53-57

[108] Thornley JHM, Cannell MGR. Dynamics of mineral N availability in grassland ecosystems under increased [CO_2]: Hypotheses evaluated using the Hurley pasture model. Plant and Soil. 2000;**224**:153-170

[109] Thornley JHM. Simulating grass-legume dynamics: A phenomenological sub-model. Annals of Botany. 2001;**88**: 905-913

[110] Eckersten H, AfGeijersstam L, Torssell B. Modelling nitrogen fixation of pea (*Pisum sativum* L.). Acta Agriculturae Scandinavica B. 2006;**56**:129-137

[111] Schwinning S, Parsons AJ. Analysis of the coexistence mechanisms for grasses and legumes in grazing systems. Journal of Ecology. 1996;**84**:799-813

[112] Schmid M, Neftel A, Riedo M, Fuhrer J. Process-based modelling of nitrous oxide emissions from different nitrogen sources in mown grassland. Nutrient Cycling in Agroecosystems. 2001;**60**:177-187

[113] Boote KJ, Jones JW, Hoogenboom G, Pickering NB. The CROPGRO model for grain legumes. In: Tsuji GY, Hoogenboom G, Thornton PK, editors. Understanding Options for Agricultural Production. 1998. pp. 99-128

[114] Sau F, Boote KJ, Ruíz-Nogueira B. Evaluation and improvement of CROPGRO-soybean model for a cool environment in Galicia, Northwest Spain. Field Crops Research. 1999;**61**:273-291

[115] Hartkamp AD, Hoogenboom G, White JW. Adaptation of the CROPGRO growth model to velvet bean (*Mucuna pruriens*): I. Model development. Field Crops Research. 2002;**78**:9-25

[116] Boote KJ, Hoogenboom G, Jones JW, Ingram KT. Modeling nitrogen fixation and its relationship to nitrogen uptake in the CROPGRO model. In: Ma L, Ahuja LR, Bruulsema TW, editors. Quantifying and Understanding Plant Nitrogen Uptake for Systems Modeling. Florence, USA: CRC Press; 2008. pp. 13-46

[117] Wu L, McGechan MB. Simulation of nitrogen uptake, fixation and leaching in a grass/white clover mixture. Grass and Forage Science. 1999;**54**:30-41

[118] Herridge DF, Turpin JE, Robertson MJ. Improving nitrogen fixation of crop legumes through breeding and agronomic management: Analysis with simulation modelling. Australian Journal of Experimental Agriculture. 2001;**41**:391-401

[119] Brisson N, Launay M, Mary B, Beaudoin N. Nitrogen transformations.

In: Brisson N, Launay M, Mary B, Beaudoin N, editors. Conceptual Basis, Formalizations and Parameterization of the STICS Crop Model. Versailles Cedex, France: Quae; 2009. pp. 141-165

[120] Corre-Hellou G, Brisson N, Launay M, Fustec J, Crozat Y. Effect of root depth penetration on soil nitrogen competitive interactions and dry matter production in pea–barley intercrops given different soil nitrogen supplies. Field Crops Research. 2007;**103**:76-85

[121] Corre-Hellou G, Faure M, Launay M, Brisson N, Crozat Y. Adaptation of the STICS intercrop model to simulate crop growth and N accumulation in pea-barley intercrops. Field Crops Research. 2009;**113**:72-81

[122] Liu Y, Wu L, Baddeley JA, Watson CA. Models of biological nitrogen fixation of legumes. A review. Agronomy for Sustainable Development. 2011;**31**:155-172

[123] Macduff JH, Jarvis SC, Davidson IA. Inhibition of N_2 fixation by white clover (*Trifolium repens* L.) at low concentrations of NO^{3-} inflowing solution culture. Plant and Soil. 1996; **180**:287-295

[124] Voisin AS, Salon C, Jeudy C, Warembourg FR. Symbiotic N_2 fixation activity in relation to C economy of *Pisum sativum* L. as a function of plant phenology. Journal of Experimental Botany. 2003;**54**:2733-2744

[125] Voisin AS, Bourion V, Duc G, Salon C. Using an ecophysiological analysis to dissect genetic variability and to propose an ideotype for nitrogen nutrition in pea. Annals of Botany. 2007;**100**:1525-1536

[126] Yu M, Gao Q, Shaffer MJ. Simulating interactive effects of symbiotic nitrogen fixation, carbon dioxide elevation and climatic change on legume growth. Journal of Environmental Quality. 2002;**31**: 634-641

[127] Zhang F, Dashti N, Hynes RK, Smith DL. Plant growth promoting rhizobacteria and soybean [*Glycine max* (L.) Merr.]nodulation and nitrogen fixation at suboptimal root zone temperatures. Annals of Botany. 2006; **77**:453-459

[128] Albrecht SL, Bennett JM, Boote KJ. Relationship of nitrogenase activity to plant water stress in field-grown soybeans. Field Crops Research. 1984;**8**:61-71

[129] Pfau T, Christian N, Masakapalli SK, Sweetlove LJ, Poolman MG, Elbenhoh O. The intertwined metabolism during symbiotic nitrogen fixation elucidated by metabolic modelling. Scientific Reports. 2018;**8**:12504

[130] Hungria M, Campo RJ, Mendes IC, Graham PH. Contribution of biological nitrogen fixation to the N nutrition of grain crops in the tropics: The success of soybean (*Glycine max* L. Merr.) in South America. In: Singh RP, Shankar N, Jaiwa PK, editors. Nitrogen Nutrition and Sustainable Plant Productivity. Houston: Studium Press; 2006a. pp. 43-93

[131] Hungria M, Franchini JC, Campo RJ, Crispino CC, Moraes JZ, Sibaldelli RNR, et al. Nitrogen nutrition of soybean in Brazil: Contributions of biological N_2 fixation and of N fertilizer to grain yield. Canadian Journal of Plant Science. 2006b; **86**:927-939. DOI: 10.4141/P05-098

[132] Bashan Y, Bashan LE. How the plant growth-promoting bacterium Azospirillum promotes plant growth: A critical assessment. Advances in Agronomy. 2010;**108**:77-136. DOI: 10.1016/S0065-2113(10)08002-8

[133] Duete RRC, Muraoka T, daSilva EC, Trivelin PCO, Ambrosano EJ. Viabilidadeeconômica de doses e parcelamentos da

adubaçãonitrogenadanacultura do milhoemlatossolovermelhoeutrófico. Acta Scientiarum Agronomy. 2009;**31**: 175-181. DOI: 10.4025/actasciagron. v31i1.6646

[134] Galindo FS, Filho MCMT, Buzetti S, Santini JMK, Alves JK, Nogueira LM et al. Corn Yield and Foliar Diagnosis Affected by Nitrogen Fertilization and Inoculation with *Azospirillum brasilense*. Revista Brasileira de Ciência Avícola. 2016;**40**:e0150364

[135] Matsunaga M, Bemelmans PF, PEN de T, Dulley RD, Okawa H, Pedroso IA. Metodologia de custo de produçãoutilizadapelo IEA. Vol. 23. Agriculturaem São Paulo: BoletimTécnicodo Instituto de EconomiaAgrícola; 1976. pp. 123-139

[136] Zarei I, Sohrabi Y, Heidari GR, Jalilian A, Mohammadi K. Effects of biofertilizers on grain yield and protein content of two soybean (*Glycine max L.*) cultivars. African Journal of Biotechnology. 2014;**11**:7028-7037

[137] Yadegari M, Rahmani HA, Noormohammadi G, Ayneband A. Evaluation of bean (*Phaseolus vulgaris*) seeds inoculation with *Rhizobium phaseoli* and plant growth promoting rhizobacteria on yield and yield components. PJBS. 2008;**11**: 1935-1939

[138] Zuffo AM, Bruzi AT, de Rezende PM, de Carvalho MLM, Zambiazzi EV, Soares IO, et al. Foliar application of *Azospirillum brasilense* in soybean and seed physiological quality. African Journal of Microbiology Research. 2016;**10**: 675-680

[139] Groppa MD, Zawoznik MS, Tomaro ML. Effect of coinoculation with *Bradyrhizobium japonicum* and *Azospirillum brasilense* on soybean plants. European Journal of Soil Biology. 1998;**34**:75-80

[140] Mahato S, Neupane S. Comparative study of impact of Azotobacter and Trichoderma with other fertilizers on maize growth. Journal of Maize Research and Development. 2017;**3**:1-16

[141] Hungria M, Nogueira MA, Araujo RS. Coinoculations of soybean and common beans with rhizobia and Azospirilla: Strategies to improve sustainability. Biology and Fertility of Soils. 2013;**49**:791-801

[142] Fatima Z, Zia M, Chaudhary MF. Interactive effect of rhizobium strains and P on soybean yield, nitrogen fixation and soil fertility. Pakistan Journal of Botany. 2007;**39**:255-264

[143] Wang X, Pan Q, Chen F, Yan X, Liao H. Effects of coinoculation with arbuscular mycorrhizal fungi and rhizobia on soybean growth as related to root architecture and availability of N and P. Mycorrhiza. 2011;**21**:173-181

[144] Hungria M, Nogueira MA, Araujo RS. Coinoculation of soybeans and common beans with rhizobia and Azospirilla: Strategies to improve sustainability. Biology and Fertility of Soils. 2013;**49**:791-801. DOI: 10.1007/ s00374-012-0771-5

[145] Igiehon NO, Babalola OO. Rhizosphere microbiome modulators: Contributions of nitrogen fixing bacteria towards sustainable agriculture. International Journal of Environmental Research and Public Health. 2018;**15**:574

[146] Kristensen L, Stopes C, Kølster P, Granstedt A. Nitrogen leaching in ecological agriculture: Summary and recommendations. Biological Agriculture and Horticulture. 1995;**11**: 331-340

9

Distribution and Characterization of the Indigenous Soybean-Nodulating Bradyrhizobia in the Philippines

Maria Luisa Tabing Mason and Yuichi Saeki

Abstract

The research about the indigenous soybean bradyrhizobia in the Philippines is scarce, and this greatly influences the improvement of soybean production in the country. Thus, soil samples were collected from 11 locations in the country and were used to isolate the indigenous bradyrhizobia in the soil. Through the use of polymerase chain reaction—restriction fragment length polymorphism (PCR-RFLP) and sequence analysis of the 16S rRNA gene, 16S-23S rRNA gene internal transcribed spacer (ITS) region and *rpoB* housekeeping gene, the most abundant and dominant indigenous bradyrhizobia in the country were identified. Then, the representative isolates of the most dominant species per location were used to test their symbiotic efficiency and N-fixation ability with the local soybean cultivars. The results showed that among all the tested indigenous strains, the *B. elkanii* IS-2 is the most effective and efficient microsymbiont of the Philippines' local soybean cultivars. This report was able to provide necessary information on the distribution of soybean bradyrhizobia in the Philippines and characterized the symbiotic performance of the indigenous strains.

Keywords: tropical rhizobia, genomic diversity, N-fixation, symbiotic performance

1. Introduction

Soybean (*Glycine max* [L.]) is a leguminous plant that can form a symbiotic relationship with the nitrogen-fixing group of bacteria living in the rhizosphere, which are generally termed as rhizobia. In the Philippines, soybean production has been limited by the poor grain yield which leads to the importation of more than 90% of the country's demand. Thus, it is essential to look for an alternative way to increase the volume of production per unit area.

The research about tropical bradyrhizobia indicated a high diversity of species and their distribution has been reported to be due to several abiotic and biotic factors such as soil acidity [1–3], alkalinity [3, 4], temperature [1, 5–11], climate [12, 13], soil water status [14, 15], soil type [2, 14, 16–18], and soil management or cultural practices [2, 14, 19–22]. In case of the Philippines, the pioneer research that was able to identify the most dominant species of bradyrhizobia in the country reported that *B. elkanii* species was the most abundant, followed by the

B. diazoefficiens, *B. japonicum*, and some yet unclassified *Bradyrhizobium* sp. [14]. In this later study, it was identified that the distribution of these indigenous species of bradyrhizobia were influenced mainly by the water status of the soil, followed by soil pH, nutrient content, and soil type.

Previous studies have reported that aside from the various agro-environmental factors, the competition with the native rhizobia is a hindrance for a successful inoculation [23, 24]. The utilization of inoculants for legumes had shown promising results for the increase in grain yield as evidenced by recent reports [25, 26]. The role of the biological nitrogen fixation (BNF) in providing the N requirement of the plant in a natural way has been deemed necessary especially these times that the soil has become more degraded due to over-fertilization. The indiscriminate use of NPK fertilizer could cause soil pollution and less crop production [27]. Therefore, it is essential to select and evaluate the symbiotic competitiveness of the indigenous strains which are native and existing in high density in the country. The use of different genetic markers to accurately identify the rhizobia for taxonomic purposes has been proposed [28] and so we have used three genetic markers such as the 16S rRNA gene, 16S-23S rRNA gene internal transcribed spacer (ITS) region, and the *rpoB* housekeeping gene.

Thus, this study was formulated with the aim to utilize the recently identified indigenous bradyrhizobia in the Philippines and characterize their symbiotic performance with the local soybean cultivars.

2. Materials and methods

2.1 Collection of soil and soybean cultivation

The soil samples were collected from 11 locations in the Philippines, where some basic information on the sites are listed in **Table 1**. The collection of soil was conducted by first removing the surface litters then, obtaining a bar of soil with a dimension of approximately 20 cm in depth and 3 cm in thickness that weighs about 1 kg. A total of 10 subsamples per location were obtained and were mixed thoroughly until a 1 kg of composite soil sample was taken. A 0.5 kg soil was air-dried for the chemical analyses while the remaining 0.5 kg of the fresh soil was used for the soybean cultivation.

Location	Rep. Isolate	pH[a]	N[a] (%)	P[a] ($mgkg^{-1}$)	K[a] ($mgkg^{-1}$)	CEC[b] (meq)
1. Isabela1 (IS)	IS-2	5.90	0.13	1.86	51.80	17.1
2. Isabela 2 (GI)	GI-4	5.52	0.17	2.30	58.60	27.4
3. Benguet (BA)	BA-24	5.22	0.24	22.22	51.00	29.3
4. Nueva Ecija1 (NE1)	NE1-6	6.21	0.13	6.74	73.90	30.1
5. Nueva Ecija 2 (NE2)	NE2-37	5.81	0.22	21.63	49.40	27.7
6. Sorsogon (SO)	SO-1	5.26	0.22	2.57	55.80	20.3
7. Leyte (LT)	LT-3	5.80	0.15	6.39	174.20	40.6
8. Negros Occ. (NR)	NR-2	5.62	0.07	20.44	74.10	14.1
9. Bohol (BO)	BO-4	5.82	0.06	2.80	47.80	9.7
10. Maguindanao (SK)	SK-5	6.64	0.19	4.53	59.60	40.5
11. South Cotabato (SC)	SC-3	5.52	0.14	31.18	47.20	9.7

[a]*Mason et al., 2018*
[b]*This study*

Table 1.
Result of the soil chemical analysis on the 11 locations in the Philippines.

The cultivation of soybean was performed using a 1-L capacity culture pots (n = 3). Each pot was filled with vermiculite and a N-free solution [29] was added at 40% (vol/vol) water content. The culture pots were sterilized by autoclaving for 20 min at 121°C. Meanwhile, the soybean seeds were surface-sterilized by soaking into a 70% EtOh for 30 s, then by a diluted sodium hypochlorite solution (0.25% available chlorine) for 3 min and followed by washing with sterile distilled water for about 6–8 times. Then, a 2–3 g of soil sample was placed on the vermiculite at a depth of about 2–3 cm, the seeds were sown on the soil and the pot was weighed and recorded. The plants were grown inside a growth chamber for 28 days at 28°C (8 h, night) and 33°C (16 h, day) then were supplied weekly with sterile distilled water until the initial weight of the pot was reached.

2.2 Isolation of soybean rhizobia and DNA extraction

After 28 days, approximately 20 random nodules were collected from the roots of each soybean plants and were sterilized with 70% EtOh and sodium hypochlorite solution as previously described [29]. Each nodule was homogenized with sterile distilled water in a microtube and streaked on to a yeast-extract mannitol agar (YMA) plate [30]. The YMA plate was incubated in the dark at 28°C for about 1 week until a single colony was formed. After then, the single colony was streaked on to a YMA plate containing a 0.002% (wt/wt) bromothymol blue (BTB) [31] and was incubated as above. Repeated streaking was done until a pure single colony was obtained which was cultured for about 3–4 days in a HEPES-MES (HM) broth culture [32, 33] at 28°C in a shaker for 120 rpm. After then, the bacteria cells were collected by centrifugation at 9000×*g* and washed with sterile distilled water. The DNA was extracted by using BL buffer as described [34] from the method reported by Hiraishi et al. [35].

2.3 PCR amplification of the 16S rRNA gene, ITS region, and rpoB gene

For the amplification of the 16S rRNA gene, the primer set: 16S-F: 5′ AGAG TTTGATCCTGGCTCAG-3′ and 16S-R2: 5′- CGGCTACCTTGTTACGACTT-3′ [36]. The PCR tubes were then placed in the PCR Thermal Cycler (TaKaRa Co. Ltd.) with the following conditions: pre-run at 94°C for 5 min; followed by 30 cycles of denaturation at 94°C for 1 min, annealing at 55°C for 1 min, and extension at 72°C for 1 min. Final extension was set at 72°C for 10 min and indefinite preservation at 4°C.

On the other hand, the PCR amplification of the ITS region was conducted using the following primer set: Bra-ITS-F: 5-GACTGGGGTGAAGTCGTAAC-3′ and Bra-ITS-R1: 5′-ACGTCCTTCATCGCC TC-3′ [6]. The PCR cycle for the ITS region was almost the same with the 16S rRNA gene except for a shorter denaturation and annealing periods which were conducted at 30 s for each step.

For the *rpoB* gene, simplification was done using the following primer sets: *rpo*B83F: 5′-CCTSATCGAGGTTCAC AGAAGGC-3′ and *rpo*B1540R: 5′-AGCTGCGAGGAACCGAAG-3′ [37]. The PCR cycle conditions were as follows: pre-run at 94°C for 5 min; followed by 30 cycles of denaturation at 94°C for 30 s, annealing at 60°C for 1 min, and extension at 72°C for 1 min. Final extension was set at 72°C for 5 min and indefinite preservation at 4°C.

2.4 Restriction fragment length polymorphism (RFLP) of the 16S rRNA gene, ITS region, and rpoB gene

The successfully amplified products were subjected to the RFLP treatment using four restriction enzymes which were *HhaI*, *HaeIII*, *MspI*, and *XspI*. For the rpoB gene, the enzymes that were used for RFLP are *HaeIII*, *MspI*, and *AluI*.

The reference strains that were used in this study are the *Bradyrhizobium* USDA strains (*B. japonicum* USDA 4, 6^T, 38, 62, 115, 122, 123, 124, 125, 127, 129, 135, *B. diazoefficiens* USDA 110^T, *B. elkanii* 31, 46, 61, 76^T, 94, 130, and *B. liaoningense* 3622^T) which were previously described [38]. This was done in a 10-μL reaction mixture containing a 2.5-μL amplified PCR product and was incubated in a 37°C for 16 h. Afterward, a 3–4% agarose gel was used in a submerged gel electrophoresis for about 60 min, stained with ethidium bromide and the patterns were visualized using a Luminiscent Image Analyzer LAS-4000 (FUJIFILM Tokyo, Japan).

2.5 Single-strain inoculation test

After the amplification and the RFLP treatment of the 16S rRNA gene, a single-strain inoculation test was conducted for all the amplified isolates that shared the same restriction enzymes' fragment patterns with the USDA *Bradyrhizobium* reference strains. This was done to confirm the strain's capability to nodulate soybean and was tested on two local varieties which are the PSB-SY2 and Collection 1 which are both commercially available across the country.

The cultivation of soybean was conducted as described above, but without soil. Each isolate was cultured in a YM broth (YMB) [30] at 28°C for about 1 week on a shaker. After then, the cultures were diluted with sterile distilled water at about 10^6 cells mL^{-1} and were inoculated on the cultivated soybean at a rate of 1.0 mL per seed. This was done with three replications. After inoculation, the weight of the pot was recorded and it was placed inside a growth chamber with a condition set to mimic the average temperature in the Philippines at 26°C (8 h, night) and 33°C (16 h, day). The same condition was used for the cultivation of an uninoculated control and a positive control pot that was inoculated with *B. diazoefficiens* USDA110. The pots were kept inside the growth chamber for 28 days and were supplied weekly with sterile distilled water until the initial weight of each pot was reached.

2.6 Sequence analysis of the 16S rRNA gene and ITS region

According to the similarities of the band patterns through the RFLP treatment, a representative of the most abundant isolates was chosen for each location. In total, there were 11 isolates that were selected to confirm the nucleotide sequence of the 16S rRNA gene and the ITS region. The sequence primers that were used were reported previously [22]. From the PCR amplified product, the samples were purified according to the protocol of the manufacturer (Nucleospin® Gel and PCR Clean-up; Macherey-Nagel, Germany). Then, the samples were sent to the company for the sequence analysis (Eurofins Genomics, Tokyo, Japan).

Then, the Basic Local Alignment Search Tool (BLAST) program in DNA Databank of Japan (DDBJ) was used to determine the nucleotide homology of the isolates. Only the sequences with a similarity of at least 99% for the 16S rRNA and 96% for the ITS region with our isolates were retrieved from the BLAST database. The alignment was performed using the ClustalW and Neighbor-Joining [21] method was used to construct the phylogenetic trees. The genetic distances were computed using the Kimura 2-parameter model [39] in the Molecular Evolutionary Genetic Analysis (MEGA v7) software [40]. Subsequently, the phylogenetic trees were bootstrapped with 1000 replications. All the nucleotide sequences determined in this study were deposited in DDBJ at http://www.ddbj.nig.ac.jp/.

3. Results

3.1 Soil analysis and characterization of the indigenous bradyrhizobia

The soil samples that were used in this study were all slightly to moderately acidic (5.22–6.64) with non-saline condition (0.05–0.20 dS/m), low nutrient status as evidenced by low amounts of NPK and CEC (**Table 1**). These values are generally typical of agricultural soils that are used for crop production all throughout the year. These results showed that the soils used in this study have low fertility status that indicated the need for soil restoration strategies.

The growth morphologies of the pure single colony for each strain of bradyrhizobia were characterized and listed in **Table 2**. All the isolates were slow growers which were able to form single colonies measuring about 2 mm between 5 and 7 days upon streaking on YMA plates and incubation in a dark room. Based on the morphology, the isolates were grouped into three. Group I include the isolates IS-2, NE1–6, NR-2, and BO-4 which were translucent and the colonies are circular

Isolate	No. of isolates	Colony morphology	Formation of acid / alkaline substances
IS-2	40	Translucent, circular, liquid, slightly-convex elevation with entire margin	Alkaline
GI-4	30	Translucent, circular, slightly-mucoid, convex elevation with entire margin	Alkaline
BA-24	31	Translucent, circular, mucoid, convex elevation with entire margin	Alkaline
NE1-6	49	Translucent, circular, liquid, slightly-convex elevation with entire margin	Alkaline
NE2-37	31	Translucent, circular, slightly-mucoid, convex elevation with entire margin	Alkaline
SO-1	44	Translucent, circular, mucoid, convex elevation with entire margin	Alkaline
LT-3	42	Translucent, circular, mucoid, convex elevation with entire margin	Alkaline
NR-2	23	Translucent, circular, liquid, slightly-convex elevation with entire margin	Alkaline
BO-4	24	Translucent, circular, liquid, slightly-convex elevation with entire margin	Alkaline
SK-5	29	Translucent, circular, mucoid, convex elevation with entire margin	Alkaline
SC-3	31	Translucent, circular, mucoid, convex elevation with entire margin	Alkaline
Total	**374**		

Table 2.
Characterization of the morphology of the indigenous bradyrhizobia isolated from Philippines' soil according to their growth on Yeast-Extract Mannitol Agar plate medium [30].

in shape with slightly convex elevation and an entire margin. When they were manipulated with a needle, the colony was liquid. Group II include the isolates BA-24, SO-1, LT-3, and SK-5 were translucent with circular colonies, convex elevation with entire margin. When manipulated with a needle, the colonies have mucoid viscosity. On the other hand, last group (III) are the isolates GI-4 and NE2-37 which have similar growth morphology with Group II except that their viscosity was intermediate between liquid and mucoid. All the isolates produced alkaline substances when grown on YMA plate with BTB which is an indication of the *Bradyrhizobium* genus.

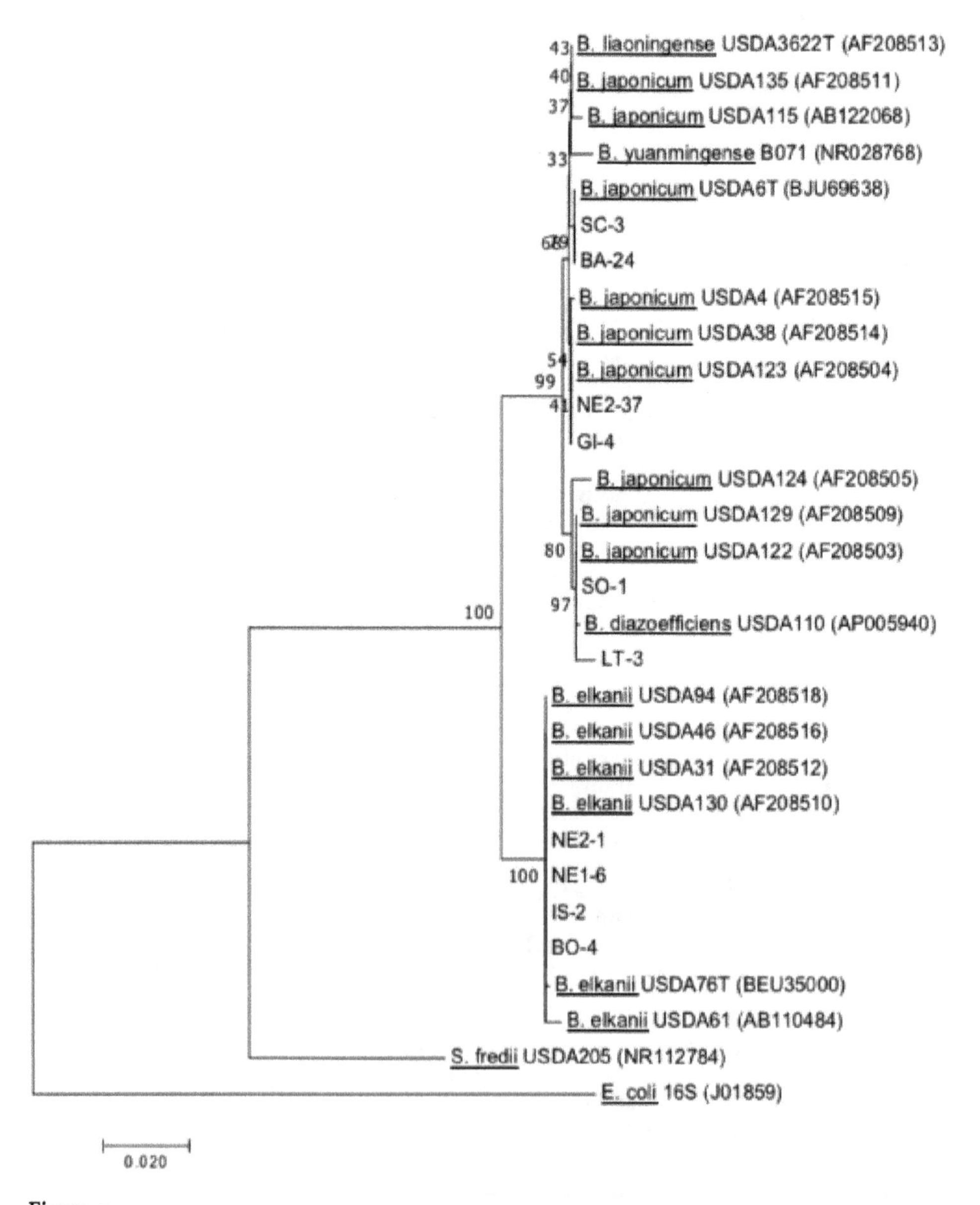

Figure 1.
Phylogenetic tree based on the sequence analysis of the 16S rRNA gene. The tree was constructed using the Neighbor-Joining method with the Kimura 2-parameter (K2P) distance correlation model and 1000 bootstrap replications in MEGA v.7 software. The accession numbers are indicated only for sequences obtained from BLAST. The isolates in this study are indicated with letters and number combinations, for example: BO-4–isolate no. 4 collected from Bohol.

3.2 Distribution of indigenous soybean bradyrhizobia

As seen in **Figure 1**, it is evident that the 11 most abundant indigenous soybean rhizobia in the Philippines are classified under the genus *Bradyrhizobium*, and are separated into its two species, *B. japonicum* and *B. elkanii*, according to the

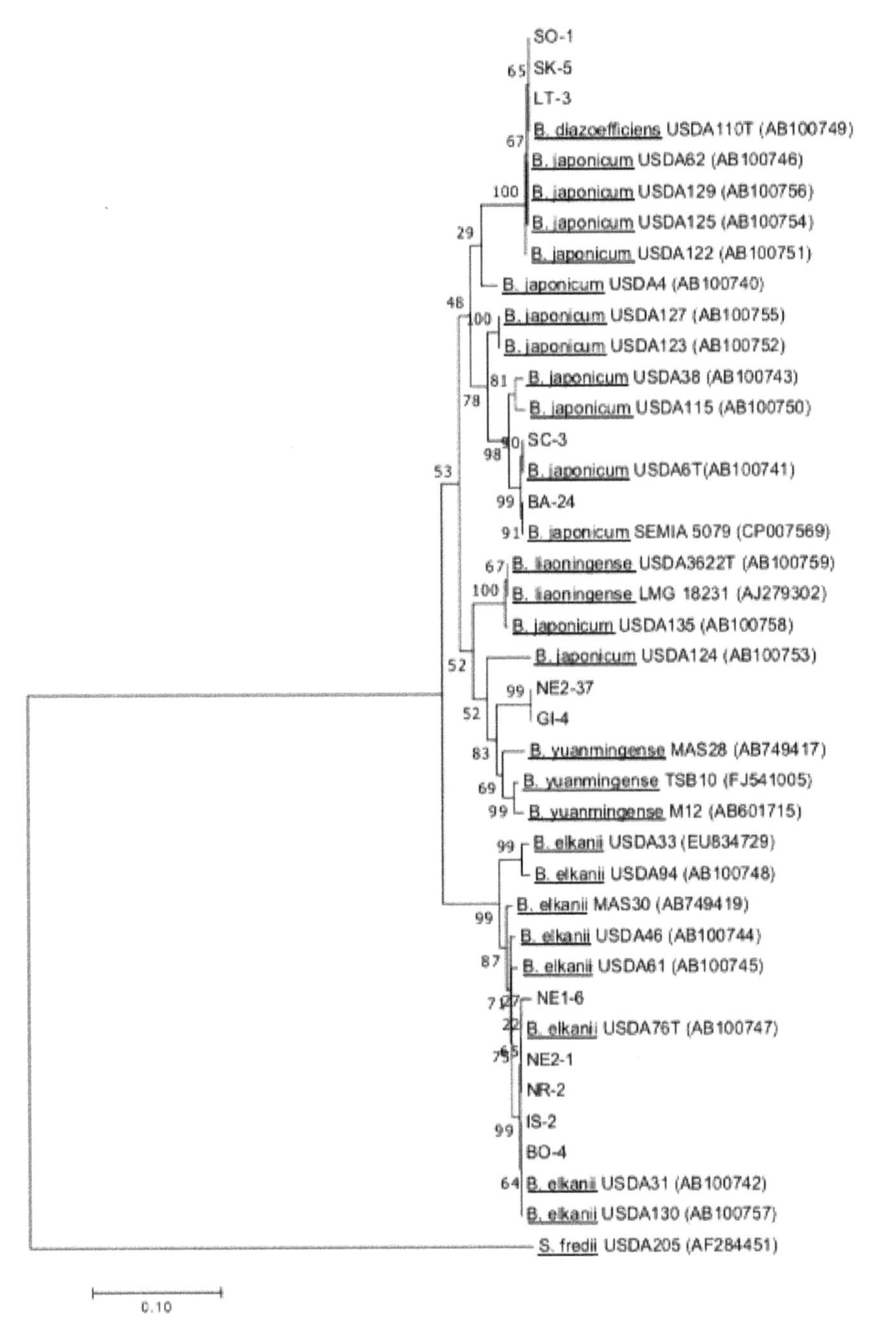

Figure 2.
Phylogenetic tree based on the sequence analysis of the 16S-23S rRNA internal transcribed spacer (ITS) region. The tree was constructed using the Neighbor-Joining method with the Kimura 2-parameter (K2P) distance correlation model and 1000 bootstrap replications in MEGA v.7 software. The accession numbers are indicated only for sequences obtained from BLAST. The isolates in this study are indicated with letters and number combinations, for example: BO-4–isolate no. 4 collected from Bohol.

phylogenetic tree from the sequence analysis of the 16S rRNA gene. To further confirm the classification of the indigenous bradyrhizobia, the phylogenetic trees constructed from the ITS region and the *rpoB* gene are presented in **Figures 2** and **3**, respectively. For the ITS region and the *rpoB* gene, the isolates were distinctly grouped into three species, *B. elkanii*, *B. japonicum*, and *B. diazoefficiens*. Additionally, an independent cluster composed of the representative isolates GI-4

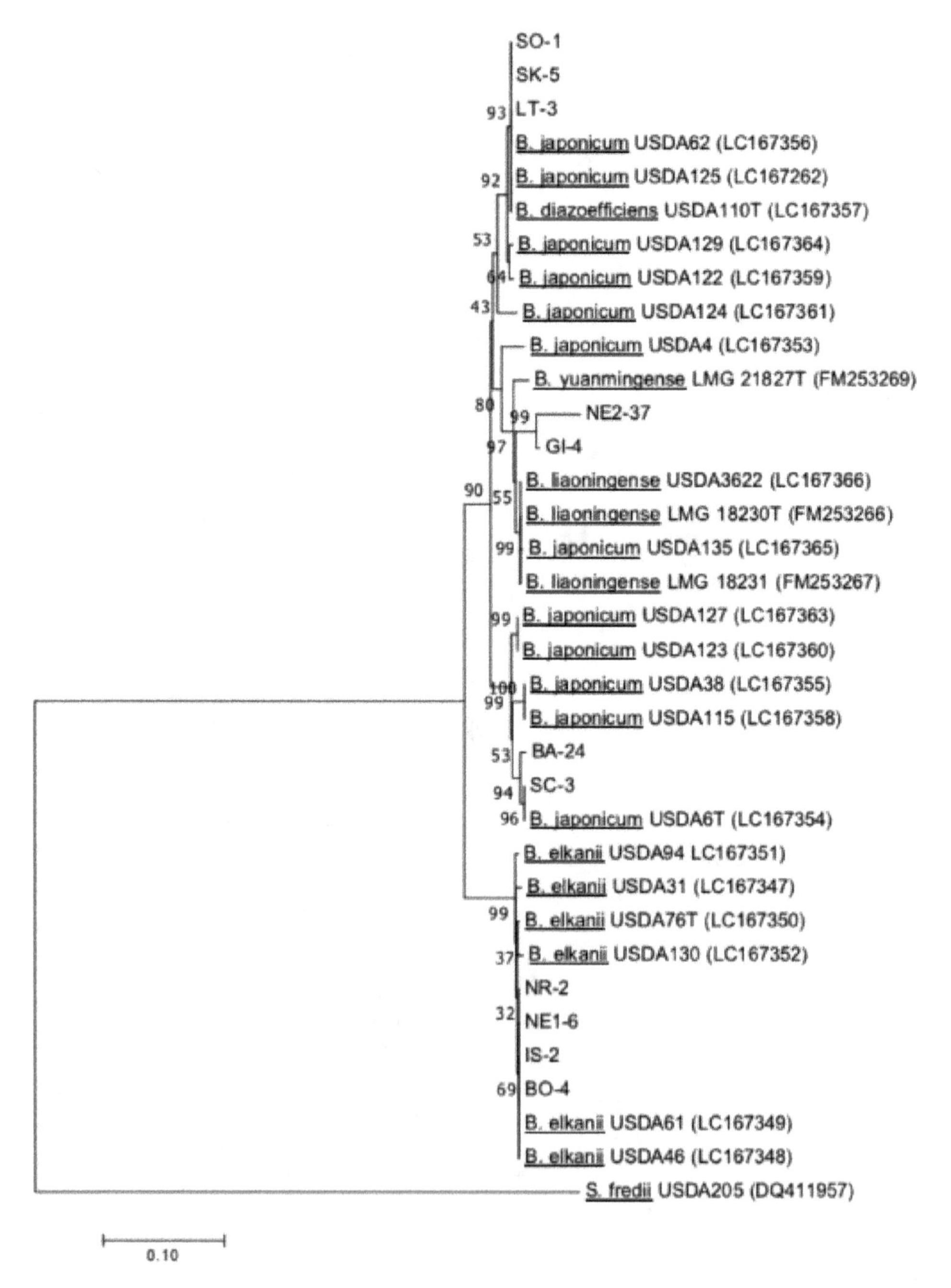

Figure 3.
Phylogenetic tree based on the sequence analysis of the rpoB housekeeping gene. The tree was constructed using the Neighbor-Joining method with the Kimura 2-parameter (K2P) distance correlation model and 1000 bootstrap replications in MEGA v.7 software. The accession numbers are indicated only for sequences obtained from BLAST. The isolates in this study are indicated with letters and number combinations, for example: BO-4–isolate no.4 collected from Bohol.

and NE2–37 that are seen in the ITS region and *rpoB* phylogenetic trees were treated as *Bradyrhizobium* sp. due to their nucleotide divergence with the known species from the BLAST engine.

Meanwhile, the distribution of the most abundant soybean bradyrhizobia in the country is shown in **Table 3**, which was classified according to the results of the sequence analysis of the three genetic markers used in this study. From here, it can be seen that 4 of the 11 locations were dominated with *B. elkanii* species (37.74%), 3 locations were dominated by the isolates under the *B. diazoefficiens* (28.54%), whereas 2 locations each were dominated by the species of *B. japonicum* (16.98% and *Bradyrhizobium* sp. (16.74%). This indicated that in the Philippines, the species of *B. elkanii* is the most prevalent in terms of population and the most widespread in terms of location as its presence was detected even in minor populations on all the locations except for one, which was Sorsogon.

3.3 Symbiotic efficiency and N-fixation ability of the indigenous bradyrhizobia

Upon classification, it is important to determine the capability of the indigenous bradyrhizobia for their symbiotic performance and N-fixation ability. As can be seen in **Figure 4A**, although USDA110 strain has the highest N-fixation ability, it should be noted that the amount of N that was fixed by *B. elkanii* IS-2 is the highest among all the indigenous bradyrhizobia isolated from the Philippines' soil on Rj_4 plants. However, the N-fixation ability of IS-2 was comparably similar with other strains (GI-4, NE2–37, and SK-5) with the non-*Rj* plants. The lowest N-fixation ability was observed from the strain LT-3 which was classified under the *B. diazoefficiens* species. This suggested that the process of biological N-fixation is a mutual relationship that is influenced by both the plant and the rhizobia and that the plant-rhizobia compatibility should be taken into consideration for inoculation strategies.

Presented in **Figure 4B** is the nodulation test performed on the strains and it can be seen for Rj_4 plants, there was not much significant difference in the nodulation

Location	*B. elkanii*	*B. diazoefficiens*	*B. japonicum*	*Bradyrhizobium* sp.	No. isolates
Isabela1 (IS)	100.0	-	-	-	40
Isabela 2 (GI)	16.7	-	-	83.3	36
Benguet (BA)	3.0	-	97.0	-	33
Nueva Ecija1 (NE1)	87.5	1.8	-	10.7	56
Nueva Ecija 2 (NE2)	8.6	-	2.9	88.6	35
Sorsogon (SO)	-	100	-	-	44
Leyte (LT)	2.3	97.7	-	-	43
Negros Occ. (NR)	74.2	3.2	22.6	-	31
Bohol (BO)	82.8	13.8	3.4	-	29
Maguindanao (SK)	23.2	67.4	-	9.3	43
South Cotabato (SC)	8.8	-	91.2	-	34
Total isolates	**160**	**121**	**72**	**71**	**424**
Percentage (%)	**37.74**	**28.54**	**16.98**	**16.74**	

Table 3.
Percentage distribution of the dominant Bradyrhizobium species in the Philippines as identified from the sequence analysis of the 16S rRNA gene, 16S-23S internal transcribed spacer (ITS) region, and rpoB housekeeping gene.

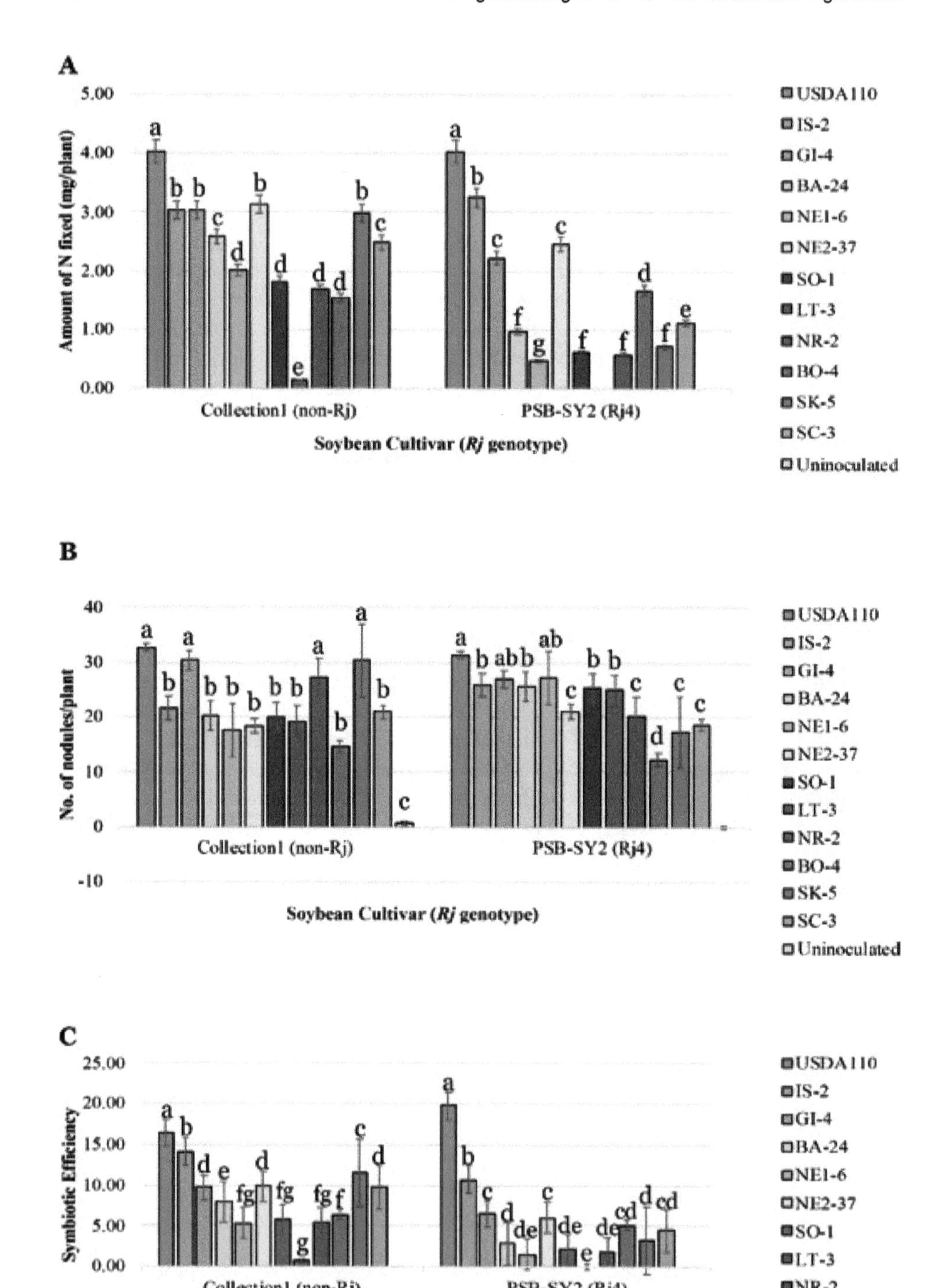

Figure 4.

Characterization of the dominant indigenous Bradyrhizobium strains isolated from the 11 locations in the Philippines based on the (A) amount of Nitrogen fixed (B) nodulation ability and (C) symbiotic efficiency as influenced by the single-strain inoculation test against the reference strain B. diazoefficiens USDA110 for the two soybean cultivars from the Philippines. Different letters indicate a significant difference by Tukey's test at $p > 0.05$, $n=3$, bar=SE.

ability of the strains, except for the low nodulation ability that was observed for the BO-4. In contrast, a significant difference in the nodulation ability was detected on the strains upon inoculation on the non-*Rj* plants. Although all the strains were able to form nodules on both soybean cultivars, the strains GI-4, NR-2, and SK-5 obtained the highest number of nodules for the non-*Rj* plants.

On the other hand, the symbiotic efficiency of the strains used in this study is presented in **Figure 4C**. Similar with the N-fixation ability, the USDA110 still possesses the highest symbiotic efficiency. But among all the indigenous bradyrhizobia, the strain IS-2 obtained the highest efficiency regardless of the *Rj* genotype of the soybean plants. As with the N-fixation, LT-3 obtained the least efficiency for symbiosis. This result indicated that the symbiotic efficiency of the rhizobia might not be directly influenced by the *Rj* genotype of the plant.

4. Discussion

4.1 Distribution and characterization of bradyrhizobia in the Philippines

The distribution of the most dominant and abundant species of soybean bradyrhizobia in the Philippines are reported in this study along with the characterization of their growth morphology. According to our earlier reports, we have elucidated that the Philippines was dominated by the soybean-nodulating bradyrhizobia that were classified under the *B. elkanii* species and the most important agro-environmental factors that affected their diversity and prevalence in the country was the similarity of soil pH, salinity, and temperature in the study locations [5, 14]. Our observation that there are abundant and high diversity of indigenous bradyrhizobia in the Philippines is similar with previous reports in other sub-tropical and tropical regions [12, 25, 41–44]. The temperate regions of Japan and USA were studied in the past and were reported to be dominated by species of *B. japonicum* and *B. diazoefficiens* [6, 9–11, 13, 45]. Our report showed that the distribution of bradyrhizobia in a tropical region like the Philippines seemed to be different from those of temperate regions.

Meanwhile, it was included in a recent report that the distribution and abundance of *B. diazoefficiens* and *B. japonicum* at specific locations were due to the longer period of flooding conditions [14]. The effect of nutrient content and soil type were also correlated with the abundance of these two species. In a report by Shiina et al. [17], it was stated that the predominance of *B. diazoefficiens* was observed on more anaerobic condition; whereas, *B. japonicum* was predominant on aerobic soils which was supported by another study [18]. Additionally, it was reported that *B. diazoefficiens* becomes predominant with enhanced flooding condition [15]. These results confirmed that our observations for the abundant of *B. diazoefficiens*, followed by *B. japonicum* and *Bradyrhizobium* sp. on flooded areas in the country which were usually used for planting rice.

4.2 Symbiotic and N-fixation ability of indigenous bradyrhizobia

In this report, the symbiotic performance, N-fixation and nodulation ability of the indigenous soybean bradyrhizobia form the Philippines were evaluated against that of the *B. diazoefficiens* USDA110 strain. The USDA110 has been extensively used in the world as a model strain for soybean inoculation due to its high ability for N-fixation and symbiotic efficiency [25, 46, 47]. Additionally, its possession of a complete set of denitrification genes that allows the release of N_2 back into the atmosphere makes it an ideal strain also for climate change mitigation studies [17, 48–50].

Therefore, we hypothesized that the indigenous isolates SO-1, LT-3, and SK-5, which were phylogenetically clustered under the USDA110 would also prove to be as effective N-fixer and efficient microsymbiont of soybean cultivars from the Philippines. However, our results indicated that the N-fixation ability and symbiotic efficiency of LT-3 and SO-1 were very low in comparison to the other indigenous isolates. For the low performance of these two isolates, it is hypothesized that the inherent ability of these strains to fix N and establish a symbiotic relationship with soybean is low. This could be explained by the fact that their nodulation ability was comparably similar with the other strains which possess higher N-fixation ability and symbiotic efficiency. In contrast, the isolate IS-2, which was clustered under the *B. elkanii* species, showed the highest symbiotic efficiency for both *Rj*-genotypes of the soybean cultivar used and the highest N-fixation ability for Rj_4 plants. In a previous report, the *Rj* genes that could restrict the nodulation of soybean by some strains of bradyrhizobia was summarized [51] but in case of our present report, all the strains in this study were not restricted by the two *Rj*-plants that were used. This led us to consider that the low N-fixation and symbiotic performance of some strains were not due to the restrictions from the *Rj*-genotypes of the plants but could be attributed to the strains' intrinsic capabilities. These observations might explain the reason for low yield of soybean in the Philippines. It was reported that many strains of *B. elkanii* were relatively inefficient microsymbionts of soybean and can induce chlorosis in soybean plants [52]. In a previous report [53], the high temperatures in tropical regions can limit the nodulation which could explain the low soybean yield.

It was expected that the strains which were classified as *B. diazoefficiens* could provide a better symbiotic performance than the other strains that were collected. However, the data showed that *B. elkanii* might establish a better symbiosis with local soybean cultivars in the Philippines. This result is crucial in order to devise strategies on how to increase the local production of soybean by inoculation with the indigenous strains.

Upon considering these results with the N-fixation and symbiotic performance ability of the strains, the number of nodules that can be formed from the single-strain inoculation does not seem to influence the amount of N that each strain can fix nor their symbiotic ability.

5. Conclusion

In this report, we have revealed that the distribution of tropical soybean bradyrhizobia seemed to be different than those of temperate bradyrhizobia in terms of population dominance of *B. elkanii* on higher temperature region like the Philippines. Additionally, it is proposed that for the Philippines, the most efficient N-fixer and symbiotically efficient species of bradyrhizobia would be *B. elkanii*. Yet, our results were made under the laboratory conditions only, so the results that were obtained here might not be as expected when done in field condition. For future research, utilization of more local soybean varieties with different soil types both in a controlled environment and on natural field condition would be beneficial to target the development of a site-specific and useful potential soybean inoculant. The data generated in this report would be beneficial for the augmentation of inoculation strategies in the country.

Acknowledgements

The authors would like to acknowledge the contributions of John Philip Tanay, Emmanuel Victor Buniao, Mary Joy Portin, and Maria Leah Sevilla of Central Luzon

State University for their help on some laboratory experiments. This work was supported by the JSPS Grant-in-Aid for Scientific Research (KAKENHI Grant Number: 18K05376).

Author details

Maria Luisa Tabing Mason[1] and Yuichi Saeki[2*]

1 College of Agriculture, Central Luzon State University, Science City of Munoz, Nueva Ecija, Philippines

2 Faculty of Agriculture, University of Miyazaki, Miyazaki, Japan

*Address all correspondence to: yt-saeki@cc.miyazaki-u.ac.jp

References

[1] Adhikari D, Kaneto M, Itoh K, Suyama K, Pokharel BB, Gaihre YK. Genetic diversity of soybean-nodulating rhizobia in Nepal in relation to climate and soil properties. Plant and Soil. 2012;**357**:131-145. DOI: 10.1007/s11104-012-1134-6

[2] Loureiro MDF, Kaschuk G, Alberton O, Hungria M. Soybean [*Glycine max* (L.) Merrill] rhizobial diversity in Brazilian oxisols under various soil, cropping, and inoculation managements. Biology and Fertility of Soils. 2007;**43**:665-674. DOI: 10.1007/s00374-006-0146-x

[3] Suzuki K, Oguro H, Yamakawa T, Yamamoto A, Akao S, Saeki Y. Diversity and distribution of indigenous soybean-nodulating rhizobia in the Okinawa islands, Japan. Soil Science and Plant Nutrition. 2008;**54**:237-246. DOI: 10.1111/j.1747-0765.2007.00236.x

[4] Saeki Y, Kaneko A, Hara T, Suzu ki K, Yamakawa T, Nguyen MT, et al. Phylogenetic analysis of soybean-nodulating rhizobia isolated from alkaline soils in Vietnam. Soil Science and Plant Nutrition. 2005;**51**: 1043-1052. DOI: 10.1111/j.1747- 0765.2005.tb00143.x

[5] Mason MLT, Matsuura S, Domingo AL, Yamamoto A, Shiro S, Sameshima-Saito R, et al. Genetic diversity of indigenous soybean-nodulating *Bradyrhizobium elkanii* from southern Japan and Nueva Ecija, Philippines. Plant and Soil. 2017;**417**:349-362. DOI: 10.1007/s11104-017-3263-4

[6] Saeki Y, Aimi N, Tsukamoto S, Yamakawa T, Nagatomo Y, Akao S. Diversity and geographical distribution of indigenous soybean-nodulating bradyrhizobia in Japan. Soil Science and Plant Nutrition. 2006;**52**:418-426. DOI: 10.1111/j.1747-0765.2006.00050.x

[7] Saeki Y, Minami M, Yamamoto A, Akao S. Estimation of the bacterial community diversity of soybean-nodulating bradyrhizobia isolated from Rj-genotype soybeans. Soil Science and Plant Nutrition. 2008;**54**:718-724. DOI: 10.1111/j.1747-0765.2008.00300.x

[8] Saeki Y, Ozumi S, Yamamoto A, Umehara Y, Hayashi M, Sigua GC. Changes in population occupancy of bradyrhizobia under different temperature regimes. Microbes and Environments. 2010;**25**:309-312. DOI: 10.1264/jsme2.ME10128

[9] Saeki Y. Characterization of soybean-nodulating rhizobia communities and diversity. In: Sudarić A, editor. Soybean—Molecular Aspects of Breeding. InTech; 2011. pp. 163-184. DOI: 10.5772/14417

[10] Shiro S, Yamamoto A, Umehara Y, Hayashi M, Yoshida N, Nishiwaki A, et al. Effect of *Rj* genotype and cultivation temperature on the community structure of soybean-nodulating bradyrhizobia. Applied and Environmental Microbiology. 2012;**78**:1243-1250. DOI: 10.1128/AEM.06239-11

[11] Shiro S, Matsuura S, Saiki R, Sigua GC, Yamamoto A, Umehara Y, et al. Genetic diversity and geographical distribution of indigenous soybean-nodulating bradyrhizobia in the United States. Applied and Environmental Microbiology. 2013;**79**:3610-3618. DOI: 10.1128/AEM.00236-13

[12] Risal CP, Yokoyama T, Ohkama-Ohtsu N, Djedidi S, Sekimoto H. Genetic diversity of native soybean bradyrhizobia from different topographical regions along the southern slopes of the Himalayan Mountains in Nepal. Systematic and Applied Microbiology. 2010;**33**:416-425. DOI: 10.1016/j.syapm.2010.06.008

[13] Saeki Y, Shiro S. Comparison of soybean-nodulating Bradyrhizobia community structures along north latitude between Japan and USA. In: Ohyama T, editor. Advances in Biology and Ecology of Nitrogen Fixation. Rijeka, Croatia: InTech; 2014. pp. 195- 224. DOI: 10.5772/57165

[14] Mason MLT, Tabing BLC, Yamamoto A, Saeki Y. Influence of flooding and soil properties on the genetic diversity and distribution of indigenous soybean-nodulating bradyrhizobia in the Philippines. Heliyon. 2018;**4**:e00921. DOI: 10.1016/j. heliyon.2018. e00921

[15] Saeki Y, Nakamura M, Mason MLT, Yano T, Shiro S, Sameshima-Saito R, et al. Effect of flooding and the *nos*Z gene in bradyrhizobia on bradyrhizobial community structure in the soil. Microbes and Environments. 2017;**32**:154-163. DOI: 10.1264/jsme2. ME16132

[16] Sessitsch A, Howieson JG, Perret X, Antoun H, Martinez-Romero E. Advances in *rhizobium* research. Critical Reviews in Plant Sciences. 2002;**21**:323-378

[17] Shiina Y, Itakura M, Choi H, Saeki Y, Hayatsu M, Minamisawa K. Relationship between soil type and N_2O reductase genotype (*nos*Z) of indigenous soybean bradyrhizobia: *nos*Z-minus populations are dominant in andosols. Microbes and Environments. 2014;**29**:420-426. DOI: 10.1264/jsme2.ME14130

[18] Siqueira AF, Minamisawa K, Sanchez C. Anaerobic reduction of nitrate to nitrous oxide is lower in *Bradyrhizobium japonicum* than in *Bradyrhizobium diazoefficiens*. Microbes and Environments. 2017;**32**:398-401. DOI: 10.1264/jsme2.ME17081

[19] Bizarro MJ, Giongo A, Vargas LK, Roesch LFW, Gano KA, Saccol de Sa' EL, et al. Genetic variability of soybean bradyrhizobia populations under different soil managements. Biology and Fertility of Soils. 2011;**47**:357-362. DOI: 10.1007/s00374-010-0512-6

[20] Grossman JM, Schipanski ME, Sooksanguan T, Seehaver S, Drinkwater LE. Diversity of rhizobia in soybean [*Glycine max* (Vinton)] nodules varies under organic and conventional management. Applied Soil Ecology. 2011;**50**:14-20. DOI: 10.1016/j.apsoil. 2011.08.003

[21] Saitou N, Nei M. The neighbour-joining method: A new method for reconstructing phylogenetic trees. Molecular Biology and Evolution. 1987;**4**:406-425. DOI: citeulike-article-id:93683

[22] Yan J, Han XZ, Ji ZJ, Li Y, Wang ET, Xie ZH, et al. Abundance and diversity of soybean-nodulating rhizobia in black soil are impacted by land use and crop management. Applied and Environmental Microbiology. 2014;**80**:5394-5402. DOI: 10.1128/AEM.01135-14

[23] Grönemeyer JL, Kulkarni A, Berkelmann D, Hurek T, Reinhold-Hurek B. Identification and characterization of rhizobia indigenous to the Okavango region in Sub-Saharan Africa. Applied and Environmental Microbiology. 2014;**80**:7244-7257. DOI: 10.1128/AEM.02417-14

[24] Yamakawa T, Saeki Y. Inoculation methods of *Bradyrhizobium japonicum* on soybean in South-West area of Japan. In: A Comprehensive Survey of International Soybean Research—Genetics, Physiology, Agronomy and Nitrogen Relationships. Rijeka, Croatia: InTech; 2012. pp. 83-114. DOI: 10.5772/52183

[25] Chibeba AM, Kyei-Boahen S, Guimarães M de F, Nogueira MA, Hungria M. Isolation, characterization and selection of indigenous

Bradyrhizobium strains with outstanding symbiotic performance to increase soybean yields in Mozambique. Agriculture, Ecosystems and Environment. 2017;**246**:291-305. DOI: 10.1016/j.agee. 2017.06.017

[26] Heerwaarden J, Baijukya F, Kyei-Boahen S, Adjei-Nsiah S, Ebanyat P, Kamai N, et al. Soyabean response to rhizobium inoculation across sub-Saharan Africa: Patterns of variation and the role of promiscuity. Agriculture, Ecosystems and Environment. 2018;**261**:211-218. DOI: 10.1016/j. agee.2017.08.016

[27] Ashraf MA, Maah MJ, Yusoff I. Soil Contamination, Risk Assessment and Remediation. London, UK: IntechOpen; 2014. 56pp. DOI: 10.5772/57287

[28] Rivas R, Martens M, de Lajudie P, Willems A. Multilocus sequence analysis of the genus *Bradyrhizobium*. Systematic and Applied Microbiology. 2009;**32**:101-110

[29] Saeki Y, Akagi I, Takaki H, Nagatomo Y. Diversity of indigenous *Bradyrhizobium* strains isolated from three different *Rj*-soybean cultivars in terms of randomly amplified polymorphic DNA and intrinsic antibiotic resistance. Soil Science and Plant Nutrition. 2000;**46**:917-926. DOI: 10.1080/00380768.2000.10409157

[30] Vincent JM. A Manual for the Practical Study of the Root-Nodule Bacteria. Oxford: Blackwell Scientific; 1970

[31] Keyser HH, Bohlool BB, Hu TS, Weber DF. Fast-growing rhizobia isolated from root nodules of soybean. Science. 1982;**215**:1631-1632

[32] Cole MA, Elkan GH. Transmissible resistance to Penicillin-G, Neomycin, and Chloramphenicol in *Rhizobium japonicum*. Antimicrobial Agents and Chemotherapy. 1973;**4**:248-253

[33] Sameshima R, Isawa T, Sadowsky MJ, Hamada T, Kasai H, Shutsrirung A, et al. Phylogeny and distribution of extra-slow-growing *Bradyrhizobium japonicum* harboring high copy numbers of RSα, RSβ and IS1631. FEMS Microbiology Ecology. 2003;**44**:191-202. DOI: 10.1016/S0168-6496(03)00009-6

[34] Minami M, Yamakawa T, Yamamoto A, Akao S, Saeki Y. Estimation of nodulation tendency among *Rj*-genotype soybeans using the bradyrhizobial community isolated from an Andosol. Soil Science and Plant Nutrition. 2009;**55**:65-72. DOI: 10.1111/j.1747-0765.2008.00333.x

[35] Hiraishi A, Kamagata Y, Nakamura K. Polymerase chain reaction amplification and restriction fragment length polymorphism analysis of 16S rRNA genes from methanogens. Journal of Fermentation and Bioengineering. 1995;**79**:523-529. DOI: 10.1016/0922-338X(95)94742-A

[36] Weisburg WG, Barns SM, Pelletier DA, Lane DJ. 16S ribosomal DNA amplification for phylogenetic study. Journal of Bacteriology. 1991;**173**:697-703

[37] Martens M, Dawyndt P, Coopman R, Gillis M, De Vos P, Willems A. Advantages of multilocus sequence analysis for taxonomic studies: A case study using 10 housekeeping genes in the genus *Ensifer* (including former *Sinorhizobium*). International Journal of Systematic and Evolutionary Microbiology. 2008;**58**:200-214

[38] Saeki Y, Aimi N, Hashimoto M, Tsukamoto S, Kaneko A, Yoshida N, et al. Grouping of *Bradyrhizobium* USDA strains by sequence analysis of 16S rDNA and 16S-23S rDNA internal transcribed spacer region. Soil Science and Plant Nutrition. 2004;**50**:517-525. DOI: 10.1080/00380768.2004.10408508

[39] Kimura M. A simple method for estimating evolutionary rates of base substitutions through comparative studies of nucleotide sequences. Journal of Molecular Evolution. 1980;**16**: 111-120. DOI: 10.1007/BF01731581

[40] Kumar S, Stecher G, Tamura K. MEGA7: Molecular evolutionary genetics analysis version 7.0 for bigger datasets. Molecular Biology and Evolution. 2016;**33**:1870-1874. DOI: 10.1093/molbev/msw054

[41] Ansari PG, Rao DLN, Pal KK. Diversity and phylogeny of soybean rhizobia in central India. Annales de Microbiologie. 2014;**64**:1553-1565. DOI: 10.1007/s13213-013-0799-2

[42] Appunu C, N'Zoue A, Laguerre G. Genetic diversity of native bradyrhizobia isolated from soybeans (*Glycine max* L.) in different agricultural-ecological-climatic regions of India. Applied and Environmental Microbiology. 2008;**74**:5991-5996. DOI: 10.1128/AEM.01320-08

[43] Guimarães AA, Florentino LA, Almeida KA, Lebbe L, Silva KB, Willems A, et al. High diversity of *bradyrhizobium* strains isolated from several legume species and land uses in Brazilian tropical ecosystems. Systematic and Applied Microbiology. 2015;**38**:433-441. DOI: 10.1016/j.syapm.2015.06.006

[44] Ribeiro PR, Santos JV, Costa EM, Lebbe L, Assis ES, Louzada MO, et al. Symbiotic efficiency and genetic diversity of soybean bradyrhizobia in Brazilian soils. Agriculture, Ecosystems and Environment. 2015;**212**:85-93

[45] Saeki Y, Shiro S, Tajima T, Yamamoto A, Sameshima-Saito R, Sato T, et al. Mathematical ecology analysis of geographical distribution of soybean-nodulating bradyrhizobia in Japan. Microbes and Environments. 2013;**28**:470-478. DOI: 10.1264/jsme2.ME13079

[46] Siqueira AF, Ormeno-Orillo E, Souza RC, Rodrigues EP, Almeida LGP, Barcellos FG, et al. Comparative genomics of *Bradyrhizobium japonicum*, CPAC 15 and *Bradyrhizobium diazoefficiens*, CPAC 7: Elite model strains for understanding symbiotic performance with soybean. BMC Genomics. 2014;**15**:420. DOI: 10.1186/1471-2164-15-420

[47] Soe KM, Bhromsiri A, Karladee D, Yamakawa T. Effects of endophytic actinomycetes and *Bradyrhizobium japonicum* strains on growth, nodulation, nitrogen fixation and seed weight of different soybean varieties. Soil Science and Plant Nutrition. 2012;**58**:319-325. DOI: 10.1080/00380768.2012.682044

[48] Akiyama H, Hoshino YT, Itakura M, Shimomura Y, Wang Y, Yamamoto A, et al. Mitigation of soil N_2O emission by inoculation with a mixed culture of indigenous *Bradyrhizobium diazoefficiens*. Scientific Reports. 2016;**6**:1-8. DOI: 10.1038/srep32869

[49] Itakura M, Uchida Y, Akiyama H, Hoshino YT, Shimomura Y, Morimoto S, et al. Mitigation of nitrous oxide emissions from soils by *Bradyrhizobium japonicum* inoculation. Nature: Climate Change. 2013;**3**:208-212. DOI: 10.1038/nclimate1734

[50] Sameshima-Saito R, Chiba K, Minamisawa K. Correlation of denitirifying capability with the existence of *nap, nir, nor* and *nos* genes in diverse strains of soybean bradyrhizobia. Microbes and Environments. 2006;(3):174-184. DOI: 10.1264/jsme2.21.174

[51] Hayashi M, Saeki Y, Haga M, Harada K, Kouchi H, Umehara Y. *Rj* (*rj*) genes involved in nitrogen-fixing root nodule formation in soybean. Breeding

Science. 2012;**61**:544-553. DOI: 10.1270/jsbbs.61.544

[52] Devine TE, Kuykendall LD, O'Neill JJ. DNA homology group and the identity of bradyrhizobial strains producing rhizobitoxine-induced foliar chlorosis on soybean. Crop Science. 1988;**28**:939-941. DOI: 10.2135/cropsci 1988.0011183X002800060014x

[53] Eaglesham ARJ, Ayanaba A. Tropical stress ecology of rhizobia, root nodulation and legume nitrogen fixation. In: Subba Rao NS, editor. Selected Topics in Biological Nitrogen Fixation. New Delhi: Oxford/IBH Publishing; 1984. pp. 1-35

10

Management Practices to Sustain Crop Production and Soil and Environmental Quality

Upendra M. Sainju, Rajan Ghimire and Gautam P. Pradhan

Abstract

Improved management practices can be used to sustain crop yields, improve soil quality, and reduce N contaminations in groundwater and the atmosphere due to N fertilization. These practices include crop rotation, cover cropping, application of manures and compost, liming, and integrated crop-livestock system. The objectives of these practices are to reduce the rate of N fertilization, enhance N-use efficiency, increase crop N uptake, promote N cycling and soil N storage, and decrease soil residual N. This chapter discusses improved management practices to reduce N fertilization rate, sustain crop yields, and improve soil and environmental quality. The adaptation of these practices by farmers, producers, and ranchers, however, depends on social, economic, soil, and environmental conditions.

Keywords: crop yields, environmental quality, management practices, nitrogen fertilizer, nitrogen-use efficiency, soil quality

1. Introduction

Legume-integrated crop rotations provide opportunity to reduce N fertilizer rates due to increased N supply by legume residues to succeeding crops compared with nonlegume monocropping [1, 2]. As little or no N fertilizer is applied to legumes during their growth, inclusion of legumes in rotation with nonlegumes helps to reduce the overall N rate for a crop rotation, which increase farm income by reducing C footprints and lowering the cost of N fertilization [1, 3]. Legumes also fix atmospheric N and release it for as long as 3 years, increasing yields of succeeding crops compared with nonlegume crops in crop rotations [4]. Crop rotations also reduce disease, pest, and weed infestations [5], improve soil structure and organic matter storage [6], increase water-use efficiency [7], and enhance soil health through microbial proliferation [8]. Crop rotation can also increase N uptake efficiency of diverse crops and reduce soil residual N compared with monocropping [2].

Cover cropping has many beneficial effects on sustaining crop yields and improving soil and environmental quality. Cover crops planted after the harvest of cash crops use soil residual N, reducing N leaching. The additional residues supplied by cover crops increase soil organic matter and fertility [9, 10]. Legume cover crops reduce N fertilization rates and enhance crop yields, but nonlegume cover crops are more effective on enhancing C sequestration [11, 12]. Similarly, integrate crop-livestock system, while reducing feed cost and supplying meat, milk, and wood, enhances N cycling and soil fertility, and control weeds [13, 14].

Continuous application of NH_4-based N fertilizers to nonlegume crops can reduce soil pH compared with legume-nonlegume crop rotations where N fertilizer is not applied to legumes [15]. After 16–28 years of management implications, soil pH was reduced by 0.22–0.42 from the original level in continuous nonlegumes compared with crop rotations containing legumes and nonlegumes [15]. Soil acidification from N fertilization to crops primarily results from (1) increased removal of basic cations, such as calcium (Ca), magnesium (Mg), potassium (K), and sodium (Na) in crop grains and stover due to increased yield; (2) leaching of soil residual NO_3-N, Ca, and Mg; and (3) microbial oxidation (or nitrification) of NH_4-based N fertilizers that release H^+ ions [16]. Alkalinity produced during plant uptake of N or conversion of inorganic N to organic form, however, can partly or wholly counter the acidity from nitrification [17]. Increased toxicity of aluminum (Al), iron (Fe), and manganese (Mn) and reduced availability of most nutrients, such as P, Ca, Mg, K, and Na, during acidification can reduce crop growth and yield [18].

Here we discuss various management strategies to reduce N fertilization rates, increase N-use efficiency, and decrease N leaching and N_2O emissions due to N fertilization. These practices will reduce the cost of N fertilization while sustaining crop production and reducing soil and environmental degradation.

2. Management practices

Management practices that reduce N fertilization rates without affecting crop yields and quality are needed to reduce soil and environmental degradation, as soil degradation is directly related to increased N rates. Some of these practices include crop rotation, cover cropping, application of manure and compost, and integrated crop-livestock system. These practices can increase N inputs, reduce N fertilization rates, conserve soil organic matter, and enhance soil health and environmental quality without affecting crop yields compared with traditional management practices. We discuss these practices as follows.

2.1 Crop rotation

Crop rotations that include legumes and nonlegumes in the rotation can substantially reduce N fertilization rates compared with nonlegume monocropping because legumes supply N to the soil due to their greater N concentration from atmospheric N fixation than nonlegumes. As no N fertilizer is applied to legumes, overall N fertilization rate is lower for the legume-nonlegume rotation than continuous nonlegumes while still maintaining crop yields. Sainju et al. [19] observed that annualized crop biomass and grain yields under rainfed condition were similar or greater with legume-based rotations that included pea, durum (*Triticum turgidum* L.), canola (*Brassica napus* L.), and flax (*Linum usitatissimum* L.) than with continuous durum (**Table 1**). Crop rotation is an effective management practice to control weeds, diseases, and pests [7]; reduce the risk of crop failure, farm inputs, and duration of fallow; and improve the economic and environmental sustainability of dryland cropping systems [20]. Diversified crop rotations can efficiently use water and N compared with monocropping [7, 21]. For instance, wheat and barley can efficiently utilize soil water in wheat-pea and barley-pea rotations than continuous wheat and barley. This is because pea uses less water than wheat and barley, resulting in more water available for succeeding crops in the rotation [7, 21].

Crop rotation†	Annualized biomass yield (Mg ha^{-1})	Annualized grain yield (Mg ha^{-1})
CD	3.32b‡	1.77a
D-C-D-P	4.02a	1.76a
D-D-C-P	3.90a	1.70a
D-F-D-P	3.39b	1.63ab
D-D-F-P	3.56b	1.54b

†Crop rotations are CD, continuous durum; D-C-D-P, durum-canola-durum-pea; D-D-C-P, durum-durum-canola-pea; D-F-D-P, durum-flax-durum-pea; and D-D-F-P, durum-durum-flax-pea.
‡Numbers followed by different letters within a column are significantly different at $P \le 0.05$ by the least square means test.

Table 1.
Effect of crop rotation on average annualized crop biomass (stems and leaves) and grain yields of durum, canola, flax, and pea from 2006 to 2011 in eastern Montana, USA (Sainju et al., 2017d).

Crop rotation can enhance or maintain soil organic C and N levels compared to monocropping. Both soil C and N stocks can be influenced by the quality and quantity of residue returned to the soil from crops involved in the rotation [12, 22]. Crop rotation can sequester C at 200 ± 120 kg C ha^{-1} $year^{-1}$, reaching equilibrium in 40–60 years compared with monocropping [23]. Sainju [24] found that soil organic C at 0–5 and 5–10 cm was similar in no-till malt barley-pea rotation (NTB-P) and no-till continuous malt barley (NTCB), both of which had greater soil organic C than no-till malt barley-fallow (NTB-F) and conventional till malt barley-fallow (CTB-F) due to greater amount of crop residue returned to the soil and reduced mineralization of soil organic matter (**Figure 1**). Similarly, Sainju et al. (2017d) found that soil total C at 0–125 cm was similar to continuous durum and rotations that included durum, canola, pea, and flax, except D-D-F-P (**Table 2**). Soil total N at 0–120 cm was greater with spring wheat-pea rotation than continuous spring wheat (**Table 3**) [25].

In an experiment evaluating the effects of crop rotation and cultural practice (traditional and ecological) on N balance in dryland agroecosystems, Sainju et al. [26, 27] observed that N fertilization rates were lower with legume-based crop rotations (D-C-D-P, D-D-C-P, D-F-D-P, and D-D-F-P) than nonlegume monocropping (CD) (**Table 4**). Traditional cultural practices included conventional till, recommended seed rate, broadcast N fertilization, and reduced stubble height and ecological practices inlcuded no-till, increased seed rate, banded N fertilization, and increased stubble height. They found that both total N input and output were greater with legume-based rotations than nonlegume monocropping due to pea N fixation and increased grain N removal. As a result, N balance was positive, indicating N surplus in legume-based rotations, and negative, indicating N deficit in nonlegume monocropping. This suggests that external N input is lower to sustain crop yields in legume-based crop rotations than nonlegume monocropping.

Legume-nonlegume rotation can also resist soil acidification compared with continuous nonlegumes. Sainju et al. [18] reported that soil pH at 0–7.5 cm after 30 years of experiment initiation was 0.13–0.44 greater and at 7.5–15.0 cm was 0.11–0.29 greater with spring wheat-barley/pea rotation (FSTW-B/P) than continuous spring wheat (NTCW, STCW, and FSTCW) (**Table 5**). They explained this as a result of lack of N fertilization to pea and reduced N fertilization rate to spring wheat following pea whose residue supplied N to spring wheat because of higher M concentration than spring wheat and barley residues. Soil residual NO_3-N, which can pollute groundwater through leaching, was lower with legume-based crop rotations containing durum, canola, pea, and flax than continuous durum (**Table 6**), suggesting that legume-based crop rotations can reduce N fertilization rate and the potential for N leaching compared with nonlegume monocropping.

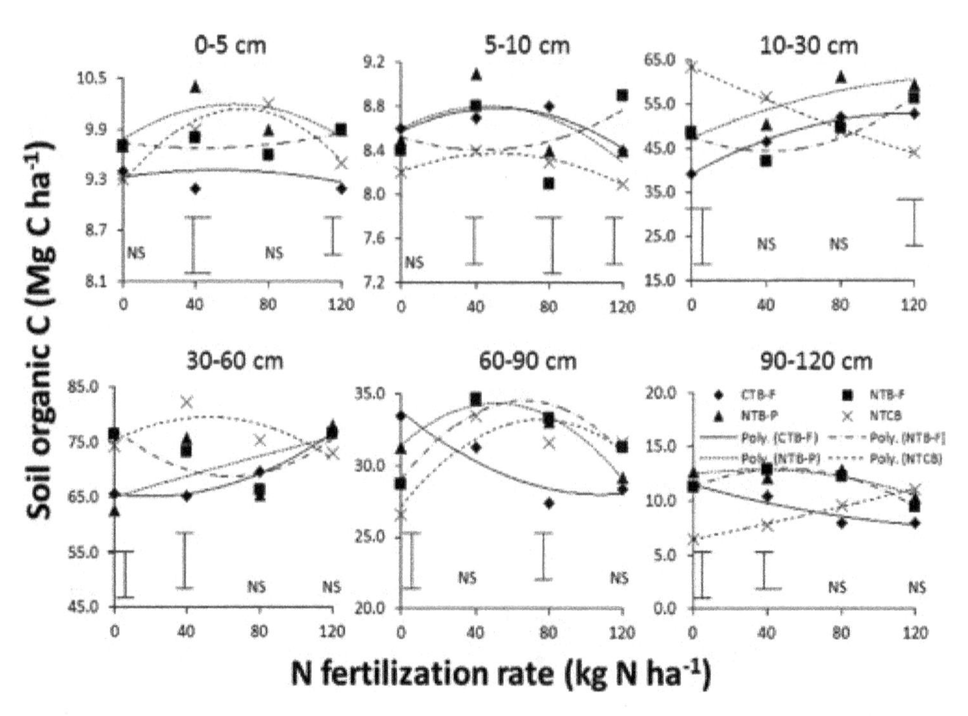

Figure 1.
Soil organic C at the 0–120 cm depth as affected by 6 years of N fertilization rates to malt barley in various cropping systems in eastern Montana, USA. CTB-F denotes conventional till malt barley-fallow; NTB-F, no-till malt barley-fallow; NTB-P, no-till malt barley-pea; and NTCB, no-till continuous malt barley. Vertical bars denote least significant difference between tillage and cropping sequence treatments within a N rate at P = 0.05 [24].

Crop rotation†	STC at 0–125 cm (Mg C ha^{-1})
CD	394.6a‡
D-C-D-P	395.4a
D-D-C-P	387.1a
D-F-D-P	395.4a
D-D-F-P	370.2b

†Crop rotations are CD, continuous durum; D-C-D-P, durum-canola-durum-pea; D-D-C-P, durum-durum-canola-pea; D-F-D-P, durum-flax-durum-pea; and D-D-F-P, durum-durum-flax-pea.
‡Numbers followed by different letters within a column are significantly different at P ≤ 0.05 by the least square means test.

Table 2.
Soil total C (STC) at the 0–125 cm depth after 6 years as affected by crop rotation in eastern Montana, USA [19].

2.2 Cover cropping

Cover crops have been grown successfully in regions with mild winter to provide vegetative cover for reducing soil erosion. Cover crops are usually grown in the fall after the harvest of summer cash crops and have many benefits for sustaining crop yields and improving soil and water quality. Winter cover crops use soil residual N that may otherwise leach into groundwater after crop harvest in the fall, thereby reducing soil profile NO_3-N content and N leaching [29, 30]. Summer cover crops are grown in the summer to replace fallow when no other crops are grown. Depending on the species, cover crops can maintain or increase soil organic C and N

Crop rotation[a]	STN (Mg N ha^{-1})							
	0–5 cm	5–10 cm	10–20 cm	20–40 cm	40–60 cm	60–90 cm	90–120 cm	0–120 cm
CW	0.82	0.91	1.46	2.34b[b]	2.11	2.29b	2.11	12.03b
W-P	0.85	0.90	1.53	2.66a	2.24	2.55a	2.23	12.96a
W-B-P	0.79	0.86	1.44	2.43ab	2.17	2..35b	2.22	12.17b
W-B-C-P	0.81	0.88	1.47	2.54a	2.26	2.51a	2.10	12.62ab

[a]*Crop rotations are CW, continuous spring wheat; W-P, spring wheat-pea; W-B-P, spring wheat-barley hay-pea; and W-B-C-P, spring wheat-barley hay-corn-pea.*
[b]*Numbers followed by different letters within a column are significantly different at* $P \leq 0.05$ *by the least square means test.*

Table 3.
Soil total N (STN) at the 0–120 cm depth after 6 years as affected by crop rotation in eastern Montana, USA [25].

by providing additional crop residue which increases biomass C and N inputs to the soil [9, 10, 12] and sequester atmospheric C and/or N, thereby reducing the rate of N fertilization to summer crops [9, 10]. Other benefits of cover crops include increased soil aggregation and water infiltration capacity [31], improved water holding capacity [32], and reduced soil erosion [33] compared with no cover crop.

Integrating legumes in crop rotations can supply N to succeeding crops and increase crop yields compared to nonlegumes or no cover crop rotations [10]. In contrast, nonlegume cover crops are effective in increasing soil organic C through increased biomass production compared with legumes or no cover crop [9, 10, 12]. Nonlegumes also reduce NO_3-N leaching from the soil profile better than legumes, or no cover crop do [29]. As none of the cover crops are effective enough to provide most of these benefits, i.e., to supply N, sustain crop yields, increase soil organic matter, and reduce N leaching, a mixture of legume and nonlegume cover crops is ideal to supply both C and N inputs in adequate amounts that help to improve soil and water quality by increasing organic matter content and the potential for reducing N leaching compared with legumes and increase crop yields compared with nonlegumes [12, 34, 35].

Sainju et al. [36] found higher biomass yield with hairy vetch/rye (*Secale cereale* L.) mixture than rye, hairy vetch, or winter weeds, and N concentration in the mixture similar to hairy vetch, except in 2001 (**Table 7**). As a result, they observed greater biomass C and N contents with hairy vetch/rye mixture than rye and winter weeds and similar to or greater than hairy vetch. The C/N ratio of cover crop biomass, which measures the decomposition rate of the residue, was similar between hairy vetch/rye mixture and hairy vetch.

Because of increased C supply, soil organic C at 0–10 and 10–30 cm was also greater with hairy vetch/rye than other cover crops (**Figure 2**). At 30–60 cm, soil organic C was greater with hairy vetch/rye than other cover crops, except hairy vetch. Soil total N at 0–15, 15–30, and 0–120 cm was also greater with hairy vetch and hairy vetch/rye mixture than other cover crops (**Figure 3**). Similarly, soil residual NO_3-N content at 0–120 cm was greater with hairy vetch than other cover crops and is slightly greater than that with 120–130 kg N ha^{-1} (**Figure 4**). Nitrogen loss at 0–120 cm during the winter fallow period from November to April was lower with hairy vetch/rye than other cover crops (**Table 8**). Nitrogen fertilizer equivalence of rye and winter weeds for cotton and sorghum ranged from −129 to 69 kg N ha^{-1}, but those of hairy vetch and hairy vetch/rye ranged from 92 to 220 kg N ha^{-1} (**Table 9**), suggesting that hairy vetch and hairy vetch/rye can increase cotton and sorghum yields similar to those by 92–220 kg N ha^{-1} [11]. These results suggest that hairy vetch/rye mixture can produce crop yields similar to hairy vetch.

Parameter	Traditional (kg N ha^{-1} year^{-1})					Ecological (kg N ha^{-1} year^{-1})				
	CD[a]	D-C-D-P[a]	D-D-C-P[a]	D-F-D-P[a]	D-D-F-P[a]	CD	D-C-D-P	D-D-C-P	D-F-D-P	D-D-F-P
N inputs										
N fertilization rate	83A[b]	62B	59B	52B	54B	87A	60B	63B	55B	56B
Pea N fixation	0C	84AB	76B	80AB	75B	0C	84AB	78B	87A	82AB
Atmospheric N deposition	14	14	14	14	14	14	14	14	14	14
N added by crop seed	3	3	3	3	3	3	3	3	3	3
Nonsymbiotic N fixation	5	5	5	5	5	5	5	5	5	5
Total N input	105B	167A	156A	154A	150A	109B	166A	162A	164A	159A
N outputs										
Grain N removal	49B	62A	57AB	54AB	55AB	52AB	65A	64A	63A	54AB
Denitrification	12	10	9	8	9	13	9	10	9	9
Ammonia volatilization	12	9	9	8	8	13	9	9	8	8
Plant senescence	5	7	6	6	6	6	7	7	7	6
N leaching	9	12	12	12	12	9	12	12	12	12
Gaseous N (NO_x) emissions	2	3	3	3	3	2	3	3	3	3
Surface runoff	1	2	1	1	1	1	2	2	2	2
Total N output	91B	105A	98AB	92B	94AB	96AB	107A	107A	103A	94AB
Changes in N level[c]	14B	62A	58A	62A	56A	13B	59A	55A	61A	65A
N sequestration rate (0–125 cm)[d]	50	45	42	46	43	52	48	46	44	40
N balance[e]	−36 (±11)B	17 (±5)A	16 (±4)A	16 (±4)A	13 (±3)A	−39 (±12)B	11 (±3)A	9 (±2)A	17 (±4)A	25 (±5)A

[a] *Crop rotation are CD, continuous durum; D-C-D-P, durum-canola-durum-pea; D-D-C-P, durum-durum-canola-pea; D-F-D-P, durum-flax-durum-pea; and D-D-F-P, durum-durum-flax-pea.*
[b] *Numbers followed by the same letter within a row are not significantly different at* $P \leq 0.05$.
[c] *Changes in N level = total N input − total N output.*
[d] *Determined from the linear regression analysis of soil total N (STN) at 0–125 cm from the year 2005 to 2011.*
[e] *N balance = changes in N levels − N sequestration rate (0–125 cm).*

Table 4.
Annual N balance due to the difference between total N inputs and outputs and N sequestration rate under dryland agroecosystems from 2005 to 2011 in eastern Montana, USA [26, 27].

Tillage and cropping sequence[a]	Soil depth					
	0–7.5 cm	7.5–15 cm	15–30 cm	30–60 cm	60–90 cm	90–120 cm
pH						
NTCW	5.33ab[b]E[c]	6.50abD	7.60C	8.35B	8.58A	8.75A
STCW	5.05bE	6.15bD	7.58C	8.25B	8.63A	8.70A
FSTCW	5.02bE	6.33bD	7.80C	8.30B	8.68AB	8.73A
FSTW-B/P	5.46aE	6.44bD	7.60C	8.15B	8.51A	8.59A
STW-F	5.73aE	7.03aD	7.65C	8.25B	8.50AB	8.66A
Contrast						
NT vs. T	0.29	0.26	−0.09	0.08	−0.08	0.04
CW vs. W-F	−0.68***	−0.88**	−0.08	0.01	0.13	0.04
CW vs. W-B/P	−0.43*	−0.11	0.20	0.15	0.16	0.14
Buffer pH						
NTCW	6.45bE	7.10abD	7.43C	7.60B	7.70AB	7.73A
STCW	6.38bE	7.00bD	7.43C	7.58B	7.68A	7.70A
FSTCW	6.43bE	7.05bD	7.45C	7.60B	7.70AB	7.73A
FSTW-B/P	6.66aD	7.13abC	7.44B	7.58B	7.69AB	7.70A
STW-F	6.80aE	7.24aD	7.44C	7.59B	7.66AB	7.72A
Contrast						
NT vs. T	0.05	0.08	−0.01	0.01	0.01	0.01
CW vs. W-F	−0.43***	−0.24**	−0.01	−0.01	0.01	−0.01
CW vs. W-B/P	−0.24*	−0.08	−0.01	0.03	0.01	0.03

**Significant at P = 0.05.*
***Significant at P = 0.01.*
****Significant at P = 0.001.*
[a]*FSTCW, fall and spring till continuous spring wheat; FSTW-B/P, fall and spring till spring wheat-barley (1994–1999) followed by spring wheat-pea (2000–2013); NTCW, no-till continuous spring wheat; STCW, spring till continuous spring wheat; and STW-F, spring till spring wheat-fallow. CW represents continuous wheat; NT, no-till; T, till; W-B/P, spring wheat-barley/pea; and W-F, spring wheat-fallow.*
[b]*Numbers followed by the same lowercase letter within a column among treatments in a set are not significantly different at P ≤ 0.05.*
[c]*Numbers followed by the same uppercase letter within a row among soil depths in a set are no significantly different at P ≤ 0.05.*

Table 5.
Effect of tillage and crop rotation combination on soil pH and buffer pH at the 0–120 cm depth after 30 years of experiment initiation in eastern Montana, USA [18].

The mixture can also increase soil organic matter and reduce N fertilization rate and the potential for N leaching compared with rye and winter weeds. Therefore, legume-nonlegume cover crop mixture can provide several benefits, such as reducing the cost of N fertilization, maintaining crop yields, enhancing soil organic matter, and reducing N leaching compared with either cover crop alone or no cover crop.

2.3 Application of manure and compost

Manure and compost are rich sources of nutrients, and their application can increase soil organic C and total N, improving soil quality and crop production compared to no fertilizer application [37, 38]. Sainju et al. [39, 40] compared soil organic C and total N after 10 years of poultry litter with inorganic N

Crop rotation[a]	NO$_3$-N content at various depths (kg N ha^{-1})						
	0–5 cm	5–10 cm	10–20 cm	20–50 cm	50–88 cm	88–125 cm	0–125 cm
CD	2.47a[b]	1.81a	2.43a	8.49a	9.37a	9.17a	33.87a
DCDP	1.82a	1.22b	1.94b	6.47a	7.77a	6.71b	26.32b
DDCP	1.86a	1.19b	1.93b	5.97a	8.07a	6.38b	25.59b
DFDP	1.90a	1.37b	2.20a	6.59a	9.62a	8.64ab	30.60a
DDFP	1.74a	1.28b	2.29a	6.27a	8.63a	6.65b	27.02b

[a]*Crop rotations are CD, continuous durum; DCDP, durum-canola-durum-pea; DDCP, durum-durum-canola-pea; DDFP, durum-durum-flax-pea; and DFDP, durum-flax-durum-pea.*
[b]*Numbers followed by different letters within a column are significantly different at $P \leq 0.05$ by the least square means test.*

Table 6.
Soil NO$_3$-N content at the 0–125 cm depth as affected by crop rotation and cultural practice averaged across years from 2006 to 2011 in eastern Montana, USA [28].

Cover crop†	Biomass yield (Mg ha^{-1})	Concentration		Content		C/N ratio
		C (g kg^{-1})	N (g kg^{-1})	C (kg ha^{-1})	N (kg ha^{-1})	
2000						
Weeds	1.65d‡	370b	15b	587d	25d	24b
Rye	6.07b	430a	15b	2670b	68c	29a
Vetch	5.10c	394ab	33a	2006c	135b	12c
Vetch/rye	8.18a	366b	38a	3512a	310a	10c
2001						
Weeds	0.75d	391b	20b	277d	15b	20c
Rye	3.81b	448a	8d	1729b	32b	57a
Vetch	2.44c	398b	32a	964c	76a	12c
Vetch/rye	5.98a	434a	14c	2693a	84a	32b
2002						
Weeds	1.25c	375b	18b	476c	23b	21b
Rye	2.28b	434a	11b	986b	25b	40a
Vetch	5.16a	361b	36a	2094a	167a	10c
Vetch/rye	5.72a	381b	33a	2260a	186a	11c

†Cover crops are rye, cereal rye; vetch, hairy vetch; vetch/rye, hairy vetch and rye biculture; and weeds, winter weeds.
‡Numbers followed by the same letter within a column of a year are not significantly different at $P \leq 0.05$.

Table 7.
Effect of cover crop species on aboveground biomass yield and C and N contents in cover crops from 2000 to 2002 in central Georgia, USA [36].

fertilizer applications, both applied at 100 kg N ha^{-1} to corn and cotton (**Tables 10** and **11**). They found that soil organic C and total N at 0–20 cm were greater with poultry litter application than inorganic N fertilization, regardless of tillage practices. As a result, poultry litter application sequestered C at 461 kg C ha^{-1} year^{-1} and N at 38 kg N ha^{-1} year^{-1} compared to 38 kg C ha^{-1} year^{-1} and 4 kg N ha^{-1} year^{-1}, respectively, with N fertilization. As poultry litter also supplied C at 1.7 Mg C ha^{-1} year^{-1} [40] and only 60% of N from poultry litter was available

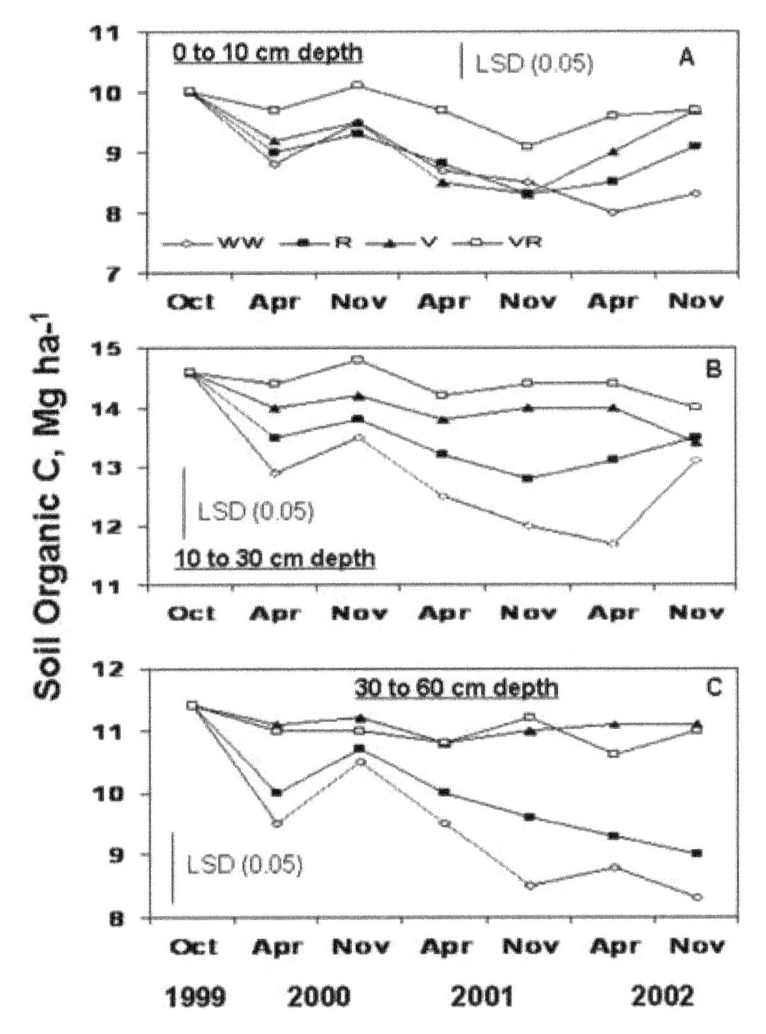

Figure 2.
Effect of cover crop on soil organic C at the (A) 0-10 cm, (B) 10-30 cm, and (C) 30-50 cm depths in a chisel-tilled system (October 1999–November 2002, central Gerogia, USA). R denotes cereal rye; V, hairy vetch; VR, hairy vetch and rye biculture; and WW, winter weeds. Vertical line with LSD (0.05) is the least significant difference between cover crops within a sampling date at P = 0.05 [12].

to crops in the first year [37], Sainju et al. [39, 40] reported that part of non-mineralized C and N from the litter converted to soil organic C and N, thereby increasing their levels with poultry litter application. In contrast, little or no C was supplied by inorganic N fertilizer, and most of N supplied by the fertilizer can either be taken up by the crop or lost to the environment through leaching, denitrification, and volatilization.

Because of lower N availability from poultry litter as a result of reduced N mineralization, total aboveground biomass and N uptake of corn, cotton, and rye cover crop were lower with poultry litter application than inorganic N fertilization (**Table 12**). Although soil health and quality can be improved with poultry litter application through organic matter enrichment, crop yields can be lower compared with N fertilization. For enhancing soil and environmental quality and sustaining crop yields, both inorganic N fertilizer and manure/compost should be applied as a mixture in balanced proportion as per crop demand after analyzing soil NO_3-N test to a depth of 60 cm. This could reduce N fertilization rate and undesirable consequences of N fertilization on soil and environmental quality.

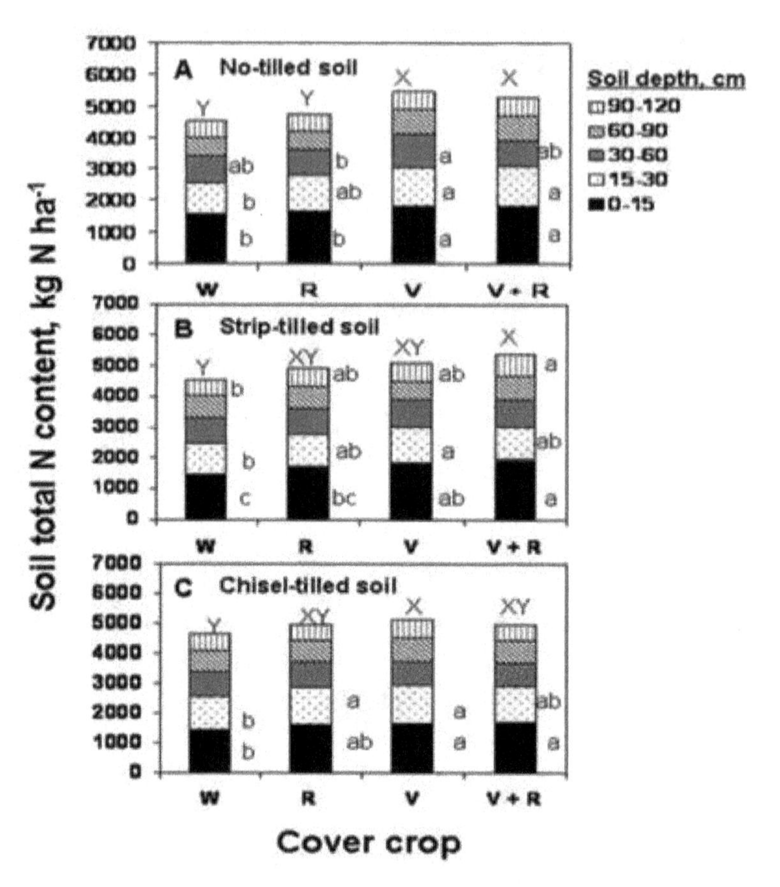

Figure 3.
Effect of cover crop on soil total N at the 0–120 cm depth in (A) no-tilled, (B) strip-tilled, and (C) chisel-tilled soils after 3 years in Central Georgia, USA. R denotes cereal rye; V, hairy vetch; V + R, hairy vetch and rye biculture; and WW, winter weeds. Bars followed by the same lowercase letter within a soil depth are not significantly different between cover crops at P = 0.05. Bars followed by the same uppercase letter at the top are not significantly different between cover crops at the 0–120 cm depth at P ≤ 0.05 [34].

2.4 Integrated crop-livestock system

Integrated crop-livestock systems were commonly used to sustain crop and livestock products throughout the world before commercial fertilizers were introduced in 1950 [41]. The system is still common among producers in developing countries, especially in Africa and Asia where fertilizers are scarce and expensive [42, 43]. The integrated crop-livestock system has the potential to improve soil quality and sustain crop yields [41, 44]. The major benefits of the system are (1) production of crops, meat, and milk, (2) production of crop residue for animal feed, (3) production of manure to apply as fertilizer, (4) use of animals as draft power for tillage, and (5) control of weeds and pests [41, 42].

Animal grazing during fallow periods in wheat-fallow systems can be used to effectively control weeds [14] and insects, such as wheat stem saw fly [*Cephus cinctus* Norton (Hymenoptera: Cephidae)] [13]. The animal usually grazes on crop residues and weeds during the fallow period. Although grazing can reduce the quantity of crop residue returned to the soil, the number of animals grazed per unit area can be adjusted in such a way that crop residue cover in the grazing treatment will be similar to that in the conservation tillage system where soil erosion is minimal [14]. Animal feces and urine returned to the soil during grazing can enrich

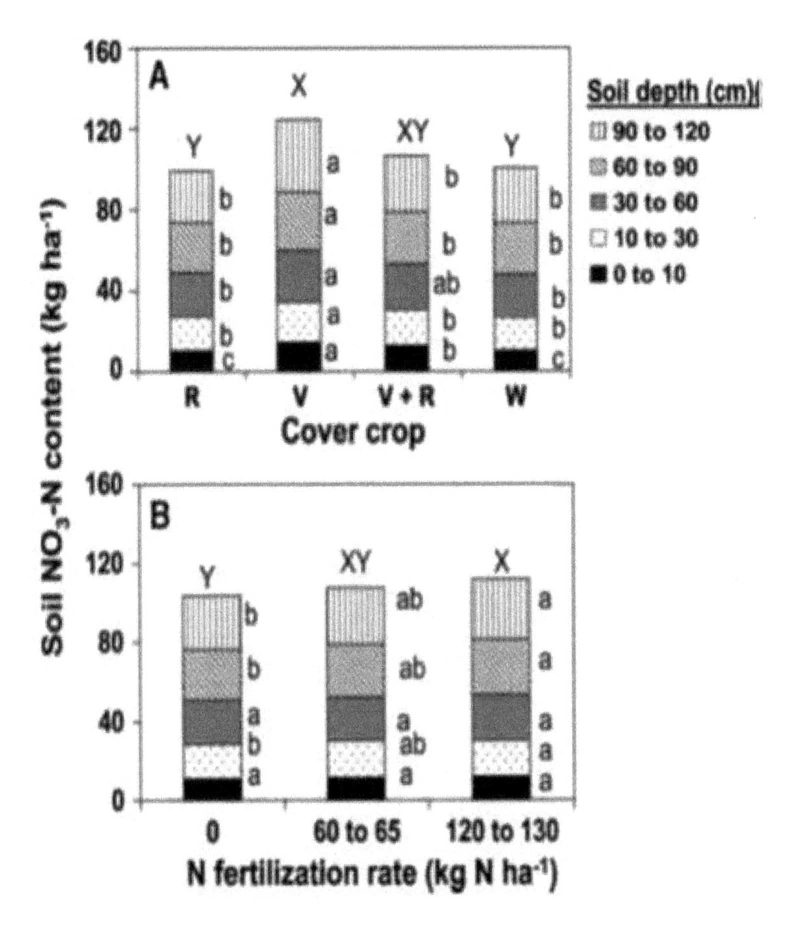

Figure 4.
Effect of (A) cover crop and (B) N fertilization rate on soil NO_3-N content at the 0–120 cm depth in Central Georgia, USA. R, denotes cereal rye; V, hairy vetch; V + R, hairy vetch and rye biculture; and W, winter weeds. Bars followed by the same lowercase letter within a soil depth are not significantly different between cover crops at P = 0.05. Bars followed by the same uppercase letter at the top are not significantly different between cover crops at the 0–120 cm depth at P ≤ 0.05 [35].

Cover crop†	Total crop residue and soil N‡ (kg N ha^{-1})			Total crop residue and soil N§ (kg N ha^{-1})		
	November 2000	April 2001	Loss	November 2001	April 2002	Loss
Rye	5057bc¶	4888b	169b	4820b	4764b	56a
Vetch	5455a	5235a	220a	5323a	5244a	79a
Vetch/rye	5249ab	5141a	108c	5222a	5182a	40a
Weeds	4869c	4709b	160b	4725b	4649b	76a

†Cover crops are rye, cereal rye; vetch, hairy vetch; vetch/rye, hairy vetch and rye biculture; and weeds, winter weeds or no cover crop.
‡Include soil NH_4-N + NO_3-N + organic N contents at 0–120 cm, and N returned to the soil from cotton biomass (stems + leaves) in November 2000 and cover crop biomass in April 2001.
§Include soil NH_4-N + NO_3-N + organic N contents at 0–120 cm, and N returned to the soil from sorghum biomass (stems + leaves) in November 2001 and cover crop biomass in April 2002.
¶Numbers followed by the same letter within a column are not significantly different at P ≤ 0.05.

Table 8.
Effect of cover crop on N loss from crop residue and soil N (NH_4-N + NO_3-N + organic N contents) at the 0–120 cm depth during the two winter seasons (from November 2000 to April 2001 and from November 2001 to April 2002) in central Georgia, USA [35].

Parameter	Cover crop				Regression analysis[a]	
	Winter weeds	Rye	Hairy vetch	Hairy vetch/rye	R^2	*P*
2000 cotton						
Lint yield	—	—	—	—	0.25	0.67
Lint N uptake	—	—	—	—	0.25	0.67
Biomass yield	−13	30	149	93	0.96	0.13
Biomass N uptake	−21	2	165	92	0.99	0.06
Soil inorganic N	−60	−190	220	140	0.64	0.40
2001 sorghum						
Lint yield	7	−64	107	179	0.96	0.12
Lint N uptake	25	−67	167	150	0.96	0.14
Biomass yield	32	−168	194	194	0.99	0.02
Biomass N uptake	69	−84	192	83	0.98	0.08
Soil inorganic N	59	12	116	71	0.86	0.25
2002 cotton						
Lint yield	—	—	—	—	0.28	0.82
Lint N uptake	—	—	—	—	0.24	0.87
Biomass yield	−21	−61	139	205	0.96	0.12
Biomass N uptake	−35	−13	134	160	0.97	0.11
Soil inorganic N	−74	5	176	160	0.70	0.37

[a]*Regression analysis of N fertilization rates versus cotton and sorghum yields and N uptake and soil inorganic N.*

Table 9.
Nitrogen fertilizer equivalence (kg N ha^{-1}) of cover crops and soil inorganic N (NH_4-N + NO_3-N) content at the 0–30 cm depth for cotton and sorghum yields and N uptake from 2000 to 2002 in central Georgia, USA [11].

soil nutrients, improve soil quality, and increase crop yields [44]. The distribution of feces and urine by animals during grazing at the soil surface can be uneven; however, distribution can be more uniform with sheep than with cattle grazing [45].

Hatfield et al. [14] reported that sheep grazing during fallow did not affect soil organic matter and nutrient levels compared to the non-grazed treatment in the North Central Montana. Sheep grazing can increase soil bulk density and extractable P and grass yields compared to cattle grazing [45]. Snyder et al. [46] found similar or greater wheat grain yields with and without animal grazing. Similarly, Quiroga et al. [47] observed that 10 years of cattle grazing did not alter soil P concentration in Argentina. In contrast, Niu et al. [48] in Australia observed greater soil P and K concentrations in sheep camping than in non-camping sites due to increased animal excreta. Cattle and sheep grazing in the pasture can increase soil P and K concentrations compared to non-grazing [45].

Sainju et al. [49] reported that annualized wheat grain and biomass yields were lower with spring wheat-fallow and winter wheat-fallow rotations than continuous spring wheat due to the absence of crops during the fallow period (**Table 13**). In

Tillage†	N source‡	SOC concentration ($g C kg^{-1}$)		SOC content ($Mg C ha^{-1}$)	Changes in SOC from 1996 to 2006 ($Mg C ha^{-1}$)	C sequestration rate ($kg C ha^{-1} year.^{-1}$)
	100 kg N ha^{-1}	0–10 cm	10–20 cm	0–20 cm	0–20 cm	0–20 cm
NT	AN	13.5	11.0	40.1	1.47	147
	PL	15.9	10.5	43.7	5.10	510
MT	AN	15.9	11.0	42.6	3.97	397
	PL	15.4	10.6	42.2	3.63	363
CT	AN	14.3	10.7	37.4	−1.20	−120
	PL	15.3	11.8	43.7	5.10	510
LSD (0.05)		—	—	3.1	3.1	310
Means	AN	14.6a§	10.9a	40.0b	1.41b	141b
	PL	15.6a	11.0a	43.2a	4.61a	461a

†Tillage is CT, conventional till; MT, mulch till; and NT, no-till.
‡N source is AN, NH_4NO_3; and PL, poultry litter.
§Numbers followed by different letters within a column in a set are significantly different at $P \leq 0.05$ by the least square means test.

Table 10.
Effect of tillage and N source on soil organic C (SOC) at the 0–20 cm depth after 10 years in Alabama, USA [40].

Tillage[a]	N source[b]	STN concentration ($g N kg^{-1}$)		STN content ($Mg N ha^{-1}$)	Change in STN from 1996 to 2006 ($Mg N ha^{-1}$)	N sequestration rate ($kg N ha^{-1} year^{-1}$)
	(100 kg N ha^{-1})	0–10 cm	10–20 cm	0–20 cm	0–20 cm	0–20 cm
NT	AN	1.23	1.03	3.44	−0.23	−23
	PL	1.52	1.02	4.19	0.49	49
MT	AN	1.42	1.01	3.84	0.15	15
	PL	1.49	0.92	3.91	0.21	21
CT	AN	1.31	0.98	3.67	−0.03	−3
	PL	1.51	1.04	4.11	0.41	41
LSD (0.05)[c]		—	—	0.24	0.24	24
Means	AN	1.55b[d]	1.59a	3.65b	−0.04b	−4b
	PL	1.65a	1.59a	4.07a	0.38a	38a

[a]Tillage is CT, conventional till; MT, mulch till; and NT, no-till.
[b]N source is AN, ammonium nitrate; and PL, poultry litter.
[c]Least significant differences between treatments at $P = 0.05$.
[d]Numbers followed by the same letter within a column in a set are not significantly different at $P \leq 0.05$.

Table 11.
Effects of tillage and N source on soil total N and N sequestration rate at the 0–20 cm depth after 10 years in Alabama, USA [39].

Cropping system	N source	Total crop biomass	Total N uptake
	100 kg N ha^{-1}	(Mg ha^{-1})	(kg N ha^{-1})
Rye/cotton-rye/cotton-corn		137.0a†	1544a†
Cotton-cotton-corn		110.2b	1247b
	NH_4NO_3	133.3a	1502a
	Poultry litter	111.8b	1289b

†Numbers followed by the same letter within a column in a set are not significantly different at $P \leq 0.05$.

Table 12.
Effects of cropping system and N source on total biomass (stems + leaves) residues of rye, cotton, and corn and N uptake from 1997 to 2005 in Alabama, USA [39, 40].

Year	Cropping sequence† (Mg ha^{-1})			Weed management‡ (Mg ha^{-1})			Mean
	CSW	SW-F	WW-F	Chem.	Mech.	Graz.	
Annualized grain yield							
2004	5.55a§A¶	2.90aC	3.53aB	3.92aA	4.01aA	4.05aA	3.99a
2005	2.68bA	1.83bB	1.15eC	1.84cA	1.92bA	1.90bA	1.89b
2006	2.57bA	1.45cB	1.70 dB	1.89cA	1.90bA	1.92bA	1.90b
2007	1.86cB	1.18cC	2.95bA	1.89cA	2.03bA	2.00bA	2.00b
2008	2.61bA	1.56bcC	2.22cB	2.09bA	2.17bA	2.14bA	2.13b
Mean	3.05A	1.78C	2.31B	2.32A	2.42A	2.40A	
Annualized biomass yield							
2004	6.60aA	3.10aC	3.57aB	3.61aAB	3.41aB	3.89aA	4.42a
2005	3.28bA	1.65bB	1.94bcB	2.52bA	2.17bcA	2.19bA	2.29b
2006	2.96cA	1.57bcB	1.64cB	1.79bB	2.51bA	1.87bcB	2.06bc
2007	2.18dA	1.55bcB	2.25bA	1.78bA	2.21bcA	2.00bA	2.00c
2008	1.92dA	1.17cB	1.49cAB	1.08cB	1.91cA	1.58cA	1.53d
Mean	2.58A	1.49C	1.83B	1.79B	2.20A	1.91B	

†Cropping sequences are CSW, continuous spring wheat; SW-F, spring wheat-fallow; and WW-F, winter wheat-fallow.
‡Weed management practices are Chem., chemical where weeds were controlled with herbicide applications; Graz., grazing where weeds were controlled with sheep grazing; and Mech., mechanical where weeds were controlled with tillage.
§Numbers followed by the same lowercase letters within a column in a set are not significantly different at $P \leq 0.05$.
¶Numbers followed by the same uppercase letters within a row in a set are not significantly different at $P \leq 0.05$.

Table 13.
Effects of cropping sequence and weed management practice on annualized wheat grain and biomass (stems + leaves) yield from 2004 to 2008 in western Montana, USA [49].

contrast, wheat grain yield was not different among weed management practices where sheep grazing was used among one of the treatments to control weeds along with herbicide application and tillage, although wheat biomass yield was lower with sheep grazing and herbicide application than tillage. Soil organic C, total N, and NO_3-N contents varied among weed management practices and soil depths, but the contents at 0–120 cm were not affected by weed management practices (**Table 14**).

Weed management†	SOC content (Mg C ha^{-1})						
	0–5 cm	5–10 cm	10–30 cm	30–60 cm	60–90 cm	90–120 cm	0–120 cm
Chem.	18.3a‡	19.2a	61.7a	38.0a	32.2a	29.1b	198.4a
Mech.	17.3a	17.4a	58.2ab	38.0a	35.8a	37.0a	203.5a
Graz.	16.9a	17.7a	54.2b	36.1a	31.2a	31.4ab	187.5a
STN content (Mg N ha^{-1})							
Chem.	1.69a	1.89a	6.48a	4.96a	3.58a	2.79a	21.40a
Mech.	1.61a	1.74b	5.91a	5.00a	3.43a	2.99a	20.55a
Graz.	1.53a	1.79ab	6.33a	5.60a	3.86a	2.87a	22.09a
NO_3-N content (kg N ha^{-1})							
Chem.	12.6a	12.4a	20.6a	16.0a	18.9b	38.0a	118.6a
Mech.	10.3a	12.0a	21.1a	14.5a	28.8a	37.6a	124.4a
Graz.	9.9a	10.9a	18.7a	17.5a	23.2ab	35.0a	115.2a

†Weed management practices are Chem., chemical where weeds were controlled with herbicide applications; Graz., grazing where weeds were controlled with sheep grazing; and Mech., mechanical where weeds were controlled with tillage.

‡Numbers followed by different letters within a column are significantly different at $P \leq 0.05$ by the least square means test.

Table 14.
Soil organic C (SOC), total N (STN), and NO_3-N contents at the 0–120 cm depth after 5 years of weed management experiment initiation in western Montana, USA [50].

Soil P, K, and SO_4-S contents at 0–30 cm were lower with sheep grazing than other weed management practices, but pH, electrical conductivity, and Ca, Mg, and Na contents were similar or greater with sheep grazing (**Table 15**). Consumption of crop residue by sheep during grazing, but little P and K inputs to the soil through urine and feces, reduced soil P and K concentrations with sheep grazing compared with other weed management practices [49]. These results suggest that sheep grazing can reduce the cost of animal feed without seriously affecting crop yields and sustain soil organic matter and nutrients compared with other weed management practices, except P and K which need to be added with inorganic fertilizers to eliminate their deficiency. As soil residual NO_3-N content was not different among weed management practices, long-term study may be needed to evaluate if animal grazing can reduce N fertilization rate for crop production. However, animal grazing can recycle nutrients and control weeds effectively compared with herbicide application and tillage, thereby saving the cost of fertilization and weed control.

Legumes in the crop rotation can supply N from its residue to succeeding crops, thereby reducing N fertilization rates to succeeding nonlegumes. Also diversified crop rotations can use N and water more efficiently and reduce weed, pest, and disease infestations, thereby enhancing crop yields compared with continuous nonlegume monocropping. Cover crops grown to replace the fallow period can reduce soil erosion, enhance soil organic matter, and help to enrich soil health and fertility. Legume covers crop supply N and reduce N fertilization rate. Application of manure and compost can also enhance soil health and quality; however, additional inorganic N fertilization at lower rate is required to sustain crop yield and quality. Similarly, integrated crop-livestock system can help to reduce N fertilization rate by returning N and other nutrients through urine and feces to the soil during animal grazing without affecting crop yields. Some additional N fertilizer, however, may be required for sustainable crop production, because animals

Chemical properties	Soil depth	Weed management (WM)†		
		Chem.	Mech.	Graz.
P content (kg ha^{-1})	0–5 cm	34.5a‡	35.7a	30.8a
	5–10 cm	30.4a	29.3a	17.8b
	10–30 cm	81.2a	80.7a	40.1b
K content (kg ha^{-1})	0–5 cm	263a	271a	222b
	5–10 cm	176a	191a	139b
	10–30 cm	792a	859a	577b
pH	0–5 cm	6.45a	6.94a	6.72a
	5–10 cm	6.31a	6.64a	6.51a
	10–30 cm	7.06a	7.34a	7.31a
EC (S m^{-1})	0–5 cm	0.035a	0.037a	0.035a
	5–10 cm	0.024a	0.024a	0.024a
	10–30 cm	0.025a	0.026a	0.27a
Ca content (Mg ha^{-1})	0–5 cm	2.05a	2.06a	2.08a
	5–10 cm	2.14b	2.31a	2.25ab
	10–30 cm	10.70b	11.70ab	12.90a
Mg content (kg ha^{-1})	0–5 cm	278a	288a	304a
	5–10 cm	362b	382ab	417a
	10–30 cm	2619a	2593a	2640a
Na content (kg ha^{-1})	0–5 cm	11.7a	12.5a	12.8a
	5–10 cm	15.2b	15.2b	18.4a
	10–30 cm	84.8ab	76.6b	95.0a
SO_4-S content (kg ha^{-1})	0–5 cm	8.5ab	10.0a	7.4b
	5–10 cm	9.0ab	10.6a	7.1b
	10–30 cm	34.0ab	40.8a	28.8b

†Weed management practices are Chem., chemical where weeds were controlled with herbicide applications; Graz., grazing where weeds were controlled with sheep grazing; and Mech., mechanical where weeds were controlled with tillage.
‡Numbers followed by the same letter within a row in a set are not significantly different at $P \leq 0.05$.

Table 15.
Effect of weed management practice on soil nutrients, pH, and electrical conductivity (EC) at the 0–30 cm depth after 5 years of experiment initiation in western Montana, USA [49].

return only a part of nutrients through urine and feces to the soil, while most of the crop residue grazed is used to increase the live weight of the animal. The choice of the management practice to reduce N fertilization rate to crops depends on soil and climatic conditions and social, cultural, and economic perspectives of the producers.

2.5 Liming

Soil acidification can be reduced by applying lime. However, lime is bulky and requires in large amount to neutralize soil acidity. The transportation cost to carry lime from manufactures to farms is also high and especially so in hilly regions where roads are few or lacking. As a result, it is expensive to apply lime and most producers in developing countries cannot afford to apply it. Furthermore, neutralization of soil acidity with lime application is only temporary in nature. This suggests that lime should be applied frequently to neutralize acidity, which increases the cost of production. The best practice to reduce soil acidity is to reduce the rate of N fertilization. Several management practices, such as legume-nonlegume crop rotation, cover cropping, application of manures and compost, and integrated crop-livestock system, can reduce N fertilization rate without affecting crop yields.

3. Conclusions

Degradation in soil and environmental quality can be mitigated, and crop yields can be sustained by reducing N fertilization rates and using novel management techniques that increase N cycling and N-use efficiency. These techniques include legume-nonlegume crop rotation, cover cropping, application of manures and compost, and integrated crop-livestock system. Soil acidity can be neutralized by lime application, but the effect is temporary. It is expensive to apply lime, and many producers in developing countries cannot afford to do so. Adaptation of these techniques to specific places depends on soil and climatic conditions and social, cultural, and economic perspectives of the producers.

Author details

Upendra M. Sainju[1*], Rajan Ghimire[2] and Gautam P. Pradhan[3]

1 Northern Plains Agricultural Research Laboratory, US Department of Agriculture, Agricultural Research Service, Sidney, Montana, USA

2 Agricultural Science Center, New Mexico State University, Clovis, New Mexico, USA

3 Williston Research and Extension Center, North Dakota State University, Williston, North Dakota, USA

*Address all correspondence to: upendra.sainju@ars.usda.gov

References

[1] MacWilliams M, Wismer SM, Kulshrestha S. Life-cycle and economic assessments of western Canadian pulse systems: The inclusion of pulses in crop rotations. Agricultural Systems. 2014; **123**:43-53

[2] Varvel GE, Peterson TA. Residual soil nitrogen as affected by continuous cropping, two-year, and four-year crop rotations. Agronomy Journal. 1990;**82**: 958-962

[3] Gan Y, Liang C, Wang X, McConkey B. Lowering carbon footprint of durum wheat by diversifying cropping systems. Field Crops Research. 2011;**122**: 199-206

[4] Lupawi NZ, Soon YK. Nitrogen-related rotational effects of legume crops on three consecutive subsequent crops. Soil Science Society of America Journal. 2016;**80**:306-316

[5] Stevenson FC, van Kessel C. A landscape-scale assessment on the nitrogen and non-nitrogen rotation benefits of pea in a crop rotation. Soil Science Society of America Journal. 1996;**60**:1797-1805

[6] Bremer E, Janzen HH, Ellert BH, McKenzie RH. Soil organic carbon after twelve years of various crop rotations in an aridic boroll. Soil Science Society of America Journal. 2007;**72**: 970-974

[7] Miller PR, McConkey B, Clayton GW, Brandt SA, Staricka JA, Johnston AM, et al. Pulse crop adaptation in the northern Great Plains. Agronomy Journal. 2002;**94**:261-272

[8] Trabelsi D, Ben-Amar H, Mengoni A, Mhandi R. Appraisal of the crop rotation effect of rhizobium inoculation on potato cropping systems in relation to soil bacterial communities. Soil Biology and Biochemistry. 2012;**54**:1-6

[9] Kuo S, Sainju UM, Jellum EJ. Winter cover crop effects on soil organic carbon and carbohydrate. Soil Science Society of America Journal. 1997a;**61**:145-152

[10] Kuo S, Sainju UM, Jellum EJ. Winter cover cropping influence on nitrogen in soil. Soil Science Society of America Journal. 1997b;**61**:1392-1399

[11] Sainju UM, Singh BP, Whitehead WF, Wang S. Tillage, cover crop, and nitrogen fertilization effect on soil nitrogen and cotton and sorghum yields. European Journal of Agronomy. 2006a; **25**:372-382

[12] Sainju UM, Singh BP, Whitehead WF, Wang S. Carbon supply and storage in tilled and non-tilled soils as influenced by cover crop and nitrogen fertilization. Journal of Environmental Quality. 2006b;**35**:1507-1517

[13] Hatfield PG, Blodgett SL, Spezzano TM, Goosey HB, Lenssen AW, Kott RW. Incorporating sheep into dryland grain production systems. I. Impact on overwintering larval populations of wheat stem sawfly, Cephus cinctus Norton (Hymenoptera: Cephidae). Small Ruminant Research. 2007a;**67**:209-215

[14] Hatfield PG, Goosey HB, Spezzano TM, Blodgett SL, Lenssen AW, Kott RW. Incorporating sheep into dryland grain production systems. III. Impact on changes in soil bulk density and soil nutrient profiles. Small Ruminant Research. 2007b;**67**:222-232

[15] Liebig MA, Varvel GE, Doran JW, Wienhold BJ. Crop sequence and nitrogen fertilization effects on soil properties in the western Corn Belt. Soil Science Society of America Journal. 2002;**66**:596-601

[16] Mahler RL, Harder RW. The influence of tillage methods, cropping

sequence, and N rates on the acidification of a northern Idaho soil. Soil Science. 1984;**137**:52-60

[17] Schroder JL, Zhang H, Girma H, Raun WR, Penn CJ, Payton ME. Soil acidification from long-term use of nitrogen fertilizers on winter wheat. Soil Science Society of America Journal. 2011;**75**:957-964

[18] Sainju UM, Allen BL, Caesar-TonThat T, Lenssen AW. Dryland soil chemical properties and crop yields affected by long-term tillage and cropping sequence. Springerplus. 2015;**4**:230. DOI: 10.1186/s40064-015-1122-4

[19] Sainju UM, Lenssen AW, Allen BL, Stevens WB, Jabro JD. Soil total carbon and crop yield affected by crop rotation and cultural practice. Agronomy Journal. 2017b;**109**:1-9

[20] Gregory PJ, Ingram JSI, Anderson R, Betts RA, Brovkin V, Chase TN, et al. Environmental consequences of alternative practices for intensifying crop production. Agriculture, Ecosystems and Environment. 2002;**88**: 279-290

[21] Lenssen AW, Johnson GD, Carlson GR. Cropping sequence and tillage system influence annual crop production and water use in semiarid Montana. Field Crops Research. 2007; **100**:32-43

[22] Campbell CA, Zentner RP, Liang BC, Roloff G, Gregorich EC, Blomert B. Organic carbon accumulation in soil over 30 year in semiarid southwestern Saskatchewan: Effect of crop rotation and fertilization. Canadian Journal of Soil Science. 2000;**80**:170-192

[23] West TO, Post WM. Soil organic carbon sequestration rates by tillage and crop rotation: A global data analysis. Soil Science Society of America Journal. 2002;**66**:1930-1946

[24] Sainju UM. Cropping sequence and nitrogen fertilization impact on surface residue, soil carbon sequestration, and crop yields. Agronomy Journal. 2014; **106**:1231-1242

[25] Sainju UM. Tillage, cropping sequence, and nitrogen fertilization influence dryland soil nitrogen. Agronomy Journal. 2013;**105**:1253-1263

[26] Sainju UM, Lenssen AW, Allen BL, Stevens WB, Jabro JD. Nitrogen balance in response to dryland crop rotations and cultural practices. Agriculture, Ecosystems and Environment. 2016;**233**:25-32

[27] Sainju UM. A global meta-analysis on the impact of management practices on net global warming potential and greenhouse intensity from cropland soils. PLoS One. 2016;**11**(2):1/26-26/26. DOI: 10.137/journal.pone 0148527

[28] Sainju UM, Lenssen AW, Allen BL, Stevens WB, Jabro JD. Soil residual nitrogen under various crop rotations and cultural practices. Journal of Plant Nutrition and Soil Science. 2017a;**180**: 187-196

[29] McCracken DV, Smith MS, Grove JH, Mackown CT, Blevins RL. Nitrate leaching as influenced by cover cropping and nitrogen source. Soil Science Society of America Journal. 1994;**58**:1476-1483

[30] Sainju UM, Singh BP, Rahman S, Reddy VR. Soil nitrate-nitrogen under tomato following tillage, cover cropping, and nitrogen fertilization. Journal of Environmental Quality. 1999;**28**: 1837-1844

[31] Robertson EB, Sarig E, Firestone MK. Cover crop management of polysaccharide mediated aggregation in an orchard soil. Soil Science Society of America Journal. 1991;**55**:734-739

[32] Smith MS, Frye WW, Varco JJ. Legume winter cover crops. Advances in Soil Science. 1987;7:95-139

[33] Langdale GW, Blevins RL, Karlens DL, McCool DK, Nearing MA, Skidmore EL, et al. Cover crop effects on soil erosion by wind and water. In: Hargrove WL, editor. Cover Crops for Clean Water. Ankeny, Iowa, USA: Soil and Water Conservation Society; 1991. pp. 15-22

[34] Sainju UM, Singh BP. Nitrogen storage with cover crops and nitrogen fertilization in tilled and non-tilled soils. Agronomy Journal. 2008;**100**:619-627

[35] Sainju UM, Singh BP, Whitehead WF, Wang S. Accumulation and crop uptake of soil mineral nitrogen as influenced by tillage, cover crop, and nitrogen fertilization. Agronomy Journal. 2007;**99**:682-691

[36] Sainju UM, Whitehead WF, Singh BP. Biculture legume-cereal cover crops for enhanced biomass yield and carbon and nitrogen. Agronomy Journal. 2005; **97**:1403-1412

[37] Keeling KA, Hero D, Rylant KE. Effectiveness of composted manure for supplying nutrients. In: Fertilizer, Ag-Lime and Pest Management Conference; 17-18 January 1995; Madison, WI. Madison, Wisconsin, USA: University of Wisconsin; 1995. pp. 77-81

[38] Rochette P, Gregorich EG. Dynamics of soil microbial biomass C, soluble organic C, and CO_2 evolution after three years of manure application. Canadian Journal of Soil Science. 1998; **78**:283-290

[39] Sainju UM, Senwo ZN, Nyakatawa EZ, Tazisong IA, Reddy KC. Poultry litter increases nitrogen cycling compared with inorganic N fertilization. Agronomy Journal. 2010b;**102**:917-925

[40] Sainju UM, Senwo ZN, Nyakatawa EZ, Tazisong IA, Reddy KC. Soil carbon and nitrogen sequestration as affected by long-term tillage, cropping systems, and nitrogen fertilizer sources. Agriculture, Ecosystems and Environment. 2008;**127**:234-240

[41] Franzluebbers AJ. Integrated crop-livestock systems in the southeastern USA. Agronomy Journal. 2007;**99**: 361-372

[42] Herrero M, Thorton PK, Notenbaert AM, Wood S, Masangi S, Freeman HA, et al. Smart investments in sustainable food productions: Revisiting mixed crop-livestock systems. Science. 2010; **327**:822-825

[43] Herrington LW, Hobbs PR, Tamang DB, Adhikari C, Gyawali BK, Pradhan G, et al. Wheat and Rice in the Hills: Farming Systems, Production Techniques and Research Issues for Rice-Whet Cropping Patterns in the Mid-Hills of Nepal. Nepal Agricultural Research Council, Khumaltar (NARC)/ International Maize and Wheat Improvement Center (CIMMYT); 1992

[44] Maughan MW, Flores JPC, Anghinoni I, Bollero G, Fernandez FG, Tracy BG. Soil quality and corn yield under crop-livestock integration in Illinois. Agronomy Journal. 2009;**101**: 1503-1510

[45] Abaye AO, Allen VG, Fontenot JP. Grazing sheep and cattle together or separately: Effects on soils and plants. Agronomy Journal. 1997;**89**:380-386

[46] Snyder EE, Goosey HB, Hatfield PG, Lenssen AW. Sheep grazing on wheat-summer fallow and the impact on soil nitrogen, moisture, and crop yield. In: Proceeding, Western Section American Society of Animal Science. Vol. 58. Champaign, IL; 2007. pp. 221-224

[47] Quiroga A, Fernandez R, Noellemeyer E. Grazing effect on soil properties in conventional and no-till systems. Soil and Tillage Research. 2009;**105**:164-170

[48] Niu Y, Li G, Li L, Chan KY, Oates A. Sheep camping influences soil properties and pasture production in an acidic soil of New South Wales, Australia. Canadian Journal of Soil Science. 2009;**89**:235-244

[49] Sainju UM, Lenssen AW, Goosey H, Snyder E, Hatfield P. Sheep grazing in the wheat-fallow system affects dryland soil properties and grain yield. Soil Science Society of America Journal. 2011;**75**:1789-1798

[50] Sainju UM, Lenssen AW, Goosey H, Snyder E, Hatfield P. Dryland soil carbon and nitrogen influenced by sheep grazing in the wheat-fallow system. Agronomy Journal. 2010a;**102**: 1553-1561

11

Impact on Crop, Soil and Environment

Upendra M. Sainju, Rajan Ghimire and Gautam P. Pradhan

Abstract

Nitrogen (N) is a major limiting nutrient to sustain crop yields and quality. As a result, N fertilizer is usually applied in large quantity to increase crop production throughout the world. Application of N fertilizers has increased crop yields and resulted in achievement of self-sufficiency in food production in many developing countries. Excessive application of N fertilizers beyond crops' demand, however, has resulted in undesirable consequences of degradation in soil, water, and air quality. These include soil acidification, N leaching in groundwater, and emissions of nitrous oxide (N_2O), a potent greenhouse gas that contributes to global warming. Long-term application of ammonia-based N fertilizers, such as urea, has increased soil acidity which rendered to soil infertility where crops fail to respond with further application of N fertilizers. Another problem is the groundwater contamination of nitrate-N (NO_3-N) which can be a health hazard to human and livestock if its concentration goes above 10 mg L^{-1} in drinking water. The third problem is emissions of N_2O gas which is 300 times more powerful than carbon dioxide in terms of global warming potential. This chapter examines the effect of N fertilization on soil and environmental quality and crop yields.

Keywords: crop yields, environmental quality, management practices, nitrogen fertilizer, nitrogen-use efficiency, soil quality

1. Introduction

Nitrogen (N) is a major limiting factor for sustainable and profitable crop production. However, excessive N application through fertilizers and manures can degrade soil and environmental quality by increasing soil acidification, N leaching, and emissions of ammonia (NH_3) and nitrogen oxide (NO, N_2O, and NO_2) gases, out of which N_2O is considered a highly potent greenhouse gas that contributes to global warming [1, 2]. Nitrogen application more than crop's need can also result in reduced yield [3]. Additional N inputs include dry and wet (snow and rain) depositions from the atmosphere, biological N fixation, and irrigation water. Because crops can remove about 40–60% of applied N, the soil residual N (nitrate-N [NO_3-N] + ammonium-N [NH_4-N]) after crop harvest can be lost to the environment through leaching, denitrification, volatilization, surface runoff, soil erosion, and N_2O emissions [3, 4]. One option to reduce soil residual N is to increase N-use efficiency. Nitrogen-use efficiency for crops, however, can be lower at high N fertilization rates [5]. Improved management practices can increase N-use efficiency, enhance soil N storage, and reduce N fertilizer application which reduce N losses to the environment [4]. An account of N inputs, outputs, and retention in the soil provides N balance and helps to identify dominant processes of N flow in the agroecosystem [4].

Economically profitable crop yields could be achieved by recommended N fertilization rates [6]. However, such a yield potential for a crop varies with soil and climatic conditions, crop species, variety, nutrient cycling, and competitions with weeds and pests [6]. Crop production can be optimized and potential for N losses minimized by adjusting N fertilization rates using soil residual and potentially mineralizable N values. Studies show that ~1–2% of soil organic N in the 0–30 cm depth is mineralized every year [6]. Measuring the actual amount of N mineralized is a time taking process. A commonly used method for measuring soil available N and determining nitrogen rates for crops in semiarid regions of northern Great Plains, USA is based on testing NO_3-N content in soils to a depth of 60 cm after crop harvest in the fall season of the previous year and deduct the value from recommended N rates for the current crop year [7, 8]. In semiarid regions such as Great Plains of USA, N losses to the environment due to N leaching, volatilization, and denitrification during the winter are considered minimal due to cold weather and limited precipitation in the region.

Nitrogen fertilizers are being increasingly applied to crops to enhance their yield and quality in South Asia, where land available for crop production is limited, the proportion of cultivated land to population is low, and the pressure to increase crop yields to meet the demand for growing population is high. Continuous application of N fertilizers to nonlegume crops and excessive application rates in some places have led to undesirable consequences, such as reduced crop yields and degraded soil and environmental quality from soil acidification, N leaching, and greenhouse gas (N_2O) emissions. In this chapter, we discuss the consequences of N fertilization to crop yields and soil and environmental quality.

2. Crop yields, nitrogen uptake, and nitrogen-use efficiency

Nitrogen fertilization can increase crop yields and N uptake compared with no N fertilization. This has been documented for malt barley (*Hordeum vulgare* L.), cotton (*Gossypium hirsutum* L.), and sorghum (*Sorghum bicolor* [L.] Moench) (**Figures 1** and **2**, **Table 1**) by various researchers in Georgia and Montana, USA [9, 10, 14]. It is not unusual to achieve higher crop yield with increased N fertilization rate due to increased soil N availability [11]. Crop yields, however, can remain at similar level or decline with further increase in N rates after reaching the maximum yield. Sainju [9] observed that annualized grain and biomass yields of barley and pea (*Pisum sativum* L.) and their C content maximized at 80 kg N ha^{-1} and then declined, as N rate increased to 120 kg N ha^{-1} (**Figure 1**). Similarly, Sainju et al. [10] reported that malt barley yield and N uptake increased from 0 to 40 kg N ha^{-1} and then declined with further increase in N rates in no-till and conventional till malt barley-fallow rotation (**Figure 2**). In no-till continuous malt barley and malt barley-pea rotation, they found that increased N rate from 0 to 120 kg N ha^{-1} continued to increase malt barley yield and N uptake. Increased soil residual N due to fallow as a result of enhanced soil N mineralization from increased soil temperature and water content resulted in a reduced response of malt barley yield and N uptake with N fertilization in no-till and conventional till malt barley-fallow rotation. A study reported a need of 27 kg of total soil and fertilizer N to produce 1 Mg of malt barley grain in irrigated no-till field in Colorado, USA [11].

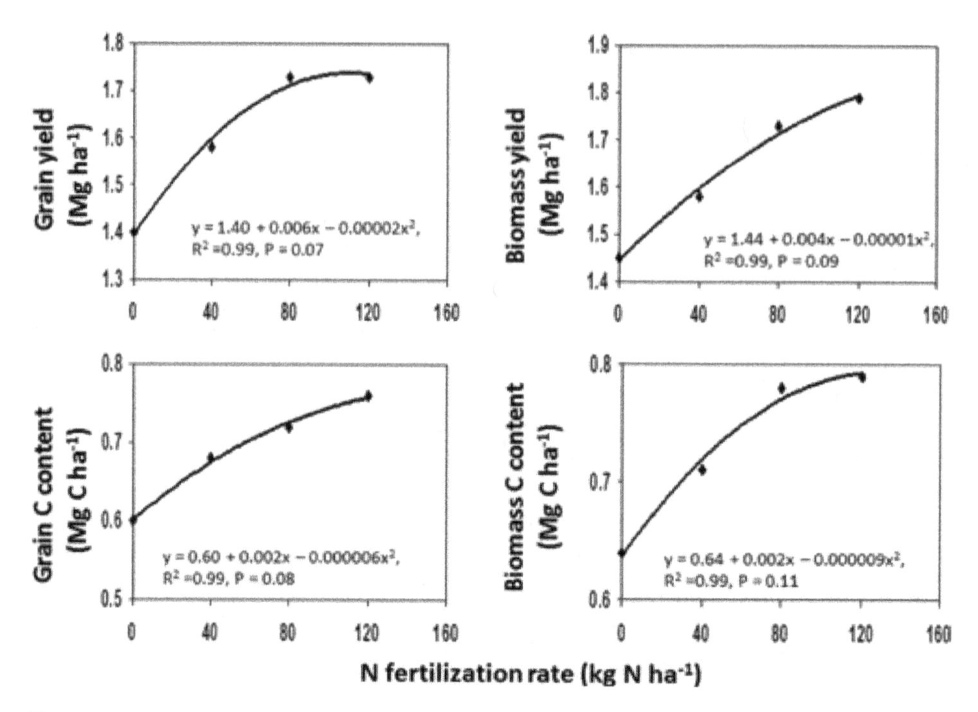

Figure 1.
Annualized grain and biomass yields of barley and pea and C content as affected by N fertilization rate in eastern Montana, USA [9].

Increased N fertilization rate can also increase grain quality, such as protein concentration [10, 11]. Increased N fertilization rates increased malt barley grain yield and protein concentration, but reduced kernel plumpness in Canada [12]. While some studies reported malt barley grain protein concentration of <130 g kg^{-1} with N rate of 168–200 kg ha^{-1} (e.g., [13]) others, observed an increase in protein concentration even with N rates <150 kg N ha^{-1} (e.g., [14]). Grain protein and kernel plumpness are important characteristics of malt barley that need to be maintained at critical levels (grain protein ≤129 g kg^{-1}, kernel plumpness ≥850 g kg^{-1}) for beer production [12]. Therefore, appropriate N fertilization rates are required to malt barley to achieve a balance between optimum grain yield, kernel plumpness, and protein concentration [15].

Sainju et al. [16] evaluated the effect of N fertilization on cotton and sorghum yields and N uptake from 2000 to 2002 in central Georgia, USA (**Table 1**). They found that cotton lint, sorghum grain, and cotton and sorghum biomass yields and N uptake increased from 0 to 60–65 kg N ha^{-1} and then remained either at a similar level or slightly increased at 120–130 kg N ha^{-1}. The response of cotton yield to N fertilization, however, depended on climatic condition, as cotton lint and biomass yields were greater in 2000 than 2002 when the growing season precipitation was below the average. The N fertilizer required for optimizing cotton and sorghum yields varied with the type of tillage and cover crop [16]. Boquet et al. [17] reported that cotton lint yield was lower with no-tillage than surface tillage without applied N, but at optimum N rate, yields were higher with no-tillage. They also found that additional N was required to optimize cotton yield following wheat (*Triticum aestivum* L.) in no-tillage and surface tillage systems without cover cropping, but no N rate was required following hairy vetch cover crop in either tillage practices. Similarly, N fertilization rates to cotton and sorghum can be reduced or eliminated

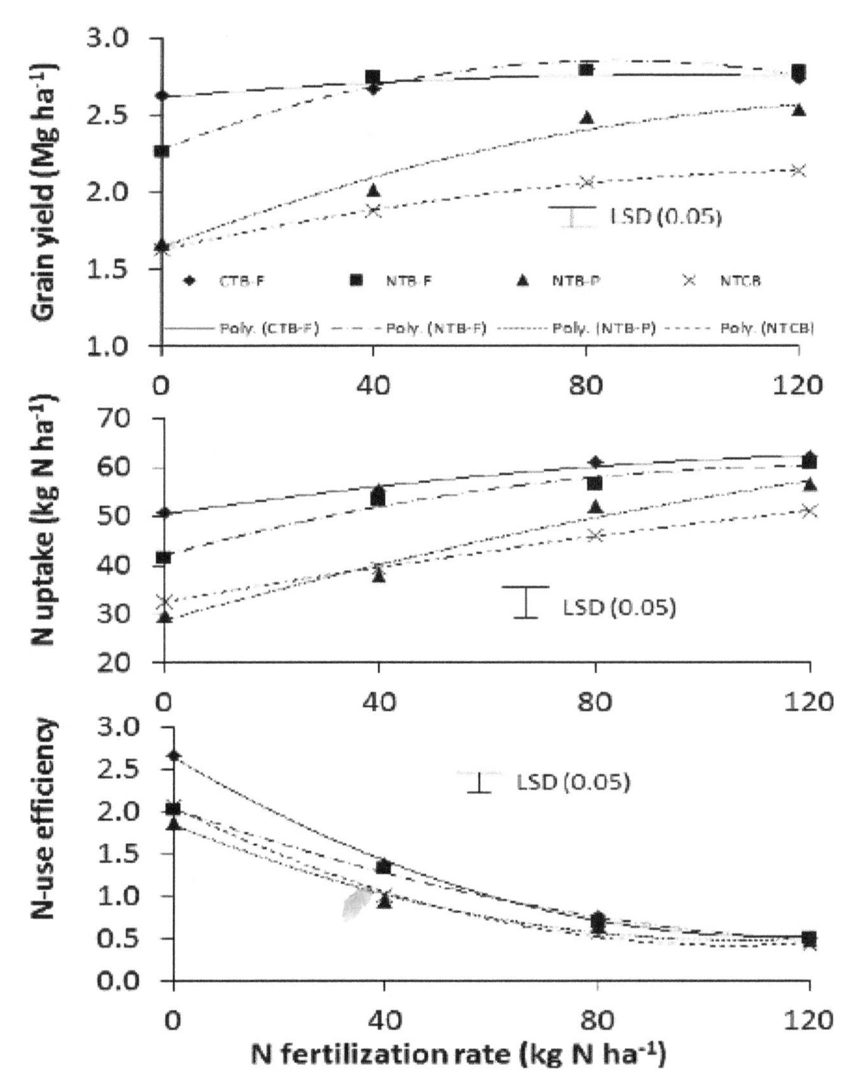

Figure 2.
Effects of cropping sequence and N fertilization rate on malt barley grain yield, N uptake, and N-use efficiency in eastern Montana, USA. CTB-F denotes conventional-till malt barley-fallow; NTB-F, no-till malt barley-fallow; NTB-P, no-till malt barley-pea; and NTCB, no-till continuous malt barley. Vertical bar with LSD (0.05) is the least significant difference between treatments at P = 0.05 [10].

by using legume cover crops, such as red clover (*Trifolium incarnatum* L.) and hairy vetch (*Vicia villosa* Roth), regardless of tillage practices [18]. The high rate of N fertilization can produce excessive vegetative growth that delays maturity and harvest and reduces cotton lint yield and N uptake [19].

Nitrogen-use efficiency, defined as crop yield or N uptake per unit applied N fertilizer, is a useful measurement of the efficiency of N fertilization to crop yields [5]. Enhancing N-use efficiency can maximize crop yield and N uptake with limited use of fertilizer N while reducing N rate and sustaining the environment [3]. Nitrogen-use efficiency, however, can decrease with increased N fertilization rate due to the inability of crops to utilize N efficiently [5]. Sainju et al. [10] found that N-use efficiency by malt barley decreased curvilinearly with increased N fertilization rate (**Figure 2**). Varvel and Peterson [5] reported that N removed by corn and sorghum grain was 50% of the applied N at low N rates and at least 20–30% at high N rates.

	2000 cotton lint (kg ha^{-1})		2000 cotton biomass (kg ha^{-1})		2001 sorghum grain (kg ha^{-1})		2001 sorghum biomass (kg ha^{-1})		2002 cotton lint (kg ha^{-1})		2002 cotton biomass (kg ha^{-1})	
Treatment	**Yield**	**N uptake**	**Yield**	**N uptake**	**Yield**	**N uptake**	**Yield**	**N uptake**	**Yield**	**N uptake**	**Yield**	**N uptake**
Cover crop[a]												
WW	699b[b]	11b	5200c	124b	2800bc	43ab	12,000ab	133ab	1091a	16a	3667a	74a
R	879a	15a	6300bc	138b	2300c	32b	9400b	81b	940ab	15a	3567a	77a
HV	660b	11b	8200a	239a	3500ab	60a	14,100a	175a	708b	13a	4067a	98a
HV/R	706b	12b	7300ab	194a	4000a	58a	14,100a	138ab	711b	14a	4233a	102a
N fertilization rate (kg N ha^{-1})												
0	736a	12a	5700b	135c	2800b	41b	11,600b	108b	1021a	17a	3700a	80b
60–65	783a	13a	7000a	178b	3100b	46b	12,400ab	135a	980a	16a	3900a	86b
120–130	689a	11a	7600a	209a	3700a	57a	13,300a	152a	587b	11b	4000a	97a

[a] *Cover crops are HV, hairy vetch; HV/R, hairy vetch/rye; R, rye; and WW, winter weeds.*
[b] *Numbers followed by the same letters within a column in a set are not significantly different at $P \leq 0.05$.*

Table 1.
Effect of cover crop and N fertilization rate on yield and N uptake by cotton lint, sorghum grain, and their biomass (stems + leaves) from 2000 to 2002 in central Georgia, USA [16].

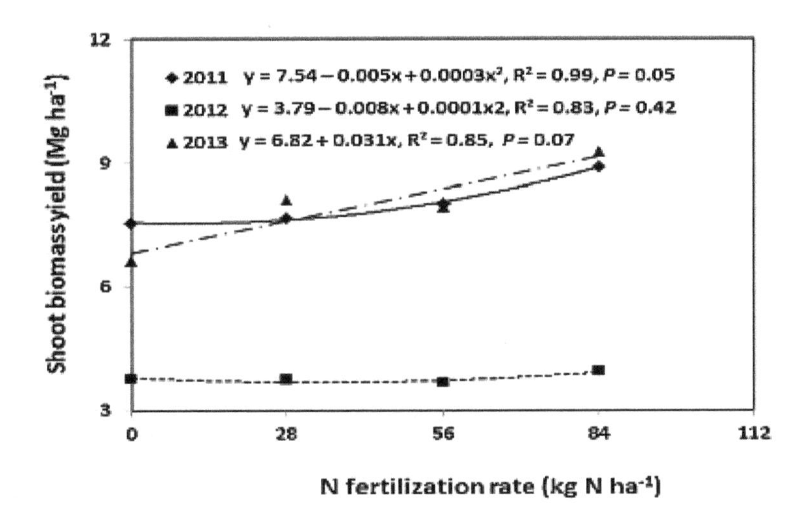

Figure 3.
Linear and quadratic responses of shoot biomass in perennial grasses with N fertilization rates from 2011 to 2013 averaged across grass species in eastern Montana, USA [20].

Nitrogen fertilization can also increase aboveground biomass yield of perennial grasses used for feedstock or bioenergy production. Sainju et al. [20] observed that yields of intermediate wheatgrass (*Thinopyrum intermedium* [Host] Barkworth and Dewey), switchgrass (*Panicum virgatum* L.), and smooth bromegrass (*Bromus inermis* L.) increased linearly or curvilinearly with increased N fertilization rate in 2011 and 2013 (**Figure 3**) when the annual precipitation was near or above the average. Biomass yield, however, did not respond to N fertilization in 2012 when the annual precipitation was below the average. Several researchers [21, 22] reported that maximum switchgrass shoot biomass yield reached at 120–140 kg N ha^{-1} in Iowa and Nebraska, USA, which had 2.5 and 2.2 times, respectively, more annual precipitation than in eastern Montana, USA. Power [23] also observed increased shoot biomass yield with increased N rate for smooth bromegrass in North Dakota, USA.

3. Soil acidification

Application of NH_4-based N fertilizers can increase soil acidity due to the release of H ions during hydrolysis [24]. Increased soil acidity following the application of N fertilizers leads to the development of infertile soils that do not respond well to crop yields with further application of N fertilizers [2, 25], thereby resulting in inefficient use of fertilizers [26]. Sainju et al. [27] reported that, after 30 years of tillage and cropping sequence, continuous application of N fertilizers reduced soil pH at the 0–7.5 cm depth from 6.30 at the initiation of the experiment to 5.73 in spring till spring wheat-fallow (STW-F) and to 5.02 in fall and spring till continuous spring wheat (FSTCW) under rainfed condition in eastern Montana, USA (**Table 2**). A similar decline in soil pH at 7.5–15.0 cm was observed from 6.75 at the initiation of the experiment to 6.15 in spring till continuous spring wheat (STCW). Buffer pH, the buffering capacity of the soil to resist changes in pH and is used to measure lime requirement, also similarly decreased with continuous N fertilization in all treatments. Both pH and buffer pH, however, did not change below 15 cm with N fertilization. Because spring wheat was grown once in 2 years in spring wheat-fallow rotation where N fertilizer was applied only to spring wheat, soil pH

Tillage and cropping sequence[a]	Soil depth					
	0–7.5 cm	7.5–15 cm	15–30 cm	30–60 cm	60–90 cm	90–120 cm
pH						
NTCW	5.33ab[b]E[c]	6.50abD	7.60C	8.35B	8.58A	8.75A
STCW	5.05bE	6.15bD	7.58C	8.25B	8.63A	8.70A
FSTCW	5.02bE	6.33bD	7.80C	8.30B	8.68AB	8.73A
FSTW-B/P	5.46aE	6.44bD	7.60C	8.15B	8.51A	8.59A
STW-F	5.73aE	7.03aD	7.65C	8.25B	8.50AB	8.66A
Contrast						
NT vs. T	0.29	0.26	−0.09	0.08	−0.08	0.04
CW vs. W-F	−0.68***	−0.88**	−0.08	0.01	0.13	0.04
CW vs. W-B/P	−0.43*	−0.11	0.20	0.15	0.16	0.14
Buffer pH						
NTCW	6.45bE	7.10abD	7.43C	7.60B	7.70AB	7.73A
STCW	6.38bE	7.00bD	7.43C	7.58B	7.68A	7.70A
FSTCW	6.43bE	7.05bD	7.45C	7.60B	7.70AB	7.73A
FSTW-B/P	6.66aD	7.13abC	7.44B	7.58B	7.69AB	7.70A
STW-F	6.80aE	7.24aD	7.44C	7.59B	7.66AB	7.72A
Contrast						
NT vs. T	0.05	0.08	−0.01	0.01	0.01	0.01
CW vs. W-F	−0.43***	−0.24**	−0.01	−0.01	0.01	−0.01
CW vs. W-B/P	−0.24*	−0.08	−0.01	0.03	0.01	0.03

Significant at P = 0.05, 0.01, and 0.001, respectively.
**Significant at P = 0.05, 0.01, and 0.001, respectively.*
***Significant at P = 0.05, 0.01, and 0.001, respectively.*
[a]*FSTCW, fall and spring till continuous spring wheat; FSTW-B/P, fall and spring till spring wheat-barley (1994–1999) followed by spring wheat-pea (2000–2013); NTCW, no-till continuous spring wheat; STCW, spring till continuous spring wheat; and STW-F, spring till spring wheat-fallow. CW represents continuous wheat; NT, no-till; T, till; W-B/P, spring wheat-barley/pea; and W-F, spring wheat-fallow.*
[b]*Numbers followed by the same lowercase letter within a column among treatments in a set are not significantly different at $P \leq 0.05$.*
[c]*Numbers followed by the same uppercase letter within a row among soil depths in a set are no significantly different at $P \leq 0.05$.*

Table 2.
Effect of tillage and crop rotation combination on soil pH and buffer pH at the 0–120 cm depth after 30 years of experiment initiation in eastern Montana, USA [27].

was less declined in this treatment than continuous spring wheat where N fertilizer was applied every year. From the same experiment, Aase et al. [28] reported an average decline of pH at 0–7.5 cm from 6.3 to 5.7 after 10 years due to continuous N fertilization.

Ghimire et al. [29] found that soil pH at 0–10 cm after 70 years of N fertilization was 5.70 with 0 kg N ha^{-1} and 5.0 with 135–180 kg N ha^{-1} under winter wheat-fallow in eastern Oregon, USA (**Figure 4**). Reduction in pH with N fertilization decreased with depth, with no significant effect below 30 cm. A study in China, where intensive farming and high rate of N fertilizer was applied for 20 years, showed that soil pH was dropped by 0.30–0.80 units from the original level [30]. In eastern Oregon, USA, application of total N fertilizer at 2.25 Mg N ha^{-1} over the 43-year period lowered soil pH by 0.60 units [31]. Liebig et al. [26] reported that, in

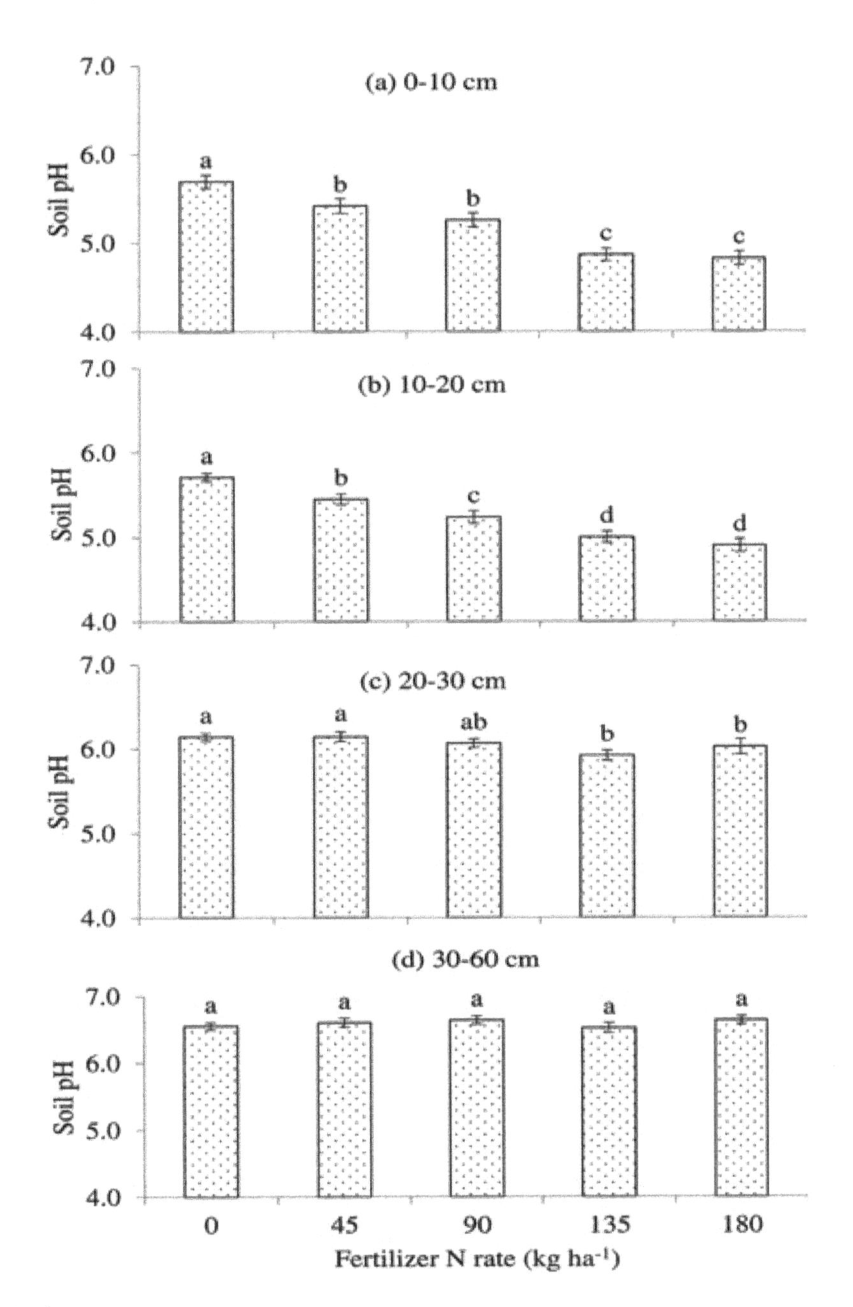

Figure 4.
Soil pH at the 0–60 cm depth from N fertilization rates to winter wheat in the winter wheat-fallow rotation after 70 years in eastern Oregon, USA. Bars with different letters at the top are significantly different at $P \leq 0.05$ *[29].*

North Dakota, USA, soil pH at 0–7.6 cm was lower under continuous corn than corn rotated with legume and other nonlegume crops because of the increased amount of N fertilizer applied. They recommended that soil samples be collected to a depth of 15 cm for measuring changes in soil pH due to N fertilization.

No-till (NT) system can increase soil acidity more than the conventional till (CT) system [32]. This is due to differences in the amount and placement of N fertilizers in the soil and removal of basic cations through grain and biomass removal between the two tillage systems [32]. Nitrogen fertilizers are usually placed at the soil surface, and N rates are usually higher in NT due to the accumulation of surface residue that partly immobilizes N than CT where fertilizers are incorporated into the soil due to tillage [33]. Because of enhanced soil water conservation, crop yields are higher in NT than CT, especially in dryland cropping systems [34]. As a result, crops remove more basic cations, resulting in increased acidity with NT compared with CT [34]. In contrast, Ghimire et al. [29] reported that soil pH decreased with increased N rate, as tillage intensity increased.

Source of N fertilizer can also have a varying effect on soil acidity. Chen et al. found that soil acidity from N fertilizer sources was in the order $(NH_4)_2SO_4$ > NH_4Cl > NH_4NO_3 > anhydrous NH_3 > urea. Similarly, Schroder et al.[25] reported that anhydrous NH_3 produce more acidity than urea. Others [35], however, observed no significant differences in acidity among $(NH_4)_2SO_4$, NH_4NO_3, anhydrous NH_3, urea, and urea-NH_4NO_3.

4. Soil organic matter

Soil organic matter refers to soil organic C and N and is a crucial component of soil health and quality [36, 37]. Nitrogen fertilization can increase soil organic C and N by increasing crop biomass yield, and the amount of residue returned to the soil [38]. Russell et al. [37], however, reported no difference in soil organic C with N fertilization rate. Sainju et al. [39] reported that 3 years of N fertilization to cotton and sorghum produced various results on soil organic C at the 0–30 cm depth in strip-tilled and chisel-tilled soils in central Georgia, USA (**Table 3**). Soil organic C at 0–10 and 10–30 cm varied with N fertilization rates in strip-tilled soil, but increased in chisel-tilled soil due to differences in tillage intensity. In strip tillage, only crop rows are tilled, leaving the area between rows undisturbed, and N fertilizer is applied in crop rows. In contrast, the land is tilled using discs in chisel tillage after N fertilizer is broadcast. Differences in N fertilization methods between tillage practices probably affected soil organic C due to N fertilization rates.

Sainju [9] observed different trends of soil organic C at the 0–120 cm depth with 6 years of N fertilization rates in various cropping systems in eastern Montana, USA (**Figure 5**). Soil organic C at 0–5 and 5–10 cm peaked at 40 kg N ha^{-1} and then declined with further increase in N rates in no-till malt barley-pea (NTB-P) and continuous no-till barley (NTCB). In no-till malt barley-fallow (NTB-F) and

N rate (kg N ha^{-1})	Soil organic C (Mg C ha^{-1})				
	0–10 cm	10–30 cm	30–60 cm	60–90 cm	90–120 cm
Strip-tilled soil					
0	10.1a[a]	16.0a	10.9	7.2	5.5
60–65	9.3b	14.4b	10.2	4.5	5.3
120–130	10.3a	14.7ab	9.8	7.3	5.8
Chisel-tilled soil					
0	8.9b	12.5b	10.1	7.4	5.9
60–65	9.6a	13.4b	10.1	7.3	5.3
120–130	9.3ab	14.8a	10.6	7.9	6.1

[a]*Numbers followed by the same letter within a column in a set are not significantly different at* $P \leq 0.05$.

Table 3.
Effect of 3 years of N fertilization rate on soil organic C at the 0–120-cm depth in strip-tilled and chisel-tilled soils under cotton and sorghum in central Georgia, USA [39].

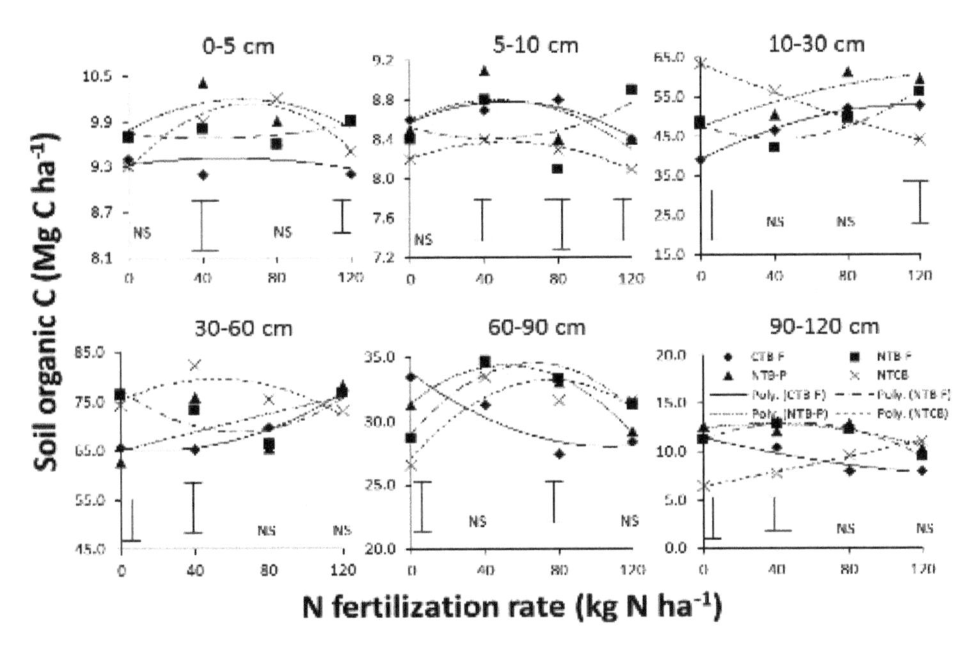

Figure 5.
Soil organic C at the 0–120 cm depth as affected by 6 years of N fertilization rates to malt barley in various cropping systems in eastern Montana, USA. CTB-F denotes conventional-till malt barley-fallow; NTB-F, no-till malt barley-fallow; NTB-P, no-till malt barley-pea; and NTCB, no-till continuous malt barley. Vertical bars denote least significant difference between tillage and cropping sequence treatments within a N rate at P = 0.05 [9].

conventional till malt barley-fallow (CTB-F), the trend of soil organic C with N rates varied at various depths. Soil organic C at these depths was greater with NTB-P and NTCB than other treatments at most N rates due to greater amount of crop residue returned to the soil. Soil organic C at 5–10, 30–60, and 60–90 cm were greater with 40 kg N ha^{-1} than other N rates. Sainju [9] also found that C sequestration rate at 0–10 cm was 83 kg C ha^{-1} $year^{-1}$ with 40 kg N ha^{-1} that was close to 94 kg C ha^{-1} $year^{-1}$ at 0–15 cm with 45 kg N ha^{-1} for dryland cropping systems in Colorado [36].

Under perennial grasses, several researchers [40, 41] did not find a significant effect of N fertilization on soil organic C at 0–30 cm after 2–5 years in Alabama and Colorado, USA. Only after 4–12 years, N fertilization increased soil organic C at 0–90 cm by 0.5–2.4 Mg C ha^{-1} $year^{-1}$ compared with no N fertilization under switchgrass in USA and Canada [42, 43]. Rice et al. [43] reported that N fertilization to cool-season grasses increased C sequestration rate at 0–30 cm by 1.6 Mg C ha^{-1} $year^{-1}$ compared with no N fertilization after 5 years in Kansas, USA. In Alberta, Canada, Bremer et al. [42] observed that N fertilization to perennial grasses increased C sequestration rate at 0–5 cm by 0.5 Mg C ha^{-1} $year^{-1}$ compared with no N fertilization after 6–12 years. In South Dakota, USA, Li et al. [44] noted C sequestration rate of 2.4 Mg C ha^{-1} $year^{-1}$ at 0–90 cm under switchgrass after 4 years. Sainju et al. [45] found increasing trend of soil total C at 30–60 cm with increased N rate under intermediate wheatgrass and smooth bromegrass and a declining trend with switchgrass after 5 years in eastern Montana (**Figure 6**). At 60–90 cm, the trend reversed with grasses. They suggested that longer than 5 years is needed to observe the effect of N fertilization on soil total C under perennial grasses.

Nitrogen fertilization has less impact on soil total N than soil organic C. Sainju and Singh [46] reported that soil total N at 0–15 cm under cotton and sorghum was greater with 60–65 than 0 kg N ha^{-1}, but not at lower depths in the

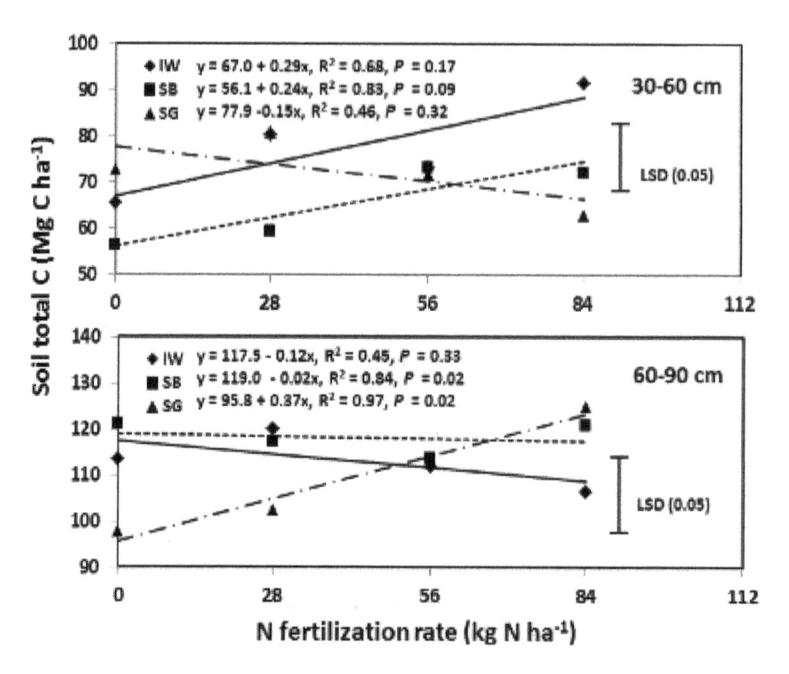

Figure 6.
Soil total C at 30–60 and 60–90 cm depths as affected by 5 years of N fertilization rates to perennial grasses in eastern Montana, USA. Perennial grasses are IW, intermediate wheatgrass; SB, smooth bromegrass, and SW, switchgrass. LSD (0.05) is least significant difference between grasses within a N rate at P = 0.05 [45].

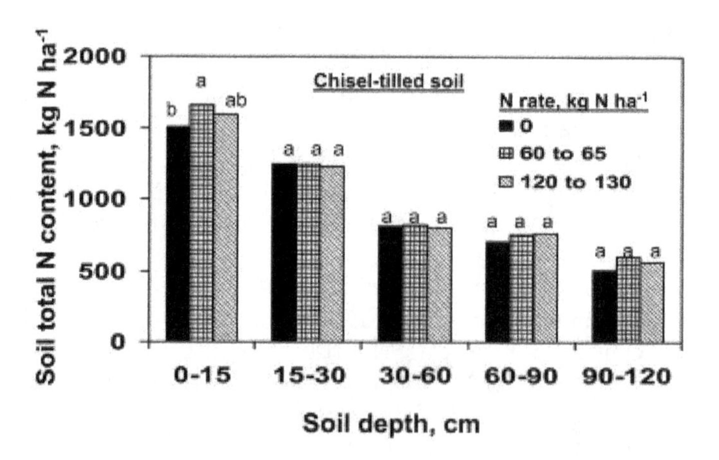

Figure 7.
Soil total N at 0–120 cm in the chisel-tilled soil as affected by 6 years of N fertilization rates to cotton and sorghum in central Georgia, USA. Bars with the same letter at the top are not significantly different among N rates at a depth at $P \leq 0.05$ [46].

chisel-tilled soil in central Georgia, USA (**Figure 7**). Ghimire et al. [29] observed that soil total N at 10–20 cm increased with increased N rates after 70 years of N fertilization to winter wheat, but the trend varied with different tillage practices at higher N rates in eastern Oregon, USA (**Figure 8**). At 0–45 kg N ha^{-1}, soil total N was greater with subsurface sweep than a moldboard plow. At 90–180 kg N ha^{-1}, soil total N was lower with disc plow than other tillage practices. Increased N substrate availability due to N fertilization along with tillage may have increased microbial activity and N mineralization and therefore reduced soil total N over time.

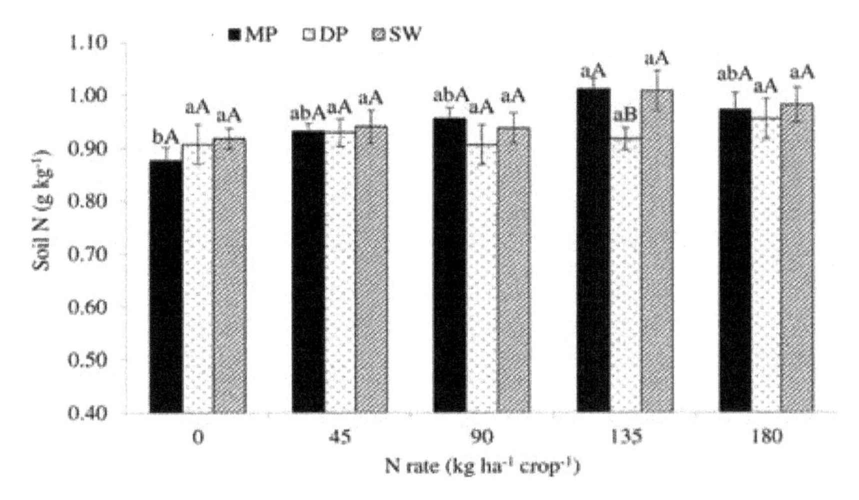

Figure 8.
Soil total N as affected by 72 years of N fertilization rates to spring wheat and tillage in eastern Oregon, USA. Tillage practices are DP, disk plow; MP, moldboard plow, and SW, subsurface sweep. Bars with different lowercase letters at the top are significantly different among tillage practices within a N rate at $P \leq 0.05$. Bars with different uppercase letters at the top are significantly different among N rates within a tillage practice at $P \leq 0.05$ [29].

5. Soil residual nitrogen and nitrogen leaching

Soil residual N refers to inorganic N (NH_4-N + NO_3-N) accumulated in the soil profile after crop harvest. This occurs because crops cannot take up all applied N fertilizer from the soil [5, 47]. Accumulation of soil NO_3-N increases with depth and is directly related to N fertilization rate [47, 48]. Deep accumulation of NO_3-N in the soil profile increases the potential for N leaching to shallow water tables [49]. Nitrogen fertilization rates that exceed crop requirement can increase NO_3-N accumulation in the soil profile and N leaching [50].

Soil inorganic N			
Treatment	**0–10 cm**	**10–30 cm**	**0–30 cm**
(kg N ha^{-1})			
Cover crop			
Winter weeds	19.6b[a]	32.9b	52.5c
Rye	19.1b	34.1b	53.2c
Hairy vetch	23.6a	38.4a	62.0a
Hairy vetch/rye	21.6a	34.8b	56.4b
N fertilization rate (kg N ha^{-1})			
0	19.6b	33.5b	53.1b
60–65	20.8b	35.3ab	56.1ab
120–130	22.5a	36.4a	59.9a

[a]*Numbers followed by the same letter within a column in a set are not significantly different at $P \leq 0.05$.*

Table 4.
Effect of cover crop and N fertilization rate on soil residual inorganic N (NH_4-N + NO_3-N) content at the 0–30 cm depth in central Georgia, USA [16].

N fertilization rate	NH_4-N content at the soil depth										
	0–5 cm	5–10 cm	10–30 cm	30–60 cm	60–90 cm	90–120 cm	0–10 cm	0–30 cm	0–60 cm	0–90 cm	0–120 cm
kg N ha^{-1}	kg N ha^{-1}										
0	2.4b[†]	2.5a	10.4a	15.8a	19.4a	23.8a	4.9b	15.3a	31.2a	50.2a	72.0a
40	2.3b	2.3a	10.6a	15.4a	19.7a	25.0a	4.7b	15.2a	30.6a	49.7a	72.7a
80	2.5b	2.5a	10.3a	15.5a	19.7a	25.1a	5.0ab	15.4a	30.8a	49.1a	72.2a
120	2.9a	2.6a	10.8a	16.2a	19.6a	25.7a	5.5a	16.1a	32.0a	50.8a	73.6a

[†]*Numbers followed by the same letters within a column are not significantly different at $P \leq 0.05$.*

Table 5.
Effect of N fertilization rate on soil residual NH_4-N content at the 0–120 cm depth from 2006 to 2011 in eastern Montana, USA [55].

N fertilization rate	NO_3-N content at the soil depth										
	0–5 cm	5–10 cm	10–30 cm	30–60 cm	60–90 cm	90–120 cm	0–10 cm	0–30 cm	0–60 cm	0–90 cm	0–120 cm
kg N ha^{-1}	kg N ha^{-1}										
0	6.7c[†]	3.7c	13.3c	15.5c	13.7c	16.7b	10.2c	23.6d	39.0d	52.7d	68.7c
40	8.1c	4.3bc	14.6c	17.5bc	17.1b	21.4ab	12.5c	27.1c	44.6c	61.6c	82.3b
80	10.1b	5.1b	16.7b	19.8b	17.7b	21.0ab	15.2b	31.9b	51.8b	69.4b	89.6b
120	12.2a	6.2a	20.0a	23.4a	21.7a	24.7a	18.3a	38.2a	61.7a	83.3a	107.0a

[†] *Numbers followed by the same letters within a column are not significantly different at $P \leq 0.05$.*

Table 6.
Effect N fertilization rate on soil residual NO_3-N content at the 0–120 cm depth from 2006 to 2011 in eastern Montana, USA [55].

One of the ways to reduce N fertilization rates to crops while maintaining yield goals is to account for N mineralized from soil organic matter during the crop growing season and soil residual N at crop planting [6]. Since the measurement of N mineralization requires a long time, N fertilization rates to dryland crops are adjusted by deducting soil NO_3 content to a depth of 60 cm after crop harvest in the previous year or at planting of the current year from recommended N rates [51]. Producers are increasingly interested in reducing the amount of N fertilizer applied to crops because of the higher cost of N fertilization and the associated environmental degradation.

Nitrogen fertilization rates to crops can be higher in the no-till than the conventional till system due to greater accumulation of surface crop residue that can enhance N immobilization [52]. On the other hand, N rates can be reduced in crop rotations containing legumes compared to monoculture nonlegume cropping systems [53]. Nonlegume monocropping can have higher soil residual NO_3-N content than legume-based crop rotations due to increased N fertilization rate [5, 27]. Increased cropping intensity can reduce soil profile NO_3-N content due to greater N immobilization, less summer fallow, and a greater amount of N removed by crops [54]. Sainju et al. [16] and Sainju [9] found that both soil NH_4-N and NO_3-N contents increased with N rates and depths (**Tables 4–6**).

It is well known that excessive N fertilizer application can increase N leaching in the groundwater, which is a major environmental concern [50]. Nitrate-N concentration >10 mg L^{-1} in the drinking water poses a serious threat to human and animal health [56]. Nitrate-N is soluble in water and moves down the soil profile with percolating water [47, 57]. Increased application of N fertilizer to crops during the last several decades has increased NO_3-N contamination of groundwater [56]. This occurs because of excessive NO_3-N accumulation in the soil profile [57] due to N fertilization rates that exceed crop requirements, accompanied by poor soil and crop management practices [56]. Nitrate-N accumulation and movement in the soil profile depend on soil properties, climatic conditions, and management practices [58]. For example, N leaching is greater in sandy than clayey soils due to the presence of a large number of macropores and leaching is higher in the humid than arid and semiarid regions due to differences in annual precipitation [56, 58]. Nitrate-N leaching occurs mostly in the fall, winter, and spring seasons in the northern hemisphere when evapotranspiration is low, crops are absent to uptake soil N, and precipitation exceeds the water holding capacity of the soil [59].

6. Greenhouse gas emissions and global warming potential

Management practices on croplands can contribute about 10–20% of global greenhouse gases (GHGs: carbon dioxide [CO_2], nitrous oxide [N_2O], and methane [CH_4]) [60]. Quantitative estimate of the impact of the GHGs to global radiative forcing is done by calculating net global warming potential (GWP) which accounts for all sources and sinks of CO_2 equivalents from farm inputs, farm operations, soil C sequestration, and N_2O and CH_4 emissions [61, 62]. The net GWP for a crop production system is expressed as kg CO_2 eq. ha^{-1} $year^{-1}$. Net GWP is also expressed as net greenhouse gas intensity (GHGI) or yield-scaled GWP, which is calculated by dividing net GWP by crop yield [61]. These values can be affected both by net GHG emissions and crop yields. Sources of GHGs in agroecosystems include N_2O and CH_4 emissions (or CH_4 uptake) as well as CO_2 emissions associated with farm machinery used for tillage, planting, harvesting, and manufacture, transportation, and applications of chemical inputs, such as fertilizers, herbicides, and pesticides, while soil C sequestration rate can be either a sink or source of CO_2

[62, 63]. In the calculations of net GWP and GHGI, emissions of N_2O and CH_4 are converted into their CO_2 equivalents of global warming potentials which are 310 and 28, respectively, for a time horizon of 100 years [60]. The balance between soil C sequestration rate, N_2O and CH_4 emissions (or CH_4 uptake), and crop yield typically controls net GWP and GHGI [61, 62].

Nitrogen fertilization typically stimulates N_2O emissions when the amount of applied N exceeds crop N demand [51, 61]. Nitrogen fertilization, however, can have a variable effect on emissions of other GHGs, such as CO_2 and CH_4 [64, 65]. Sainju et al. [65] found that the application of 80 kg N ha^{-1} to dryland malt barley increased CO_2 emissions, but not N_2O and CH_4 emissions (**Table 7**). Because N_2O emissions has a large effect on net GWP and GHGI, practices that can reduce N fertilization rates without influencing crop yields can substantially reduce net GHG emissions [61, 62]. Other factors that can influence N_2O emissions are the type, placement, time, and method of application of N fertilizers. Applying N fertilizer in the spring compared with autumn and using split application compared with one single application at planting can reduce N_2O emissions in some cases [66]. Applying N fertilizer at various depths can have a variable effect on N_2O emissions [67]. Anhydrous ammonia can increase N_2O emissions compared with urea [67, 68]. Similarly, chemical additives to reduce nitrification from N fertilizers, such as polymer-coated urea and nitrification inhibitors, can substantially reduce N_2O emissions compared with ordinary urea and non-nitrification inhibiting fertilizers [69]. Some N fertilizers, such as urea, emit both CO_2 and N_2O. Nitrogen fertilizers also indirectly emit N_2O through NH_3 volatilization and NO_3-N leaching [68].

Increased N fertilization rate can enhance net GWP and GHGI due to increased N_2O and CO_2 emissions associated with the manufacture, transport, and application of N fertilizers, regardless of cropping systems and calculation methods [61, 70]. In a meta-analysis of 12 experiments, Sainju [71], after accounting for all sources and sinks of CO_2 emissions, reported that net GWP decreased from 0 to $\leq$45 kg N ha^{-1} and net GHGI from 0 to $\leq$145 kg N ha^{-1} and then increased with increased N fertilization rate (**Figure 9**). Using partial accounting, net GWP decreased from 0 to 88 kg N ha^{-1} and net GHGI from 0 to $\leq$213 kg N ha^{-1} and then increased with increased N rate. These N rates probably corresponded to crop N demand when crops used most of the soil available N. The cropping systems that left little residual N in the soil reduced N_2O emissions, and therefore net GWP and GHGI, whereas net GWP and GHGI increased linearly with increase in N application rates that exceeded crop N demand, suggesting that excessive N fertilizer applications can induce global warming. Similar results have been reported by Li et al. [44]. Therefore, N fertilizers should be applied at optimum rates to reduce net GWP and GHGI while sustaining crop yields. The optimum N rates, however, depended on net GWP measured either per unit area or per unit crop yield.

N fertilization	**CO_2 flux**	**N_2O flux**	**CH_4 flux**
kg N ha^{-1}	**Mg C ha^{-1}**	**g N ha^{-1}**	**g C ha^{-1}**
0	1.15b[†]	308a	−314a
80	1.23a	329a	−291a

[†]Numbers followed by different letters within a column are significantly different at $P \leq 0.05$ by the least square means test.

Table 7.
Effect of N fertilization on total soil surface greenhouse gas fluxes (from March to November) averaged across years from 2008 to 2011 under rainfed malt barley in eastern Montana, USA [65].

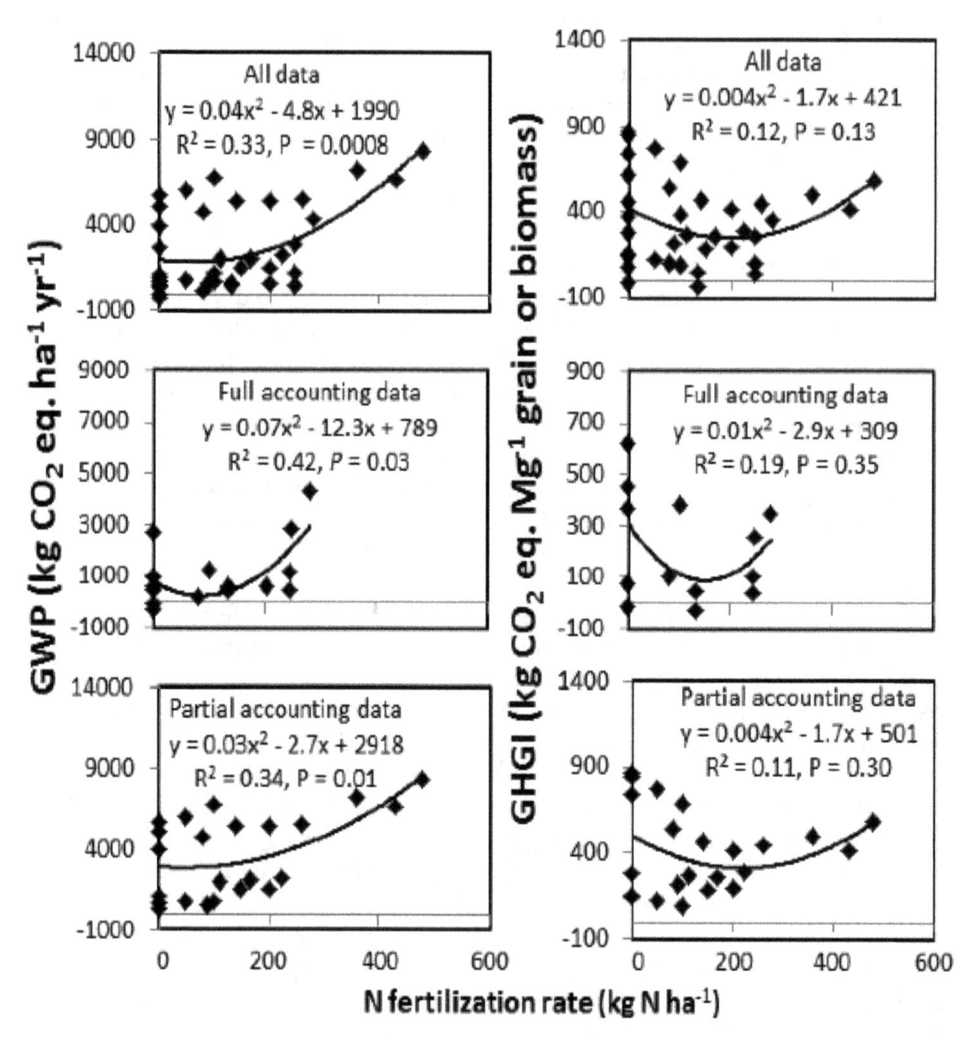

Figure 9.

The relationship between N fertilization rate and net global warming potential (GWP) and greenhouse gas intensity (GHGI). Full accounting data denote calculations of GWP and GHGI by accounting all sources and sinks of CO_2 (N_2O and CH_4 emissions, farm inputs, operations, and soil C sequestration). Partial accounting data denotes partial accounting of sources and sinks (N_2O and CH_4 emissions and/or soil C sequestration). All data denotes inclusions of full and partial accounting data [71].

Sainju [71] observed that the relationships between net GWP, net GHGI, and N rate were further improved when the duration of the experiment and soil and climatic conditions were taken into account in the multiple linear regressions. Duration of experiment and annual precipitation had positive effects, but air temperature and soil texture had negative effects on net GWP when all sources and sinks of CO_2 emissions were accounted for. With partial accounting, only air temperature had a positive effect on net GWP, but other factors had negative effects. For net GHGI, the factors having negative effects were air temperature using the complete accounting of CO_2 emissions and annual precipitation and soil texture using the partial accounting. Sainju et al. [70] reported that net GWP and GHGI calculated from soil respiration and soil C sequestration methods were lower with 80 than 0 kg N ha^{-1} (**Table 8**). They noted that, although CO_2 equivalents from N fertilization and soil respiration were higher with 80 kg N ha^{-1}, the amount of plant residue returned to the soil, soil C sequestration rate, and grain yields were greater

Cropping sequence[a]	N rate	Farm operation (A)	N fertilizer (B)[b]	Soil respiration (C)	N_2O flux (D)	CH_4 flux (E)	Annualized crop residue (F)[c]	SOC (G)[d]	GWP_R (H)[e]	GWP_C (I)[f]	Annualized grain yield (J)	$GHGI_R$ (K)[g]	$GHGI_C$ (L)[h]
	$kg\ N\ ha^{-1}$	$kg\ CO_2$ equivalent $ha^{-1}\ year^{-1}$									$kg\ ha^{-1}$	$kg\ CO_2\ kg^{-1}$ grain yield	
CTB-F		182	77	2722b[i]	425a	−16a	3476b	−114c	−89a	778a	1408b	−0.06a	0.55a
NTB-P		124	91	3303a	469a	−16a	5980a	554a	−2005c	115b	1649a	−1.22c	0.07b
NTCB		124	103	3547a	394a	−15a	5411a	268b	−1259b	337b	1683a	−0.75b	0.20b
	0	143	0	3093b	416a	−16a	4421b	−94b	−787a	635a	1399b	−0.56a	0.45a
	80	143	180	3288a	443a	−15a	5487a	566a	−1448b	185b	1761a	−0.82b	0.11b

[a] *Cropping sequences are CTB-F, conventional-till malt barley-fallow; NTB-P, no-till malt barley-pea; and NTCB, no-till continuous malt barley.*
[b] *Total CO_2 equivalents from direct and indirect sources of N fertilization.*
[c] *Total above- and below-ground crop residue.*
[d] *Carbon sequestration rate calculated from linear regression of change in soil organic C at the 0–10 cm depth from 2006 to 2011.*
[e] *Column (H) = Column (A) + Column (B) + Column (C) + Column (D) + Column (E) − Column (F) [61]. Negative values indicate GHG sink.*
[f] *Column (I) = Column (A) + Column (B) + Column (D) + Column (E) − Column (G) [61, 62]. Negative values indicate GHG sink.*
[g] *Column (K) = Column (H)/Column (J) [61]. Negative values indicate GHG sink.*
[h] *Column (L) = Column (I)/Column (J) [61]. Negative values indicate GHG sink.*
[i] *Numbers followed by the same letters within a column in a set are not significantly different at $P \leq 0.05$.*

Table 8.
Net global warming potential (GWP_R and GWP_C) and greenhouse gas intensity ($GHGI_R$ and $GHGI_C$) based on soil respiration and organic C (SOC) methods as influenced by cropping sequence and N fertilization rate in eastern Montana, USA [70].

with 80 than 0 kg N ha^{-1}, thereby resulting in lower net GWP and GHGI with N fertilization than without, regardless of the method used for calculation.

7. Conclusions

Nitrogen fertilization is one of the most commonly used practice to increase crop yields throughout the world because of abundant availability of N fertilizers and their great effectiveness to increase yields compared with other organic fertilizers, such as manure and compost. Excessive application of N fertilizers in the last several decades, however, has resulted in undesirable consequences of soil and environmental degradations, such as soil acidification, N leaching to the ground-water, and greenhouse gas (N_2O) emissions. Crop yields have declined in places where soil acidification is high due to unavailability of major nutrients and basic cations and toxic effect of acidic cations. Other disadvantages of excessive N fertilization include increased cost of fertilization, reduced N-use efficiency, and negative impact on human and livestock health. To reduce excessive N fertilization, composited soil sample to a depth of 60 cm should be conducted for NO_3-N test prior to crop planting and N fertilization rate be adjusted by deducting soil NO_3-N content from the desirable N rate.

Author details

Upendra M. Sainju[1*], Rajan Ghimire[2] and Gautam P. Pradhan[3]

1 Northern Plains Agricultural Research Laboratory, US Department of Agriculture, Agricultural Research Service, Sidney, Montana, USA

2 Agricultural Science Center, New Mexico State University, Clovis, New Mexico, USA

3 Williston Research and Extension Center, North Dakota State University, Williston, North Dakota, USA

*Address all correspondence to: upendra.sainju@ars.usda.gov

References

[1] Franzluebbers AJ. Integrated crop-livestock systems in the southeastern USA. Agronomy Journal. 2007;**99**: 361-372

[2] Herrero M, Thorton PK, Notenbaert AM, Wood S, Masangi S, Freeman HA, et al. Smart investments in sustainable food productions: Revisiting mixed crop-livestock systems. Science. 2010; **327**:822-825

[3] Eickhout B, Bouwman AP, van Zeijts H. The role of nitrogen in world food production and environmental sustainability. Agriculture, Ecosystems and Environment. 2006;**116**:4-14

[4] Ross SM, Izaurralde RC, Janzen HH, Robertson JA, McGill WB. The nitrogen balance of three long-term agroecosystems on a boreal soil in western Canada. Agriculture, Ecosystems and Environment. 2008;**127**: 241-250

[5] Varvel GE, Peterson TA. Residual soil nitrogen as affected by continuous cropping, two-year, and four-year crop rotations. Agronomy Journal. 1990;**82**: 958-962

[6] Schepers JS, Mosier AR. Accounting for nitrogen in nonequilibrium soil-crop systems. In: Follett RF, editor. Managing Nitrogen for Groundwater Quality and Farm Profitability. Madison, WI: Soil Science Society of America; 1991. pp. 125-137

[7] Lenssen AW, Johnson GD, Carlson GR. Cropping sequence and tillage system influence annual crop production and water use in semiarid Montana. Field Crops Research. 2007; **100**:32-43

[8] Miller PR, McConkey B, Clayton GW, Brandt SA, Staricka JA, Johnston AM, et al. Pulse crop adaptation in the northern Great Plains. Agronomy Journal. 2002;**94**:261-272

[9] Sainju UM. Cropping sequence and nitrogen fertilization impact on surface residue, soil carbon sequestration, and crop yields. Agronomy Journal. 2014; **106**:1231-1242

[10] Sainju UM, Lenssen AW, Barsotti JL. Dryland malt barley yield and quality affected by tillage, cropping sequence, and nitrogen fertilization. Agronomy Journal. 2013;**105**:329-340

[11] Halvorson AD, Reule CA. Irrigated, no-till corn and barley response to nitrogen in northern Colorado. Agronomy Journal. 2007;**99**:1521-1529

[12] O' Donovan JT, Turkington TK, Edney MJ, Clayton GW, McKenzie RH, Juskiew PE, et al. Seeding rate, nitrogen rate, and cultivar effects on malting barley production. Agronomy Journal. 2011;**103**:709-716

[13] Birch CJ, Long KE. Effect of nitrogen on the growth, yield, and grain protein content of barley. Australian Journal of Experimental Agriculture. 1990;**30**:237-242

[14] Weston DT, Horsley RD, Schwarz PB, Goos RJ. Nitrogen and planting effects on low-protein spring barley. Agronomy Jpornal. 1993;**85**:1170-1174

[15] Thompson TL, Ottman MJ, Riley-Saxton E. Basal stem nitrate tests for irrigating malt barley. Agronomy Journal. 2004;**86**:516-524

[16] Sainju UM, Singh BP, Whitehead WF, Wang S. Tillage, cover crop, and nitrogen fertilization effect on soil nitrogen and cotton and sorghum yields. European Journal of Agronomy. 2006; **25**:372-382

[17] Boquet DJ, Hutchinson RL, Breitenbeck GA. Long-term tillage, cover crop, and nitrogen rate effects on cotton: Yield and fiber properties. Agronomy Journal. 2004;**96**:1436-1442

[18] Sweeney DW, Moyer JL. In-season nitrogen uptake by grain sorghum following legume green manures in conservation tillage systems. Agronomy Journal. 2004;**96**:510-515

[19] Howard DD, Gwathmey CO, Essington ME, Roberts RK, Mullen MD. Nitrogen fertilization of no-till corn on loess-derived soils. Agronomy Journal. 2001;**93**:157-163

[20] Sainju UM, Allen BL, Lenssen AW, Ghimire RP. Root biomass, root/shoot ratio, and soil water content under perennial grasses with different nitrogen rates. Field Crops Research. 2017;**210**:183-191

[21] Vogel KP, Brejda JJ, Walters DT, Buxton DW. Switchgrass biomass production in the Midwest USA: Harvest and nitrogen management. Agronomy Journal. 2002;**94**:413-420

[22] Heggenstaller AH, Moore KJ, Liebman M, Anex RP. Nitrogen influences biomass and nutrient partitioning by perennial warm-season grasses. Agronomy Journal. 2009;**101**:1363-1371

[23] Power JF. Seasonal changes in smooth bromegrass top and root growth and fate of fertilizer nitrogen. Agronomy Journal. 1988;**80**:740-745

[24] Mahler RL, Harder RW. The influence of tillage methods, cropping sequence, and N rates on the acidification of a northern Idaho soil. Soil Science. 1984;**137**:52-60

[25] Schroder JL, Zhang H, Girma H, Raun WR, Penn CJ, Payton ME. Soil acidification from long-term use of nitrogen fertilizers on winter wheat. Soil Science Society of America Journal. 2011;**75**:957-964

[26] Liebig MA, Varvel GE, Doran JW, Wienhold BJ. Crop sequence and nitrogen fertilization effects on soil properties in the western Corn Belt. Soil Science Society of America Journal. 2002;**66**:596-601

[27] Sainju UM, Allen BL, Caesar-TonThat T, Lenssen AW. Dryland soil chemical properties and crop yields affected by long-term tillage and cropping sequence. Springerplus. 2015;**4**:230. DOI: 10.1186/s40064-015-1122-4

[28] Aase JK, Aase JK, Pikul JL Jr. Crop and soil responses to long-term tillage practices in the northern Great Plains. Agronomy Journal. 1995;**87**: 652-656

[29] Ghimire R, Machado S, Bista P. Soil pH, soil organic matter, and crop yields in winter wheat-summer fallow systems. Agronomy Journal. 2017;**109**:706-717

[30] Guo JH, Liu XJ, Zhang Y, Shen JL, Han WX, Zhang WF. Significant acidification in major Chinese croplands. Science. 2010;**327**:1008-1010

[31] Rasmussen PE, Rhode CR. Soil acidification from ammonium-nitrogen fertilization in moldboard plow and stubble mulch tillage in wheat-fallow system. Soil Science Society of America Journal. 1989;**53**:119-122

[32] Lilienfein J, Wilcke W, Vilele L, Lima SD, Thomas R, Zech W. Effect of no-till and conventional tillage systems on the chemical composition of soils solid phase and soil solution of Brazilian Savanna soils. Journal of Plant Nutrition and Soil Science. 2000;**163**:411-419

[33] Zibilske LM, Bradford JM, Smart JR. Conservation tillage-induced changes in

organic carbon, total nitrogen, and available phosphorus in a semi-arid alkaline subtropical soil. Soil and Tillage Research. 2002;**66**:153-163

[34] Tarkalson DD, Hergert GW, Cassman KG. Long-term effects of tillage on soil chemical properties and grain yields of a dryland winter wheat-sorghum/corn-fallow rotation in the Great Plains. Agronomy Journal. 2006; **98**:26-33

[35] Darusman L, Stone R, Whitney DA, Janssen KA, Long JH. Soil properties after twenty years of fertilization with different nitrogen sources. Soil Science Society of America Journal. 1991;**55**: 1097-1100

[36] Halvorson AD, Black AL, Krupinsky JM, Merrill SD. Dryland winter wheat response to tillage and nitrogen within an annual cropping system. Agronomy Journal. 1999;**91**:702-707

[37] Russell AE, Laird DA, Parkin TB, Mallarino AP. Impact of nitrogen fertilization and cropping system on carbon sequestration in Midwestern mollisols. Soil Science Society of America Journal. 2005;**69**:413-422

[38] Omay AB, Rice CW, Maddux LD, Gordon WB. Changes in soil microbial and chemical properties under long-term crop rotation and fertilization. Soil Science Society of America Journal. 1997;**61**:1672-1678

[39] Sainju UM, Whitehead WF, Singh BP. Carbon accumulation in cotton, sorghum, and underlying soil as influenced by tillage, cover crops, and nitrogen fertilization. Plant and Soil. 2005;**273**:219-234

[40] Ma Z, Wood C, Bransby DI. Impacts of soil management on root characteristics of switchgrass. Biomass and Bioenergy. 2000;**18**:105-112

[41] Santori F, Lal R, Ebinger MH, Parrish DJ. Potential soil carbon sequestration and CO_2 offset by dedicated energy crops in the USA. Critical Reviews in Plant Sciences. 2006; **25**:441-472

[42] Bremer E, Janzen HH, Ellert BH, McKenzie RH. Soil organic carbon after twelve years of various crop rotations in an aridic boroll. Soil Science Society of America Journal. 2007;**72**:970-974

[43] Rice CW. Soil organic carbon and nitrogen in rangeland soil under elevated carbon dioxide and land management. In: Proceedings of the Advances in Terrestrial Ecosystem and Carbon Inventory, Measurement, and Monitoring; 3–5 October 2000; Washington, DC. Beltsville. Maryland. USA: USDA-ARS; 2000

[44] Li B, Fan CH, Zhang H, Chen ZZ, Sun LY, Xiong ZQ. Combined effects of nitrogen fertilization and biochar on the net global warming potential, greenhouse gas intensity, and net ecosystem budget in intensive vegetable agriculture in southeastern China. Agriculture, Ecosystems and Environment. 2015;**100**:10-19

[45] Sainju UM, Allen BL, Lenssen AW, Mikha M. Root and soil total carbon and nitrogen under bioenergy perennial grasses with various nitrogen rates. Biomass and Bioenergy. 2017;**107**: 326-334

[46] Sainju UM, Singh BP. Nitrogen storage with cover crops and nitrogen fertilization in tilled and non-tilled soils. Agronomy Journal. 2008;**100**:619-627

[47] Sainju UM, Singh BP, Rahman S, Reddy VR. Soil nitrate-nitrogen under tomato following tillage, cover cropping, and nitrogen fertilization. Journal of Environmental Quality. 1999; **28**:1837-1844

[48] Liang BC, McKenzie AF. Changes of soil nitrate-nitrogen and denitrification as affected by nitrogen fertilizer on two

Quebec soils. Journal of Environmental Quality. 1994;**23**:521-525

[49] Keeney DR, Follett RF. Overview and introduction. In: Follett RF, editor. Managing Nitrogen for Groundwater Quality and Farm Profitability. Soil Science Society of America, Madison. USA: Wisconsin; 1991. pp. 1-3

[50] Yadav SN. Formulation and estimation of nitrate-nitrogen leaching from corn cultivation. Journal of Environmental Quality. 1997;**26**: 808-814

[51] Sainju UM, Lenssen AW, Allen BL, Stevens WB, Jabro JD. Nitrogen balance in response to dryland crop rotations and cultural practices. Agriculture, Ecosystems and Environment. 2016;**233**: 25-32

[52] Zibilske LM, Bradford JM, Smart JR. Conservation tillage-induced changes in organic carbon, total nitrogen, and available phosphorus in a semi-arid alkaline subtropical soil. Soil and Tillage Research. 2002;**66**:153-163

[53] Heichel GH, Barnes DK. Opportunities for meeting crop nitrogen needs from symbiotic nitrogen fixation. In: Bezdicek DF, editor. Organic Farming: Current Technology and its Role in Sustainable Agriculture (Special Publication 46). Madison, WI: Soil Science Society of America; 1984. pp. 49-59

[54] Wood CW, Westfall DG, Peterson GA, Burke IC. Impacts of cropping intensity on carbon and nitrogen mineralization under no-till agroecosystems. Agronomy Journal. 1990;**82**:1115-1120

[55] Sainju UM. Tillage, cropping sequence, and nitrogen fertilization influence dryland soil nitrogen. Agronomy Journal. 2013;**105**:1253-1263

[56] Hallberg GR. Nitrate in groundwater in the United States. In: Follett RF, editor. Nitrogen Management and Groundwater Protection. Amsterdam: Elsevier; 1989. pp. 35-74

[57] Timmons DR, Dylla AS. Nitrogen leaching as influenced by nitrogen management and supplemental irrigation level. Journal of Environmental Quality. 1981;**10**:421-426

[58] Pang XP, Gupta SC, Moncrief JF, Rosen CJ, Cheng HH. Evaluation of nitrate leaching potential on Minnesota glacial outwash soils using the CERES-maize model. Journal of Environmental Quality. 1998;**27**:75-85

[59] Meisinger JJ, Hargrove WL, Mikkelsen RI Jr, Williams JR, Benson VE. Effects of cover crops on groundwater quality. In: Hargrove WL, editor. Cover crops for clean water. Ankeny, Iowa, USA: Soil and Water Conservation Society; 1991. pp. 57-68

[60] Intergovernment Panel on Climate Change (IPCC). Climate change 2014: Synthesis report. Contribution of working groups I, II and III to the fifth assessment report of the Intergovernmental Panel on Climate Change. Geneva, Switzerland: IPCC; 2014

[61] Mosier AR, Halvorson AD, Reule CA, Liu XJ. Net global warming potential and greenhouse gas intensity in irrigated cropping systems in northeastern Colorado. Journal of Environmental Quality. 2006;**35**: 1584-1598

[62] Robertson GP, Paul E, Harwood R. Greenhouse gases in intensive agriculture: Contribution of individual gases to the radiative forcing of the atmosphere. Science. 2000;**289**: 1922-1925

[63] West TO, Marland G. A synthesis of carbon sequestration, carbon emissions, and net carbon flux in agriculture:

Comparing tillage practices in the United States. Agriculture, Ecosystems and Environment. 2002;**91**:217-232

[64] Bronson KF, Mosier AR. Suppression of methane oxidation in aerobic soil by nitrogen fertilizers, nitrification inhibitors, and urease inhibitors. Biology and Fertility of Soils. 1994;**17**:263-268

[65] Sainju UM, Caesar-TonThat T, Lenssen AW, Barsotti JL. Dryland soil greenhouse gas emissions affected by cropping sequence and nitrogen fertilization. Soil Science Society of America Journal. 2012;**76**:1741-1757

[66] Phillips RL, Tanaka DL, Archer DW, Hanson JD. Fertilizer application timing influences greenhouse gas fluxes over a growing season. Journal of Environmental Quality. 2009;**38**: 1569-1579

[67] Fujinuma R, Venterea RT, Rosen C. Broadcast urea reduces N_2O but increases NO emissions compared with conventional and shallow applied anhydrous ammonia in a coarse-textured soil. Journal of Environmental Quality. 2011;**40**:1806-1815

[68] Venterea RT, Burger M, Spokas KA. Nitrogen oxide and methane emissions under varying tillage and fertilizer management. Journal of Environmental Quality. 2005;**34**:1467-1477

[69] Halvorson AD, Del Grosso SJ, Alluvione F. Nitrogen source effects on nitrous oxide emissions from irrigated no-till corn. Journal of Environmental Quality. 2010;**39**:1554-1562

[70] Sainju UM, Wang J, Barsotti JL. Net global warming potential and greenhouse gas intensity affected by cropping sequence and nitrogen fertilization. Soil Science Society of America Journal. 2013;**78**:248-261

[71] Sainju UM. A global meta-analysis on the impact of management practices on net global warming potential and greenhouse intensity from cropland soils. PLoS One. 2016;**11**(2):1/26-26/26. DOI: 10.137/journal.pone 0148527

12

Promotion of Nitrogen Assimilation by Plant Growth-Promoting Rhizobacteria

Gabriel Monteiro, Glauco Nogueira, Cândido Neto, Vitor Nascimento and Joze Freitas

Abstract

Nitrogen fertilizers are one of the highest expenses in agricultural systems and usually a limitation to the productions of many agricultural crops worldwide. The intensive use of this element in modern agriculture represents a potential environmental threat, one of the many tools for the sustainable use of this resource without losing productivity is the use of plant growth-promoting rhizobacteria, especially nitrogen-fixing bacteria. However, in considering the competitiveness of the market, studies are still needed to determine the most efficient way to use this resource and if the nitrogen mineral fertilization is indeed substitutable. As a result, this study aims to deepen the scientific knowledge of the plant-microbe interactions by addressing their main characteristics and functionalities for plant growth and development and efficiency in the use of nitrogen. For this we reviewed relevant information from scientific works that address these issues.

Keywords: biochemistry, nitrogen-fixation, growth, nitrogen fertilizers, nitrogen use efficiency

1. Introduction

Nitrogen (N) is a key component of most proteins, secondary metabolites and signaling molecules [1]. It is one of the most important macronutrients for plant development and usually one of the most limiting factor to plant production [2].

The use of N-fertilizers has produced a significant increase in food production in recent decades [3], and its consumption has grown from 11.3 Tg N year^{-1} in 1961 to 107.6 Tg N year^{-1} in 2013 [4]. However, less than 50% of the added N is effectively absorbed by most cultivated plants [5, 6], and even N effectively converted to biomass, eventually returns to the environment [7]. In the soil, N is available to plants in the form of nitrate (NO_3^-), ammonium (NH_4^+) and organic compounds (usually amino acids), being the NO_3, the most abundant [8]. In its ionic form, NO_3^- has a negative charge and high water solubility, being susceptible to leaching and runoff [9]. It can also be volatilized by denitrifying microorganisms [10], and lost to the atmosphere in the form of nitrous oxide (N_2O, a greenhouse gas 296-fold more potent than a unit of CO_2). Leaching of N causes eutrophication of water bodies and contamination of groundwater [11].

N can also be lost to the atmosphere in other forms such as reactive gases (NO_x; NH_3), aggravating the greenhouse effect [12], and is also related to the acidification of soils through the formation of acid rain, and depletion of exchangeable basic soil cations [13].

To minimize the loss of N in the soil, and consequently the total amount of N necessary for a high-quality production, several strategies have been used. One is the use of urease inhibitors such as N-(n-butyl)-thiophosphoric triamide (NBPT) to delay the hydrolysis of urea, thus reducing losses to the atmosphere and microbial transformations [14]. Increased N use efficiency (NUE) has also been the subject of research, through the selection of genotypes with a higher NUE [15] or through biological nitrogen fixation [16]. Biological nitrogen fixation (BNF) occurs through the conversion of atmospheric N_2 to ammonium by free-living or symbiotic diazotrophic bacteria [17]. Plant growth-promoting rhizobacteria (PGPR) can increase the N absorption capacity through BNF [18], phytohormone production [19], stimulate the production and enzymatic assimilation of NH4+ [16], as well as the transport and partition of N [20].

The study of plant nutrition related to the use of BNF as an alternative to increase efficiency and sustainability in the use of N in agriculture is essential, given the complex nature of the interactions between soil, plant, and microorganisms. Therefore, the objective of this review is to deepen the scientific knowledge of these interactions, addressing their main characteristics and functionalities for plant growth and development.

2. Mechanisms of biological nitrogen fixation

One of the largest N reservoirs is the atmosphere, second only to the lithosphere in absolute amount of N [21]. N makes up about 78% of the atmosphere [22] and is mainly in the form of molecular N (N_2). The atoms in the N_2 molecule have low-energy orbitals and the bond between the two N molecules is relatively short (1,098 Å) and stable, with a bonding energy of 930 kJ/mol [23]. This set of characteristics gives low reaction potential to the molecule. Alternatively, N_2 can be reduced to NH_3 naturally by microorganisms through the BNF process. The BNF reaction follows the following stoichiometry:

The BNF process is catalyzed exclusively by an enzyme complex called the nitrogenase complex. The nitrogenase complex is composed of dinitrogenase reductase (Iron-protein) and dinitrogenase (Molybdenum-Iron-protein) (**Figure 1a**). Dinitrogenase-reductase is a dimer of approximately 60 kDa, composed of two identical and symmetrical subunits, which coordinate a redox center 4Fe-4S (**Figure 2**). This enzyme also has sites for the binding of ATP/ADP, one in each subunit, being able to couple the hydrolysis of ATP to fuel the transfer of electrons to the dinitrogenase [26]. Dinitrogenase is a heterotetramer $\alpha_2\beta_2$ with approximately 240 kDa.

Dinitrogenase has two cofactors containing iron (**Figure 1b**), being the group P and the FeMo cofactor [27]. Group P contains a pair of 4Fe-4S centers, which share a sulfur, forming an 8Fe-7S center. The FeMo cofactor is a variation of the iron–sulfur groups, such as the P group, but differs greatly from the other metallic sites within this family. Its structure is composed of [Mo: 7Fe: 9S: C]: Homocitrate [26, 28]. Some microorganisms also have alternative forms of the MoFe cofactor, where Molybdenum (Mo) is replaced by atoms of Vanadium (V) or Iron (Fe) depending on metal availability [29].

In the N_2 reduction reaction (Eq. 1), the reduced dinitrogenase reductase couples the hydrolysis of ATP with the transfer of electrons to the dinitrogenase. The oxidized dinitrogenase reductase detaches from the dinitrogenase, only to be reduced again (by ferredoxins or flavodoxins). Again, there is the coupling between

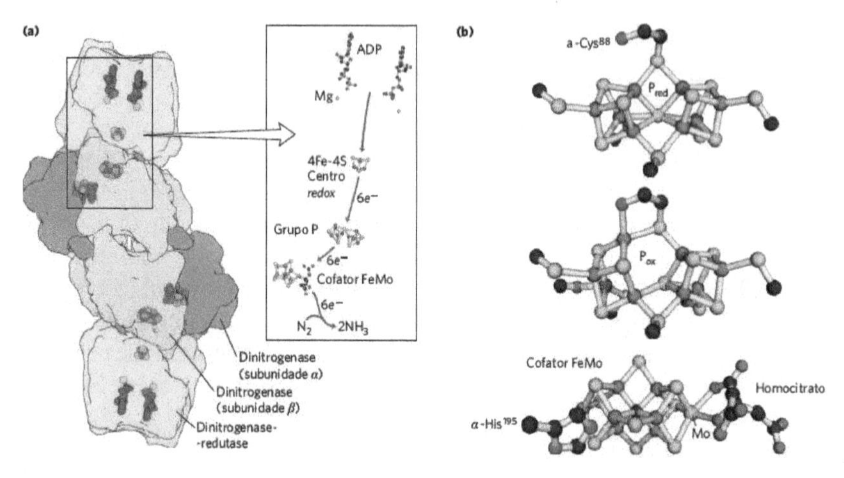

Figure 1.
Enzymes and cofactors of the nitrogenase complex. (a) The enzyme consists of two symmetrical dinitrogenase reductase molecules (in green), each with a 4Fe-4S redox center and binding sites for ATP, and two identical dinitrogenase heterodimers (in purple and blue), each one with a P group and a FeMo cofactor. (b) The cofactors for the electron transfer. A group P is shown here in its reduced (upper part) and oxidized (middle) and the cofactor FeMo is showed at the bottom. Source: Taiz et al. [24].

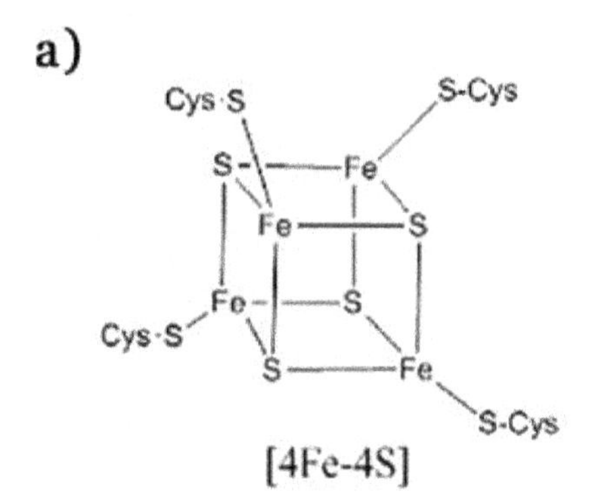

Figure 2.
Structure of 4Fe-4S clusters present in nitrogenase complex. (a) The 4Fe-4S binding site between the Dinitrogenase-reductase and Dinitrogenase contains a cubane-like structure where four iron ions and four sulfide ions are placed at the vertices. The Fe centers are typically coordinated by cysteine residuals. Source: [25].

both enzymes followed by the consumption of ATP to transfer the electrons to the dinitrogenase. This coupling-detaching cycle followed by the electron transfer is repeated until the dinitrogenase is reduced enough to reduce its substrate, which in the case of BNF, is N_2 [30].

Because it deals with large amounts of energy, enough to break the triple bond of the N_2 molecule, the nitrogenase complex is not only inactivated by the presence of oxygen (O_2) but can have its expression reduced [31]. However, diazotrophic bacte-ria are able to combine N fixation with their aerobic metabolism in different ways to avoid O_2 deactivation [32]. Notably, one of the most advanced means of controlling O_2 concentration is expressed by rhizobia-leguminous symbiosis.

3. Brazil: a success of BNF in legumes

Brazil is one of the best examples in the efficient research and use of the BNF in legumes [33]. The use of FBN is interesting from both an economic and an environ-mental point of view, since once this process is established, nitrogen fertilizers can be dispensed with in whole or in part, thus contributing to enable reforestation and minimize possible environmental impacts resulting of uof more than US$ 13 billion with the total or partial substitution of nitrogen fertilizers in legumes in Brazil [33].

Rhizobium-leguminous symbiosis is the most important symbiotic system between microorganisms and plants thanks to the efficiency of the N2 fixation process, and one of the justifications is in the amplitude and geographical distribution of the hosts and the economic impact it causes in agriculture, and one of the main sources of N for the biosphere [35]. The leguminosae family comprises almost 20 thousand species, including tree, herbaceous species used as fodder, producers of raw materials or directly in human food [36].

The most successful case here in Brazil according to Hungria et al. [18], is the symbiosis of Bradyrhizobium spp. with soybean (*Glycine max* (L.) Merrill). The main leguminous species produced in Brazil does not require nitrogen-based fertilization, due to the biological nitrogen fixation which supplies the required N for crop development. To get a sense of how important this symbiotic relationship is in Brazil, in the 2018/2019 harvest, over 35 million hectares were sown. According to the National Supply Company [37], soybean production for this same harvest was 115 million tons, resulting in an average productivity of 3,208 kg ha^{-1}, with almost zero N-fertilizer input. On the other hand, the biological nitrogen fixation with other important legumes, such as beans and peanuts, and non-legumes like sugar-cane cannot fully supply the demand for N like in soybeans.

Alfalfa is another plant species with a high potential for biological nitrogen fixation. As a legume, alfalfa is capable of symbiotically associating with rhizobial bacteria, with N inputs to reaching up to 470 kg of N ha-1 [38]. The main symbiotic alfalfa bacteria belong to the genus Ensifer, having Synorhizobium as a synonym, but previously classified as Rhizobium. Sinorhizobium meliloti and Sinorhizobium mediace are the main symbionts reported in several countries [39]. In Brazil, there are three strains of rhizobia that are used in commercial inoculants for alfalfa, which have been validated for more than two decades and with rare tests conducted with the same [39].

4. Growth promotion by associative and free-living diazotrophic bacteria and stimulating N metabolism

World production is dominated by the production of four grasses (FAO Stats 2019), namely: Sugarcane (*Saccharum officinarum* L. 1.95 Gt year-1), Maize (*Zea mays* L. 1.15 Gt year-1), Wheat (*Triticum aestivum* L. 0.76 Gt year-1) and Rice (*Oryza sativa* L. 0.75 Gt year-1). Scientific studies have shown that biological nitrogen fixation is not limited only to legumes, with a potentially important group of diazotrophic bacteria capable of forming associations through root colonization and the internal tissues of grasses [40]. These bacterial-grass associations also have different mechanisms of action than legumes, and in addition to fixing N, increase the absorption and assimilation of N by modulating the architecture and development of the root system through the production of phytohormones, such as indole-3-acetic acid [41] and gibberellins [42]. Cases such as diazotrophic bacteria such as those of the genus Azospirillum sp., Herbaspirillum sp. and Glucanobacter sp. are evidence of the influence of PGPR on N metabolism, inducing physiological and morphological changes that are associated with greater NUE.

Grasses are of great interest for the development of biological nitrogen fixation aiming at greater efficiency, in view of its relatively low NUE [43, 44]. The increase in NUE is related to several characteristics, Iqbal et al., (2020) working with different cotton genotypes analyzed the biochemical and morphological responses of the accessions according to different concentrations of NO_3^- and the root architecture and efficiency of the enzymes of the assimilation of N were essential for the increase in traction related to NUE. The inoculation of diazotrophic bacteria seems to achieve the same results in different cultures of importance to the global market such as sugarcane (*Saccharum officinarum* L.), corn (*Zea mays* L.), wheat (*Triticum aestivum* L.,) and rice (*Oryza sativa* L.). In sugarcane, the effects vary from the increase in the speed of bud sprouting and the emission of roots in sugarcane stalks used for planting [45], increases in the biomass production of the thatch [46], until the increase in the number of tillers [47]. In maize, inoculation with Azospirillum brasilense increased the transcription of the genes encoding Nitrate reductase (ZmNR), Glutamina sintase (ZmGln1–3), and the intensity of assimilation of the N [48, 49]. In wheat, the inoculation of Azospirillum brasilense was able to modify the N metabolism, resulting in an increase in growth [16].

Modulation of root architecture induced by PGPR is also a morphological trait essential to the increase in NUE. The increase in the area explored by the root triggered by the increase in the volume of the roots caused by these bacteria directly influences the interception of nutrients, among them the N. Besides the influence on the morphological characteristics, the PGPR inoculation has an impact on the metabolism activity of the N. The increase in these characteristics makes these bacteria a potential solution for increasing NUE.

$$N_2 + 8H^+ + 8e^- + ATP \xrightarrow{Nitrogenase} 2NH_3 + H_2 + 16\,ADP + 16\,Pi \qquad (1)$$

5. Conclusions

Since the discovery of the Haber-bosch process in the early 20th century, the levels of N added to the biosphere continue to increase each year and its excessive use of this element is a source of numerous environmental problems.

Therefore, biological nitrogen fixation process is an essential tool in the current economic, agricultural, and environmental context of many countries, this can be seen through studies and government data that show a reduction in financial expenses in the order of millions, this technological tool. it is a reality in vegetable crops with high potentials in the agricultural network of emerging powers worldwide, in addition to contributing to the reduction of potentially harmful agents for the worsening of the greenhouse effect.

This technology has been extensively scientifically explored, aiming to expand its possibility in other promising cultures in the agricultural and forestry world, as well as other associative and free-living diazotrophic microorganisms, and their ability to promote plant growth. However, there are still major gaps in knowledge about the diversity and mechanisms of PGPR action and further research is needed to establish the use of these new bacteria as a sustainable agricultural practice.

Acknowledgements

To Laboratory of Biodiversity Studies of Upper Plants in Federal Rural University of Amazonia.

Appendices and nomenclature

CONAB	National Supply Company
NUE	Nitrogen use efficiency
NBPT	N-(n-butyl)-thiophosphoric triamide
PGPR	Plant growth-promoting rhizobacteria
BNF	Biologic nitrogen fixation

Author details

Gabriel Monteiro[1], Glauco Nogueira[1*], Cândido Neto[2], Vitor Nascimento[3] and Joze Freitas[2]

1 Federal Rural University of Amazon (UFRA), Belém, Pará, Brazil

2 Institute of Agrarian Sciences, Federal Rural University of Amazon (UFRA), Belém, Pará, Brazil

3 Rede BIONORTE/UFPA, Belém, Pará, Brazil

*Address all correspondence to: glauand@yahoo.com.br

References

[1] Krapp, A. (2015). Plant nitrogen assimilation and its regulation: A complex puzzle with missing pieces. Current Opinion in Plant Biology, *25*, 115-122. https://doi.org/10.1016/j.pbi.2015.05.010

[2] Teixeira, E. I., George, M., Herreman, T., Brown, H., Fletcher, A., Chakwizira, E., de Ruiter, J., Maley, S., & Noble, A. (2014). The impact of water and nitrogen limitation on maize biomass and resource-use efficiencies for radiation, water and nitrogen. Field Crops Research, *168*, 109-118. https://doi.org/10.1016/j.fcr.2014.08.002

[3] Gojon, A. (2017). Nitrogen nutrition in plants: Rapid progress and new challenges. Journal of Experimental Botany, *68*(10), 2457-2462. https://doi.org/10.1093/jxb/erx171

[4] Lu, C., & Tian, H. (2017). Global nitrogen and phosphorus fertilizer use for agriculture production in the past half century: Shifted hot spots and nutrient imbalance. Earth System Science Data, *9*(1), 181-192. https://doi. org/10.5194/essd-9-181-2017

[5] Kant, S. (2018). Understanding nitrate uptake, signaling and remobilisation for improving plant nitrogen use efficiency. Seminars in Cell and Developmental Biology, *74*, 89-96. https://doi.org/10.1016/j.semcdb.2017.08.034

[6] Plett, D. C., Holtham, L. R., Okamoto, M., & Garnett, T. P. (2018). Nitrate uptake and its regulation in relation to improving nitrogen use efficiency in cereals. Seminars in Cell and Developmental Biology, *74*, 97-104. https://doi.org/10.1016/j.semcdb.2017.08.027

[7] Robertson, G. P., & Vitousek, P. M. (2009). Nitrogen in agriculture: Balancing the cost of an essential resource. Annual Review of Environment and Resources, *34*, 97-125. https://doi.org/10.1146/annurev. environ.032108.105046

[8] Miller, A. J., Fan, X., Orsel, M., Smith, S. J., & Wells, D. M. (2007). Nitrate transport and signalling. Journal of Experimental Botany, *58*(9), 2297-2306. https://doi.org/10.1093/jxb/erm066

[9] Good, A. G., & Beatty, P. H. (2011). Fertilizing nature: A tragedy of excess in the commons. PLoS Biology, *9*(8), 1-9. https://doi.org/10.1371/journal.pbio.1001124

[10] Kuypers, M. M. M., Marchant, H. K., & Kartal, B. (2018). The microbial nitrogen-cycling network. Nature Reviews Microbiology, *16*(5), 263-276. https://doi.org/10.1038/nrmicro.2018.9

[11] Sinha, E., Michalak, A. M., & Balaji, V. (2017). Eutrophication will increase during the 21st century as a result of precipitation changes. Science, *357*(6349), 1-5. https://doi.org/10.1126/science.aan2409

[12] Xu, G., Fan, X., & Miller, A. J. (2012). Plant nitrogen assimilation and use efficiency. Annual Review of Plant Biology, *63*, 153-182. https://doi.org/10.1146/annurev-arplant-042811-105532

[13] Horswill, P., O'Sullivan, O., Phoenix, G. K., Lee, J. A., & Leake, J. R. (2008). Base cation depletion, eutrophication and acidification of species-rich grasslands in response to long-term simulated nitrogen deposition. Environmental Pollution, *155*(2), 336-349. https://doi.org/10.1016/j.envpol.2007.11.006

[14] Silva, A. G. B., Sequeira, C. H., Sermarini, R. A., & Otto, R. (2017). Urease inhibitor NBPT on ammonia

volatilization and crop productivity: A meta-analysis. Agronomy Journal, *109*(1), 1-13. https://doi.org/10.2134/agronj2016.04.0200

[15] Iqbal, A., Qiang, D., Alamzeb, M., Xiangru, W., Huiping, G., Hengheng, Z., Nianchang, P., Xiling, Z., & Meizhen, S. (2020). Untangling the molecular mechanisms and functions of nitrate to improve nitrogen use efficiency. Journal of the Science of Food and Agriculture, *100*(3), 904-914. https://doi.org/10.1002/jsfa.10085

[16] Silveira, A. P. D. da, Sala, V. M. R., Cardoso, E. J. B. N., Labanca, E. G., & Cipriano, M. A. P. (2016). Nitrogen metabolism and growth of wheat plant under diazotrophic endophytic bacteria inoculation. Applied Soil Ecology, *107*, 313-319. https://doi.org/10.1016/j.apsoil.2016.07.005

[17] Franche, C., Lindström, K., & Elmerich, C. (2009). Nitrogen-fixing bacteria associated with leguminous and non-leguminous plants. Plant and Soil, *321*(1-2), 35-59. https://doi.org/10.1007/s11104-008-9833-8

[18] Hungria, M., Franchini, J. C., Campo, R. J., Crispino, C. C., Moraes, J. Z., Sibaldelli, R. N. R., Mendes, I. C., & Arihara, J. (2006). *Nitrogen nutrition of soybean in Brazil: Contributions of biological N 2 fixation and N fertilizer to grain yield*.

[19] Glick, B. R. (2012). Plant Growth-Promoting Bacteria : Mechanisms and Applications. *2012*.

[20] Mantelin, S., Desbrosses, G., Larcher, M., Tranbarger, T. J., Cleyet-Marel, J. C., & Touraine, B. (2006). Nitrate-dependent control of root architecture and N nutrition are altered by a plant growth-promoting Phyllobacterium sp. Planta, *223*(3), 591-603. https://doi.org/10.1007/s00425-005-0106-y

[21] Stevenson, F. J. (2015). *Origin and Distribution of Nitrogen in Soil* (Issue 22, pp. 1-42). https://doi.org/10.2134/agronmonogr22.c1

[22] Williams, D. R. (2016). Earth Fact Sheet. NASA Fact Sheets, 1. https://nssdc.gsfc.nasa.gov/planetary/factsheet/earthfact.html

[23] Fernandes, M. S., & Pereyra Rossiello, R. O. (1995). Mineral Nitrogen in Plant Physiology and Plant Nutrition. Critical Reviews in Plant Sciences, *14*(2), 111-148. https://doi.org/10.1080/07352689509701924

[24] Taiz, L., Zeiger, E., Møller, I. M., & Murphy, A. (2017). *Fisiologia e desenvolvimento vegetal*. Artmed Editora.

[25] Cammack, R. (2012). Iron–sulfur proteins. The Biochemist, *34*(5), 14-17. https://doi.org/10.1042/BIO03405014

[26] Einsle, O., & Rees, D. C. (2020). Structural Enzymology of Nitrogenase Enzymes. Chemical Reviews, *120*(12), 4969-5004. https://doi.org/10.1021/acs.chemrev.0c00067

[27] Morrison, C. N., Hoy, J. A., Zhang, L., Einsle, O., & Rees, D. C. (2015). Substrate Pathways in the nitrogenase MoFe protein by experimental identification of small molecule binding sites. Biochemistry, *54*(11), 2052-2060. https://doi.org/10.1021/bi501313k

[28] Einsle, O. (2014). Nitrogenase FeMo cofactor: An atomic structure in three simple steps. Journal of Biological Inorganic Chemistry, *19*(6), 737-745. https://doi.org/10.1007/s00775-014-1116-7

[29] Eady, R. R. (1996). Structure-function relationships of alternative nitrogenases. Chemical Reviews, *96*(7), 3013-3030. https://doi.org/10.1021/cr950057h

[30] Rees, D. C., Akif Tezcan, F., Haynes, C. A., Walton, M. Y., Andrade, S.,

Einsle, O., & Howard, J. B. (2005). Structural basis of biological nitrogen fixation. *Philosophical Transactions of the Royal Society A: Mathematical, Physical and Engineering Sciences*, 363(1829), 971-984. https://doi.org/10.1098/rsta.2004.1539

[31] Gallon, J. R. (1981). The oxygen sensitivity of nitrogenase: a problem for biochemists and micro-organisms. Trends in Biochemical Sciences, *6*(C), 19-23. https://doi.org/10.1016/0968-0004(81)90008-6

[32] Goldberg, I., Nadler, V., & Hochman, A. (1987). Mechanism of nitrogenase switch-off by oxygen. Journal of Bacteriology, *169*(2), 874-879. https://doi.org/10.1128/jb.169.2.874-879.1987

[33] Auras, N. É., Zilli, J. É., Soares, L. H. de B., & Fontana, J. (2018). Recomendação de uso de estirpes fixadoras de nitrogênio em leguminosas de importância agronômica e florestal. Embrapa Agrobiologia-Documentos (INFOTECA-E).https://www.infoteca.cnptia.embrapa.br/infoteca/bitstream/doc/1099443/1/recomendacaodeusodeestirpesfixadoras.pdf

[34] Barbieri, A.; Carneiro, M.A.C.; Moreira, F.M.S.; Siqueira, J.O. (1998). Nodulação em leguminosas florestais em viveiros no sul de minas gerais. CERNE, v.4, n.1, p.145-153. https://www.researchgate.net/publication/242531593

[35] Galloway, J. N., Dentener, F. J., Capone, D. G., Boyer, E. W., Howarth, R. W., Seitzinger, S. P., Asner, G. P., Cleveland, C. C., Green, P. A., Holland, E. A., Karl, D. M., Michaels, A. F., Porter, J. H., Townsend, A. R., & Vöosmarty, C. J. (2004). Nitrogen Cycles: Past, Present, and Future. Biogeochemistry, *70*(2), 153-226. https://doi.org/10.1007/s10533-004-0370-0

[36] Cantarella, H. (2007). Nitrogênio. Fertilidade Do Solo, *2*, 375-470.

[37] CONAB. (2019). *Boletim da Safra de Grãos*. Acompanhamento Da Safra 2018/19 Brasileira de Grãos - 12° Levantamento. https://www.conab.gov.br/info-agro/safras/graos/boletim-da-safra-de-graos

[38] Ormeño-Orrillo, E., Hungria, M., & Martinez-Romero, E. (2013). Dinitrogen-fixing prokaryotes. In *The Prokaryotes: Prokaryotic Physiology and Biochemistry* (Vol. 9783642301414, pp. 427-451). Springer-Verlag Berlin Heidelberg. https://doi.org/10.1007/978-3-642-30141-4_72

[39] de Soares, L.H. B, Michel, D. C., & Zilli, J. É. (2020). Fixação biológica do nitrogênio. In Ministério da Agricultura, Pecuária e Abastecimento (Ed.), *Alfafa: do cultivo aos múltiplos usos* (p. 173). MAPA/AECS.

[40] Rosenblueth, M., Ormeño-Orrillo, E., López-López, A., Rogel, M. A., Reyes-Hernández, B. J., Martínez-Romero, J. C., Reddy, P. M., & Martínez-Romero, E. (2018). Nitrogen fixation in cereals. *Frontiers in Microbiology*, *9*(AUG), 1-13. https://doi.org/10.3389/fmicb.2018.01794

[41] Spaepen, S., Vanderleyden, J., & Remans, R. (2007). Indole-3-acetic acid in microbial and microorganism-plant signaling. FEMS Microbiology Reviews, *31*(4), 425-448. https://doi. org/10.1111/j.1574-6976.2007.00072.x

[42] Bottini, R., Cassán, F., & Piccoli, P. (2004). Gibberellin production by bacteria and its involvement in plant growth promotion and yield increase. Applied Microbiology and Biotechnology, *65*(5), 497-503. https://doi.org/10.1007/s00253-004-1696-1

[43] Herrera, J. M., Rubio, G., Häner, L. L., Delgado, J. A., Lucho-Constantino, C. A., Islas-Valdez, S., & Pellet, D.

(2016). Emerging and established technologies to increase nitrogen use efficiency of cereals. Agronomy, *6*(2), 11-18. https://doi.org/10.3390/agronomy6020025

[44] Tilman, D., Cassman, K. G., Matson, P. A., Naylor, R., & Polasky, S. (2002). Agricultural sustainability and intensive production practices. Nature, *418*(6898), 671-677. https://doi.org/10.1038/nature01014

[45] Landell M.G. de A, Campana, M. P., Figueiredo, P., Xavier, M. A., Anjos, I. A. dos, Dinardo-Miranda, L. L., Scarpari, M. S., Garcia, J. C., Bidóia, M. A. P., & Silva, D. N. da. (2012). Sistema de multiplicação de cana-de-açúcar com uso de mudas pré-brotadas (MPB), oriundas de gemas individualizadas. *Ribeirão Preto: Instituto Agronômico de Campinas*, *17*.

[46] Gírio LA da S, Dias, F. L. F., Reis, V. M., Urquiaga, S., Schultz, N., Bolonhezi, D., & Mutton, M. A. (2015). Plant growth-promoting bacteria and nitrogen fertilization effect on the initial growth of sugarcane from pre-sprouted seedlings. Pesquisa Agropecuaria Brasileira, *50*(1), 33-43. https://doi. org/10.1590/s0100-204x2015000100004

[47] Oliveira, A. R. De, & Simões, W. L. (2016). Cultivares de cana-de-açúcar inoculadas com bactérias. Energia Na Agricultura, *31*, 154-161.

[48] da Fonseca Breda, F. A., da Silva, T. F. R., dos Santos, S. G., Alves, G. C., & Reis, V. M. (2019). Modulation of nitrogen metabolism of maize plants inoculated with Azospirillum brasilense and Herbaspirillum seropedicae. Archives of Microbiology, *201*(4), 547-558. https://doi.org/10.1007/s00203-018-1594-z

[49] Pereira-Defilippi, L., Pereira, E. M., Silva, F. M., & Moro, G. V. (2017). Expressed sequence tags related to nitrogen metabolism in maize inoculated with Azospirillum brasilense. Genetics and Molecular Research, *16*(2), 1-14. https://doi.org/10.4238/gmr16029682

13

Nitrogen Fixation in Soyabean Nodules Affects Seed Protein and Oil Contents: The Suggested Mechanism from the Coordinated Changes of Seed Chemical Compositions and Phosphoenol-pyruvate Carboxylase Activity Caused by Different Types of Nitrogen Fertilizer

Toshio Sugimoto, Naoki Yamamoto and Takehiro Masumura

Abstract

The contents of seed storage compounds, protein and oil, determine the best use of soybean seeds, namely materials for food processing and oil production. Genetic and environmental factors could affect the chemical compositions of soybean seeds. However, the mechanisms of how the accumulation of these primary seed com-pounds is regulated are mostly unclear. In this chapter, we describe the different effects of nodulation on the protein and oil contents in soybean seeds and the cru-cial role of phospho*enol*pyruvate carboxylase (PEPC) in the protein accumulation of soybean seeds. Based on our previous studies on soybean seeds, we introduce five manners deduced; (1) protein accumulation is independent of oil accumulation, (2) nitrogen fixation results in decreasing oil amount per seed and decreased seed oil content, (3) a high pseudo negative correlation between protein and oil contents in seeds is likely to be observed under less nitrogen supply from the soil, (4) nitrogen absorbed from soil during the late growth stage promote seed production, (5) plant-type PEPC, ex. Gmppc2 in soybean could play a role in amino acid biosynthesis for storage protein accumulation in seeds during the late maturation period.

Keywords: carbon metabolism in immature seeds, Gmppc2, plant-type, principal component analysis, role of PEPC, seed yield, slow-release N fertilizer

1. Introduction

Soybean seeds contain about 40% protein, 20% oil, 35% carbohydrate, and 5% minerals on a weight basis [1]. Contents of these compounds vary among cultivars

(CVs) and environments of plant growth. Seeds having higher protein content are preferable for food material, and which is called food bean. On the other hand, those having higher oil content are for vegetable oil production, and which is called oil bean. Germplasm stock of USDA exhibits a protein concentration from no less than 35% to more than 50% with an oil concentration of 7% to 28% [2]. Contents of these storage compounds were negatively correlated with each other, implying com-petition of the synthesis of these storage compounds during the seed maturation period. For the production of soybean seeds to have better quality, it was necessary to clarify the mechanism of how these storage compounds contents were genetically controlled and affected by environmental conditions.

One of the characteristics of soybean plants is N_2 fixation ability concerted with symbiotic microorganisms located in nodules on roots. Soybean plants supply photosynthate, sucrose, to nodules as the nutrient. Symbiotic microorganisms in nodules utilize sucrose for the source of energy required for N_2 fixation and the carbon (C) skeleton of nitrogenous compounds, ureides (allantoin and allantoic acid) [3]. Ureides are suitable forms for the transportation of nitrogen (N) in the plant body. Soybean plants supply less amount of sucrose to nodules when the soil offers enough amount of N (as a form of nitrate) [4]. It is well known that a high concentration of soil nitrate depresses the formation of nodules and N_2 fixation [3].

We did experiments to clarify two possible factors to affect seed protein content, the supply of N to seeds and C metabolism in immature soybean seeds. Firstly, we describe that N_2 fixation in nodules affects seed protein content among soybean plants having different N_2 fixation activity grown on soils with different types of coated urea slow-release N fertilizers (CUSLNFs) in Section 2. Secondly, we describe the role of a CO_2 fixing enzyme, phosopho*enol*pyruvate carboxylase (PEPC), on the protein accumulation in maturing seeds in Section 3. Here we introduce our results in the experiments on plants cultivated in the field of the Faculty of Agriculture, Kobe University, where the soil was granite-based udorthents.

2. Association of nodulation and N_2 fixation activity with the accumulation of protein and oil in soybean seeds

2.1 Soybean plants with different nodulation status reveals the effect of nodulation and N_2 fixation activity on the seed protein and oil contents

2.1.1 Estimation of N derived from nitrogen fixation (Ndfa), N derived from N fertilizer (Ndff) and N derived from soil (Ndfs) in seeds by monitoring δ^{15}N

We used the most popular cultivar in Japan, Enrei, and its two near-isogenic lines (NIL) to evaluate the effect of nodule's N_2 fixation activity on seed protein and oil contents. One of the NILs [5, 6], En1282, was the no nodulating isoline [7], and another NIL, En-b0–1, was the hyper nodulating isoline [8]. As high soil nitrate levels inhibit the nodule formation and N_2 fixation activity of soybean plants [3], we applied different types of coated urea slow-release nitrogen fertilizers (CUSLNFs) having different lifespan to inhibit the N_2 fixation activity of plants for a certain period. The used CUSLNFs continuously emit N contained in them for 0 to 30 days for M5, 60 to 120 days for MS9, and 0 to 100 days for M15, respectively. We designated four types of experimental plots with different expected soil N levels (H, high or L, low) during the early and late periods of plant growth: the plots 'L-L', 'H-L', 'L-H', and 'H-H' were with no CUSLNF, M5, MS9, and M15, respectively. N_2 fixation of Enrei and En-b0–1 were expected to be suppressed during their working period. **Table 1** shows δ^{15}N values of matured seeds of the three NILs grown on the

Genotype	N Treatment			
	L-L	H-L	L-H	H-H
En1282	3.36	2.57	1.22	1.07
	(0.60)	(0.51)	(0.66)	(0.14)
Enrei	−0.20	2.12	0.33	1.88
	(0.36)	(1.36)	(0.14)	(0.36)
En-b0–1	−0.61	1.46	1.06	1.69
	(0.22)	(0.02)	(0.09)	(0.22)

The $\sigma^{15}N$ value of an urea coated slow-released nitrogen fertilizer (Meister 15) was −1.41.
$\sigma^{15}N$ values: Averages of duplicated samples. Parentheses indicate the SD (n = 2).

Table 1.
The $\sigma^{15}N$ values of mature seeds from plants of three genotypes of soybean.

four types of plots. We discriminated N derived from N_2 fixation, soil and CUSLNFs using the $\delta^{15}N$ values of seeds based on the authorized method [9]. Namely, ratios of seed N from the three origins in the NILs were estimated by the equations described in the footnotes of **Tables 2** and **3**. Since the En1282 plants utilized the mineral N that was available in the experimental fields, the $\delta^{15}N$ value of En1282 seeds of the L-L plot (3.36) was of the N supplied from the soil and the basal compound fertil-izer. We distinguished the CULSNF-N from the N from the soil and compound fertilizer by using the $\delta^{15}N$ values of En1282 seeds of H-L, L-H, and H-H N plots then estimated the ratios of CULSNF-N as 16.6%, 44.8%, and 48.0%, respectively (**Table 2**). In the case of En-b0–1 plants grown on the L-L plot, the N from both the soil and the compound fertilizer was assumed to be negligible as %Ndfa of the plants was estimated to 118.1%. We presumed that the $\delta^{15}N$ value (−0.61) of En-b0–1seeds of L-L N plot was of the N from N_2 fixation. The contribution rates of the N_2 fixation in Enrei grown on the L-L, H-L, L-H, and H-H plots were 89.9%, 14.2%, 48.9%, and – 47.9%, respectively, and those of En-b0–1 were 100%, 35.0%, 8.9%, and – 36.5%, respectively (**Table 3**). The contribution rates of fixed-N ratio in the seeds from the plants of these nodulating genotypes grown on the H-H plot were negative values because these $\delta^{15}N$ values were higher than those of the En1282 plants. The contribution to the total N from the N_2 fixation in both Enrei and En-b0–1 was assumed to be zero. The number of nodules in a nodulating soybean

N	Ratio, %		Amount, gN/plant		
Treat.	Soil + BF	CUSLNF	Soil + BF	CUSLNF	Sum
L-L	100	0	0.63	0	0.63
H-L$	83.4	16.6	1.50	0.30	1.80
L-H$	55.2	44.8	1.20	0.98	2.18
H-H	52.0	48.0	1.32	1.22	2.53

The values were calculated by the equation described below under the assumption that the two types of CUSLNF, M-5 and MS-9, had the same $\sigma^{15}N$ values with that of Meister 15. The ratios of these N origins of seeds from plants of En1282 grown on the other N plots were estimated by using these $\sigma^{15}N$ values.
$\sigma^{15}N(En1282_{i-ii}) = (3.36) \times X_{i-ii}/100 + (-1.41) \times Y_{i-ii}/100$ ($X_{i-ii} + Y_{i-ii} = 100$), where X indicates the percentage of the sum of soil N and compound fertilizer N against the total N, Y represents the percentage of CUSLNF-N against the total N, and i–ii is the estimated N levels of the early and late periods of plant growth (H or L levels of N).

Table 2.
The ratios and amounts of N originated from the two origins, soil + compound fertilizer and CUSLNFs, in the total N in seeds from plants of non-nodulating genotype, En1282.

N	Genotype	Ratio, %	Amount		gN/plant			
Treat.		Soil + BF	SLNF	N_2 fix.	Soil + BF	SLNF	N_2 fix.	Sum
L-L	Enrei	10.1	0	89.9	0.28	0	2.48	2.75
	En-b0–1[†]	0	0	100	0	0	0.53	0.53
H-L[§]	Enrei	71.6	14.2	14.2	1.21	0.24	0.24	1.70
	En-b0–1	54.2	10.8	35.0	0.40	0.08	0.26	0.73
L-H[§]	Enrei	28.2	22.9	48.9	0.97	0.79	1.68	3.43
	En-b0–1	50.3	40.8	8.9	0.53	0.43	0.09	1.06
H-H	Enrei[‡]	76.9	71.0	−47.9	1.67	1.54	−1.04	2.17
		(52.0)	(48.0)	(0)	(1.13)	(1.04)	(0)	(2.17)
	En-b0–1[‡]	71.0	65.5	−36.5	0.95	0.87	−0.49	1.33
		(52.0)	(48.0)	(0)	(0.64)	(0.69)	(0)	(1.33)

[†]*En-b0–1 plants grown on L-L soil; it was assumed that N was only from N_2 fixation.*
[‡]*The numbers of each fraction of nitrogen origins in parentheses were calculated by the equation described in the text of Ref [5] in the cases in which N_2 fixation did not work.*
[§]*The values were calculated by the equation described in the text of Ref. [5] under the assumption that the two types of CUSLNF had the same $\sigma^{15}N$ values with that of Meister 15.*

Table 3.
The ratios and amounts of N originated from the three origins, soil + compound fertilizer, CUSLNFs, and N_2 fixation, in total N in seeds from plants of the two nodulating genotypes, Enrei and En-bo–1.

line T202 per plant during the seed maturation stages was affected by the type of applied CUSLNF (**Figure 1**) [10]. The same effects of CUSLNF was observed in Enrei (data not shown). Pot experiments under the conditions corresponded to the L-L and H-H plots showed that N_2 fixation activities of Enrei and En-b0–1 worked on L-L but did not work well on the H-H plot, respectively, judging from the changes in ureides' concentrations in xylem saps [6]. Thus, nodulation in Enrei was almost inhibited in the H-H plot supported the estimation that the amount of N from N_2 fixation was very low in this experiment. We also observed that the number

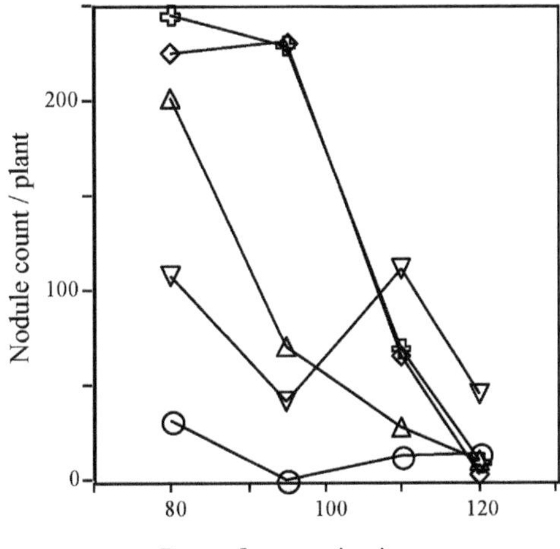

Figure 1.
Effects of N fertilizations on the changes in nodule counts per plant during seed maturation of cv T202. Symbols indicate soil type (fertilizers applied before planting) as follows circle.M-15; triangle, M-5; inverted triangle, MS-9; rhombus, urea; cross, no fertilizer.

of nodules in T202 grown in the H-L plot was rapidly decreased during the seed maturation period (**Figure 1**) [10]. The number of nodules of H-L was less than half of that of L-L at 95 days after germination, and we speculated that nodules in Enrei of the H-L plot did not develop well at the early stage of vegetative growth, resulted in much less N stored in nodules.

2.1.2 Coated urea slow-release N fertilizers differently affected seed chemical compositions and seed yields in the three NILs

Seed yields per plant of the 3 NILs are shown in **Figure 2**. Differences in seed yields of En1282 certified that fertilizers affected seed yield as expected and that the seed yields were proportional to the amount of N excreted from fertilizers. Seed yield of Enrei grown on the L-L plot was almost the same with that grown on the H-H plot, implying that the amount of N_2 fixation-derived N in seeds of the L-L plot was almost same with the amounts of fertilizer-derived N in seeds of the H-H plot. Seed yields of En-b0–1 were proportional to the amount of N applied from fertilizers, and this suggested N_2 fixation activity in En-b0–1 did not work well on this growth condition. This result is inconsistent with the low value of $\delta^{15}N$ value in seeds of the L-L plot and high ureides concentration in xylem saps from plants grown on the L-L condition in pot experiment.

The estimated amounts of N from the two origins (the soil with the compound fertilizer and the SLNF) in En1282, grown with the 4 N treatments, are listed in **Table 2**. Those from the three origins (the soil, the compound fertilizer, and N_2 fixation) in the two nodulating genotypes grown with the 4 of N conditions are listed in **Table 3**. In En1282, the amounts of N/plant from both the soil and compound fertilizer were similar among H-L, L-H, and H-H, and the estimated

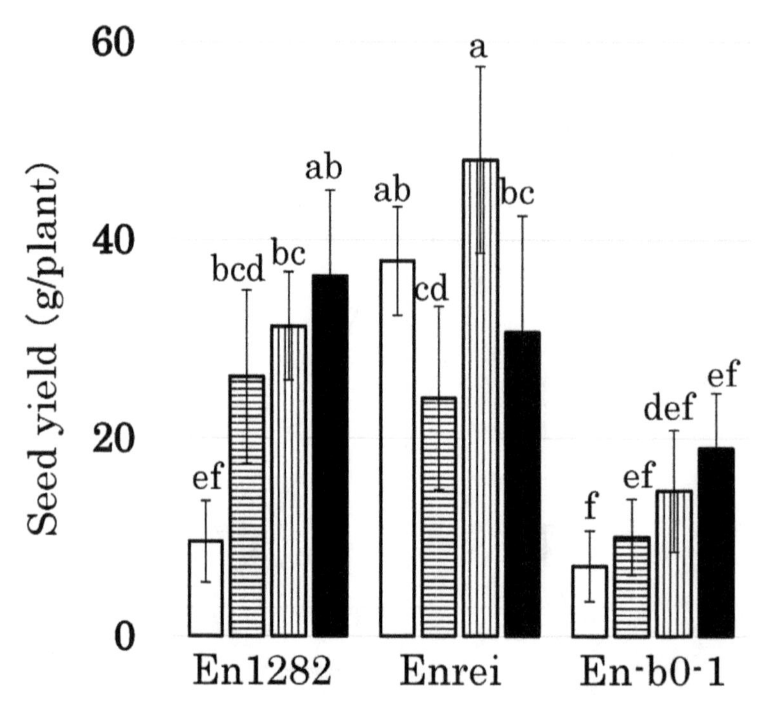

Figure 2.
Effects of the N fertilizations on the seed yields of three genotypes of soybean. Symbols indicate the types of N treatment where plants were grown: white, L-L; horizontal stripe, H-L; vertical stripe, L-H; black, H-H. Different letters indicate significant differences between genotypes and N treatments.

amounts of N/plant from the CUSLNF were 0.30, 0.98, and 1.22 g for the plants grown on H-L, L-H, and H-H, respectively. In Enrei, the amount of N/plant from N_2 fixation of L-H was 1.68 g, which was two-thirds of that on L-L. In contrast, the amount of N/plant from N_2 fixation in En-b0–1of L-H was 0.09 g/plant, which was one-fifth of the amount from the plants grown on L-L.

In the Enrei and En-b0–1 plants grown on H-H, the amount of N from N_2 fixation was negligible, and the amount of N from the CUSLNF was the highest among the N amounts from the 4 of N conditions.

For Enrei, the N supply from the soil had no apparent effect on the seed yield or seed weight (SW). It means that N derived from N_2 fixation compensates for the low supply of soil N for plant growth. Among the Enrei plants grown on the 4 types of N conditions, we observed a lower yield, a smaller SW, and a smaller number of seeds on H-L (**Figure 2**). Our observation that the $\delta^{15}N$ values of mature seeds of H-L was the highest among the Enrei plants grown on the 4 types of N conditions (**Table 1**), which implies that the ratio of N from N_2 fixation to the total N of the plants grown on H-L was less than that of the Enrei plants grown in the other condi-tions. Enrei plants develop nodules during the early stage of plant growth, and soil nitrate inhibited nodule growth of soybean plants [11]. The $\delta^{15}N$ value of seeds from plants grown on L-H (where changes in the amount of N released from the CUSLNF was opposite to the changes in the plants grown on H-L) was slightly higher than that of the plants grown on L-L. Considering this observation and the result that En1282 assimilated N well in the late growth stage (**Table 2**), the importance of N assimilation at the late plant growth was suggested. Moreira *et al* reported that foliar N application at the pod formation period increased seed yield under a certain environmental condition [12]. The work of Takahashi *et al* (1991) showed that the effectiveness of N fertilization at the late period of plant growth of soybean by the deep placement of N fertilizers that improved seed yields [13].

The inter-relationships among the seed protein content, seed oil content, and SW are illustrated for the three genotypes described in 2.1.1. (**Figures 3** and **4**). The protein content was proportional to the SW in En1282 and Enrei (**Figure 3A**). There was no such a relationship in En-b0–1. The protein content inversely correlated with the seed oil content in En-b0–1 (**Figure 3C**). The seed oil content uncorrelated with the SW and the protein content in En1282 and Enrei (**Figure 3B,C**). En-b0–1 seeds on L-L showed the highest protein content and the lowest oil content among the three genotypes (**Figure 3C**). The results of En1282 of L-L was opposite to those (**Figure 3C**).

The effects of the N treatments on the interrelationship between protein and oil contents in seeds differed among the three genotypes (**Figure 4**). The seed protein contents and seed oil contents of En1282 of L-H and H-H were higher and lower than those of L-L and H-L, respectively. The reverse was true for En-b0–1 on the same types of N conditions. The seed oil contents in Enrei of the 4 of N conditions were almost the same. The seed protein content of Enrei of L-L was the highest among the 4 N conditions, and the protein and oil contents in seeds of the three genotypes grown on H-H were almost the same.

We calculated the amounts of seed protein and seed oil by multiplying the contents of protein or oil by the SW (**Figure 5**). In En1282, the amount of oil per seed was proportional to the amount of protein per seed irrespective of the N treatment types (the r of the coefficient line was 0.998). The relationship between the amounts of oil and protein per seed seemed to be dependent on the N treatments in the two nodulating genotypes. The trends of the oil and protein amounts in Enrei and En-b0–1 seeds of L-H and H-H were similar to those of H-L and L-L. In other word, higher N_2 fixation activity of Enrei plants decreased amount of oil accumulation in seeds.

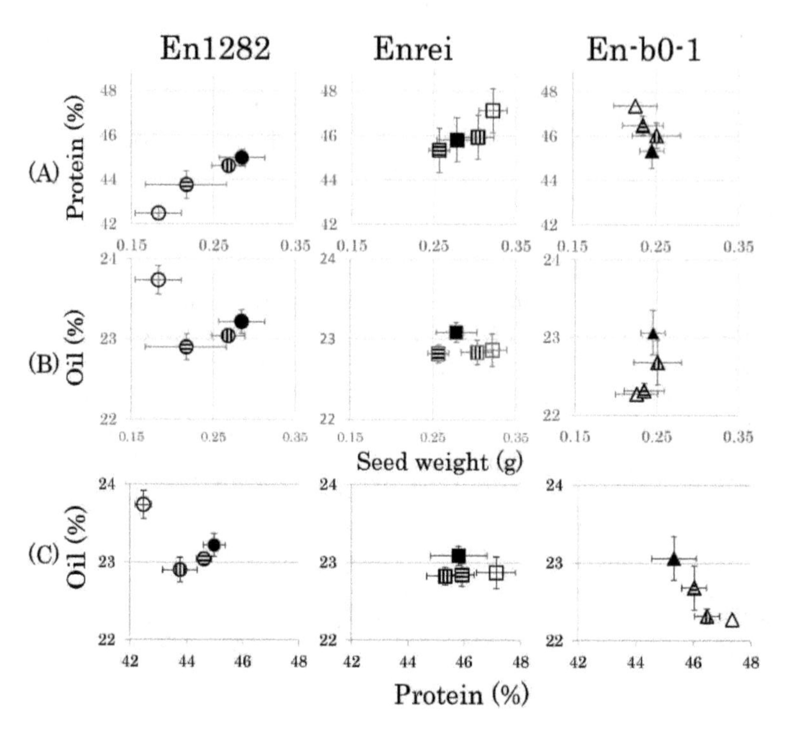

Figure 3.
Effects of N fertilizations on the contents of protein and oil in seeds from plants of three genotypes of soybean. (A) Relationships between protein content and seed weight. (B) Relationships between oil content and seed weight. (C) Relationships between oil and protein content. Shapes of symbols, circles, squares and triangles, indicate the CVs, En1282, Enrei and En-bo–1, respectively. Patterns inside symbols indicate types of soil where plants were grown as follows: white, L-L; vertical stripe, H-L; horizontal stripe, L-H; black, H-H. Correlation coefficients between protein content and seed weight from plants of En1282 and Enrei were 0.982 and 0.907, respectively. Significance levels were 0.02 and 0.10 for En1282 and Enrei, respectively. Correlation coefficient between protein content and oil content from plants of En-bo–1 was 0.924, and it's significance level was 0.10.

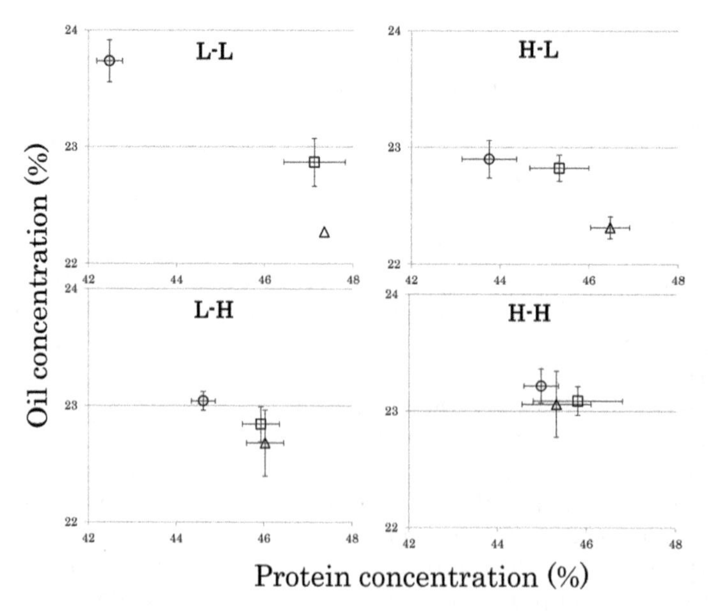

Figure 4.
Interrelationships between the protein content and oil content in seeds from plants of the three soybean genotypes. The four panels indicate the N plots as follows. Left-upper panel: L-L, right-upper panel: H-L, left-lower panel: L-H, right-lower panel: H-H. shapes of symbols indicate the same genotype as those described in the ***Fig.*** *2 legend.*

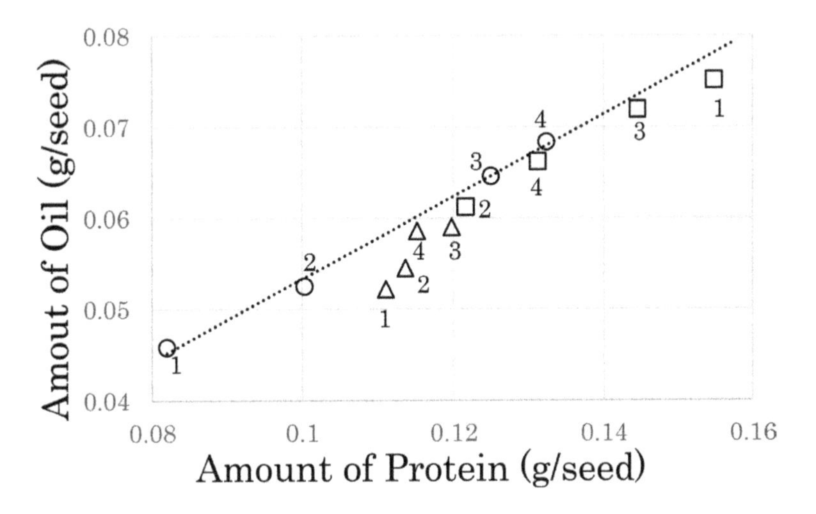

Figure 5.
Effects of the N fertilizations on the amounts of protein and oil per seed from plants of the three soybean genotypes. Shapes of symbols indicate the same genotype as those described in the ***Fig. 1*** *legend. Numbers beside marks indicate types of N plots where plants were grown as follows; 1:L-L, 2:H-L, 3:L-H, and 4: H-H. the* dotted line *indicates the coefficient line between the amount of oil per seed and the amount of protein per seed from En1282 plants (p = 0.01).*

2.2 Effect of different types of N fertilizers on seed chemical composition suggested independently regulated accumulation of protein and oil in soybean seeds

T202 plants were grown on 4 types of N fertilization conditions, where soil N levels were changed in different manners [14]. The 4 N fertilization types were (1) no N fertilizer, (2) urea, (3) M-5, and (4) M-15. The N fertilization condition (1), (3), and (4) were similar to L-L, H-L, and H-H described in 2.1, respectively. Plants of T201, a non-nodulating NIL of T202, were also grown on the 2 types of N fertilization conditions (3) and (4). Firstly, we investigated on the relationship between seed protein con-tent and seed oil content per individual plant basis (**Figure 6**). T202 and T201 exhibited very similar protein and oil contents under the N condition (4), being due to less N_2 fixation in T202. T202 under the N condition (2) exhibited similar protein contents and slightly less oil contents with those of T202 and T201 of (4). This would be due to the different characteristics of M-15 and urea as N fertilizers. T202 and T201 of (1) and (3), both of which offer low nitrogenous conditions at the seed maturation stages, exhibited negative correlations between protein contents and oil contents in each condition, implying these correlations are dependent on the availability of N from soil. Next, we investigated on the relationship between amounts of protein and oil per seed under the 4 nitrogenous conditions (**Figure 7**). Notably, observed relationships between amounts of protein and oil were exactly opposite to those of seed protein and seed oil contents of individual plants. Namely, each plant samples of (2) and (4) showed positive correlation relationships between amounts of protein and oil (**Figure 7**). In addition, a weaker positive correlation between amounts of protein and oil per seed was observed in T202 grown on the N condition (3). T201 of the N condition (4) exhibited a positive correlation between amounts of protein and oil per seed; it was similar to that of T202 grown on the same N condition. T202 of the N condition (1) did not show any correlation between amounts of protein and oil per seed, being contrast to that of protein and oil contents among individual plants. This could be due to the low variation of the amounts of protein and oil per seed in these samples. Hence, these results suggested that the observed negative

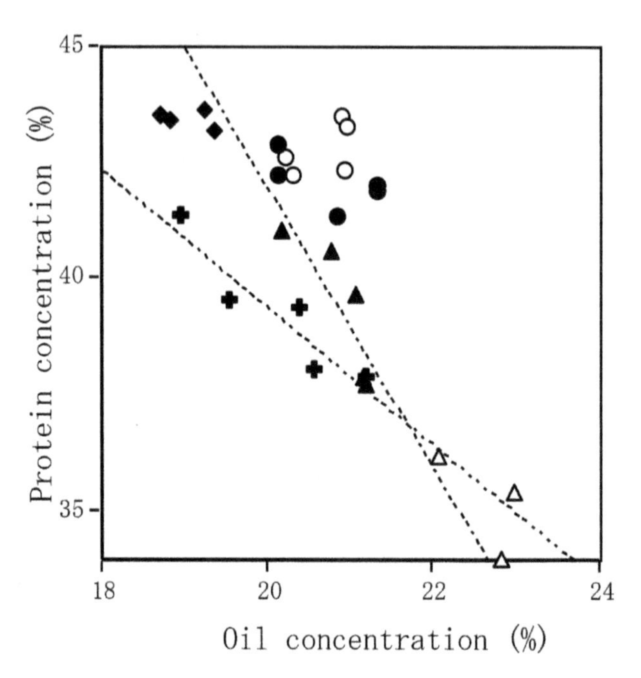

Figure 6.
Effects of soil N level and nodulation on concentrations of protein and oil in soybean seeds. Symbols indicate soil type as follows: circle, M-15; triangle, M-5; rhombus, urea; cross, no fertilizer. Solid and open symbols indicate nodulated cv, T202 and non-nodulated cv T201, respectively. Equations of correlation lines were as follows: line a, Y = -3.01X + 102.19 (r^2 = 0.709) for seeds from T202 on M-5 soil and line b, Y = -1.42X + 68.93 (r^2 = 0.864) for those from L soil, respectively. Y and X denote protein and oil concentrations, respectively.

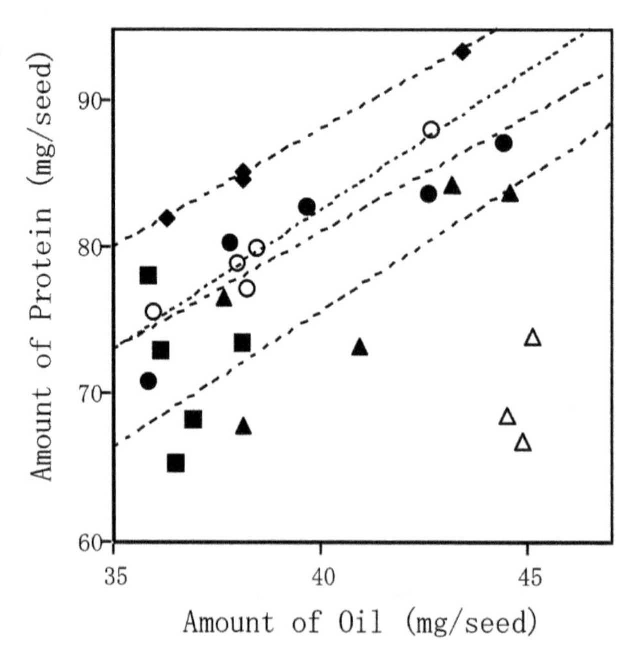

Figure 7.
Effects of soil N level and nodulation on amounts of protein and oil per seed in soybean seeds. Symbols indicate soil type as follows: Circle, M-15; triangle, M-5; rhombus, urea; cross, no fertilizer. Equations and coefficients of correlation lines were as follows: Line a, Y = 1.59X + 17.34 (r^2 = 0.817) for seeds from cv T202 on M-15 soil; line b, Y = 1.84X + 1.1.78 (r^2 = 0.632) for those on M-5 soil; line c, Y = 1.62X + 23.39 (r^2 = 0.997) for those on urea-soil, and line d, Y = 1.91X +6.21 (r^2 = 0.946) for those of cv T201 on M-15 soil, respectively. Y and X denote seed contents of protein and oil, respectively.

correlations of protein and oil contents in the N condition (1) was caused by different seed protein amounts per seed, which were made by receiving insufficient N to immature seeds. The observation that T201 of the N condition (3) (where offers low N supply during seed maturation) exhibited different protein contents and similar oil contents among individual plants implied that insufficient N supply did not affect oil accumulation in seeds.

The regression lines between amounts of protein and oil per seed in T202 of (2), (3), and (4) had similar slopes. This result implied that carbon transported into immature seeds was distributed to the biosynthesis of protein and oil in the same ratio among these plants. In addition, the Y-intercepts of the regression line of T202 under the N conditions (2) and (3) were larger and smaller than those of (4), respectively. Since the amount of N supplied to immature seeds from vegetative tissues would be increased in the order of (1), (3), (4), and (2), the differed Y-intercepts implicate another mechanism which is independent from carbon partitioning for the biosynthesis of amino acids and fatty acids. As mentioned in 2.3, seed oil accumulation ceased before seed maturation, but seed protein accumulation continued till seed maturation. The difference in the Y-intercept may be due to the different amount of accumulated proteins after the oil accumulation has ceased.

2.3 Inhibited N_2 fixation by N application at the flowering stage did not promote the protein accumulation but did the oil accumulation and dehydration in soybean seeds

2.3.1 Ammonium sulfate and NaCl applications at the flowering stage differently affects the protein and oil contents in soybean seeds

Soybean plants were grown on soil, which allows developing nodules well, and high amount of ammonium sulfate (AS) was applied (10 g of solid AS was spread around a plant) to T202 at the flowering stage [14]. We observed that the AS application did not change the protein content per seed but increased the oil content although seed protein content in AS-applied T202 individuals was lower than that in control (**Table 4**) [11]. The ureide concentrations in xylem saps from the NaCl and AS applied plants were lower than untreated ones in the same extent with each other till 25 days after the applications, which implied that plants of both treat-ments were suppressed their N_2 fixation activities to the same level as each other (**Figure 8**). Oil contents per seed of both treatments were more than those of the no treatment (**Table 4**). On the other hand, protein contents per seed from the AS treated plants were comparable to those from the no treated plants, which was quite different from those from the NaCl treated plants, of which protein content was

	Concentration, mg/g DM		Content, mg/seed		Seed weight
	Protein	Oil	Protein	Oil	g/seed
No	411	230	69.0	38.6	0.168 (0.022)
AS	391	239	69.6	42.5	0.178 (0.020)
NaCl	388	239	65.6	40.4	0.169 (0.020)

In the seed weight column, the standard deviations of seed weight from 70 grains are indicated in parentheses. Mature seeds were harvested 72 days after application. Protein and oil contents per seed were calculated by multiplying their concentrations by seed weight, respectively.

Table 4.
Effects of application of ammonium sulfate (AS) and NaCl at flowering stage on seed storage composition of soybean plants (cv T202).

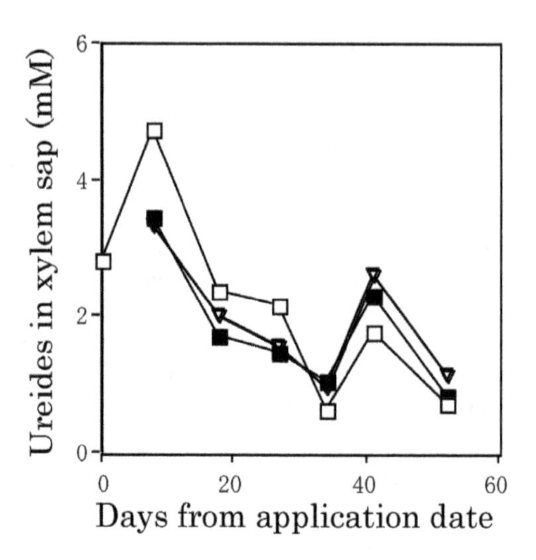

Figure 8
Effects of application of ammonium sulfate and NaCl at the flowering stage on concentration of ureides in xylem saps from nodulated soybean plants (cv T202).Symbols indicate chemicals applied as follows: Open square, no treatment; solid square, ammonium sulfate; and rectangular triangle, NaCl, respectively. Values of ureides in the xylem saps were given as means of those from two plants from each treatment groups.

lower than those from the no treated plants (**Table 4**). These results suggested that applied AS compensated the decrease in the amount of N from N_2 fixation. That suppression of N_2 fixation activity caused an increase in the amount of oil in seeds.

2.3.2 Different effects of ammonium sulfate application at the flowering stage on the accumulation of protein and oil in soybean seeds

Application of AS at flowering stage to Enrei and Tamahomare decreased protein and increased oil contents in matured seeds of both CVs [15]. Averaged protein and oil contents in seeds from the AS dressed plants were 2% lower and higher than those from the undressed plants, respectively, in Enrei. Accumulation profiles of protein and oil per seed during their ripening period were quite different from each other (**Figures 9** and **10**). The regression curve of the increase in the amount of protein per seed during seed maturation was almost identical between plants that received N at the flowering stage and those that did not. Contrary to this, the regression curves of oil content per seed showed seeds from N applied plants accumulate oil faster than those of control plants did. Seeds of both N treatments stopped in increasing oil content at the late period of seed maturation. N applied plants showed a faster seed dehydration rate than the control plants on the results of changes in the water contents of seeds during seed maturation (**Figure 11**). Matured seeds from N applied plants had less protein content than those from N unapplied ones. An increase in seed weight was higher in N applied plants than those in the control plants. These results implied that a high amount of N application at the flowering stage suppressed N_2 fixation activities of nodules, causing a decrease in sucrose consumption by nodules. The dehydration rate of seeds from N applied plants was faster than those from N unapplied plants. The fact implied that the accumulation rate of storage compounds, protein and oil, in seeds from N applied plants was faster than those from N unapplied plants. In other words, a higher amount of C (sucrose) was supposed to supply to maturing seeds in the case of N applied plants than in the case of N unapplied ones. Sudden suppression of N_2 fixation of nodules by ammonium sulfate application to plants was supposed to cause the decreased consumption of sucrose in nodules and the increase in the amount of sucrose imported into maturing seeds. Thus seeds increased the oil content resultantly.

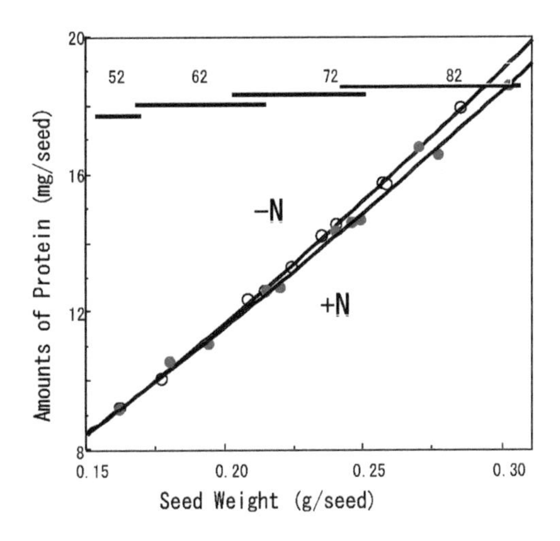

Figure 9.
Effects of N application at the flowering stage on the increase of protein content in soybean seeds during maturation. Open and solid symbols show the protein content per seed from the N undressed and dressed plants, respectively. Symbols indicated the days from the dressed day as follows: Circle, 52; square, 62; triangle, 72; inverted triangle, 82, respectively. Regression curves for values of the each treatment are shown. Equations for the curves are $Y = 56.90X^2 + 45.40X + 0.35$ $(r^2 = 0.999)$ *and* $Y = 47.96X^2 + 44.90X + 0.69$ $(r^2 = 0.995)$ *for the protein contents of undressed and dressed plants, respectively. Y and X means amount of proteinous N (mg) in a seedand seed dry weight (g), respectively.*

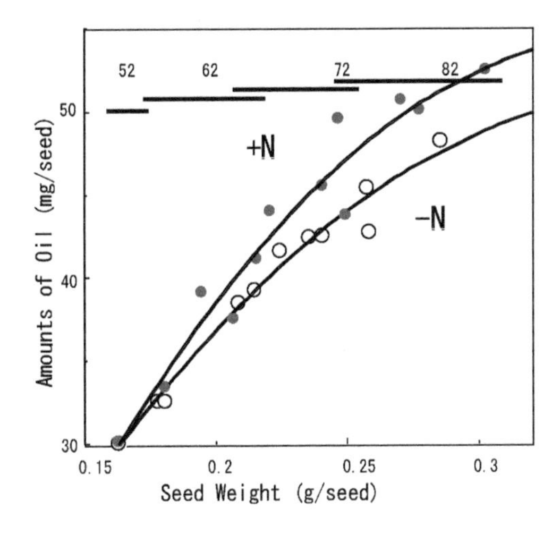

Figure 10.
Effects of N application at the flowering stage on the increase of oil content in soybean seeds during maturation. Symbols were the same as those in ***Figure** 8*. *Regression curves for values of the each treatment are shown. Equations for the curves are* $Y = -0.495X^2 + 0.366X - 0.0163$ $(r^2 = 0.977)$ *and* $Y = -0.657X^2 + .467X - 0.0284$ $(r^2 = 0.941)$ *for the oil contents of undressed and dressed plants, respectively. Y and X means amount of oil (mg) in a seed and seed dry weight (g), respectively.*

2.4 Different effects of CUSLNFs on seed protein concentrations of plants producing high and low protein content seeds – Accumulated proteins in nodules may be a factor to affect seed protein content

Plants of 13 CVs producing low-, medium- and high- protein content seeds were grown on similar conditions to the L-L, H-L, and H-H plots described in 2.1.1, and the protein and oil contents of harvested seeds were compared (**Figure 12**) [10].

Values subtracted seed protein contents from plants grown on H-H soil from those grown on L-L soil were compared among 13 CVs based on the protein concentrations

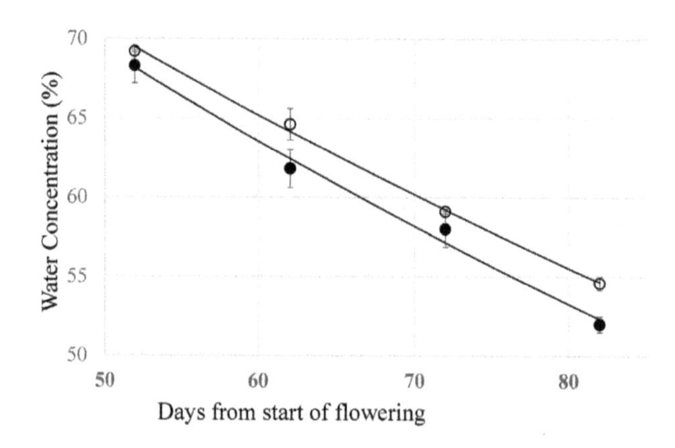

Figure 11.
Effects of N application at the flowering stage on the changes in water content of soybean seeds during maturation. Open and solid symbols show the water contents from the N undressed and dressed plants, respectively.

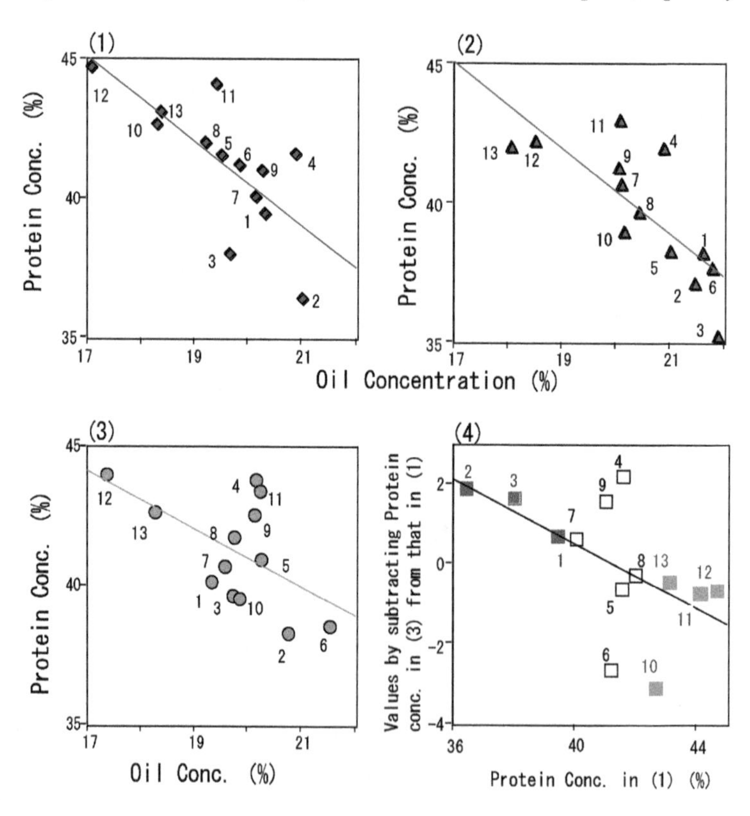

Figure12.
Different effects of CUSLNFs on the contents of protein and oil in mature seeds of 13 cultivars of soybean. Left-upper panel, (1) interrelationships between the protein content and oil content in seeds from plants grown on L-L soil. Right-upper panel, (2) interrelationships between the protein content and oil content in seeds from plants grown on H-L soil. Left-lower panel, (3) interrelationships between the protein content and oil content in seeds from plants. Right-lower panel, (4) interrelationship between the value of protein content of seeds from plants grown on H-H soil (3) subtracted from that on L-L (1) soil and protein content of seeds from plants grown on L-L soil (1). Numbers in the figure indicate the cultiver names as follows: 1, Akishirome; 2 Ginrei; 3, Tamahomare; 4, Ohtsuru; 5, Kyu-kei 273; 6, Tamamasari; 7, Nishimusume; 8, Asagoao; 9, Mizukuguriokute; 10. Toyoshirome; 11, Fukuyutaka; 12, Miyagiaosho; and 13, Bunjyocha. Cvs of which number 1 to 3, 4 to 9 and 10 to 13 were classified to those producing low (less than 40%), medium (40 to 42%) and high (more than 42%) protein content seeds, respectively, based on the results shown in panel (1). Colors of symbols in panel (4) indicate the class of seed protein content of CV as follows: brown, low; white, medium; green, high. The X-axis and Y-axis coefficient lines for each panel and their coefficients of determination are as follows: (1) $Y = -1.519X + 70.886$ ($r^2 = 0.524$), (2) Y—1.5144X + 71.301 ($r^2 = 0.583$), (3) $Y = -1.038X + 61.712$ ($r^2 = 0.312$), and (4) $Y = -0.403X + 16.604$ ($r^2 = 0.320$).

of seeds from plants grown on L-L soil (**Figure 12D**). A negative correlation between the subtract values and seed protein contents of L-L soil was observed. Seed protein contents in 3 low-protein CVs grown on L-L soil were higher than those of H-H soil. The reverse was true for four high-protein CVs. Differences in seed protein contents between plants producing low and high protein seeds were possibly ascribed to the amount of N that immature seeds received from nodules and soil.

Protein contents correlated with oil ones in all cases, and coefficients of determination were 0.724, 0.770, and 0.559 for seeds grown on the L-L, H-L, and H-H conditions, respectively. The coefficient between protein and oil concentrations from plants grown on H-H soil was lowest, and that of those grown on H-L soil was highest among plants grown on the three types of soils. This result suggests a higher amount of N supply from nodules or soil during seed maturation promotes protein accumulation but does not oil accumulation because plants grown on the H-L soil might have fewer nodules than those on L-L, and they absorb less soil N than those on the H-H soil during reproduction stage.

Different seed weight distributions between a low protein cultivar Tamahomare (TH) and those of a high one Fukuyutaka (FU) supported this idea. Plants of TH grown on L-L had smaller seeds than those on H-H. The reverse was true for FU (**Figure 13**). Seeds of CV FU grown on H-L had similar seed weight distribution to that on L-L. On the other hand, seeds of CV TH grown on H-L were smaller than those on L-L, judging from their seed weight distribution pro-files. These results suggested that immature seeds did not receive enough amount of N from vegetative tissues including nodules for the demand of seeds in the case of CVs producing low protein content seeds grown on L-L. Such plants grown on H-H absorbed and metabolized fertilizer-derived N from soil enough for imma-ture seeds' demand. These results suggested that plants of CVs producing high protein content seeds grown on L-L had enough amount of N in the vegetative tis-sues for the need of immature seeds. These results also implied that those grown on H-H where nodulation and N_2 fixation activity of plants were suppressed by soil N did not have enough amount of N in the vegetative tissues for the demand of immature seeds.

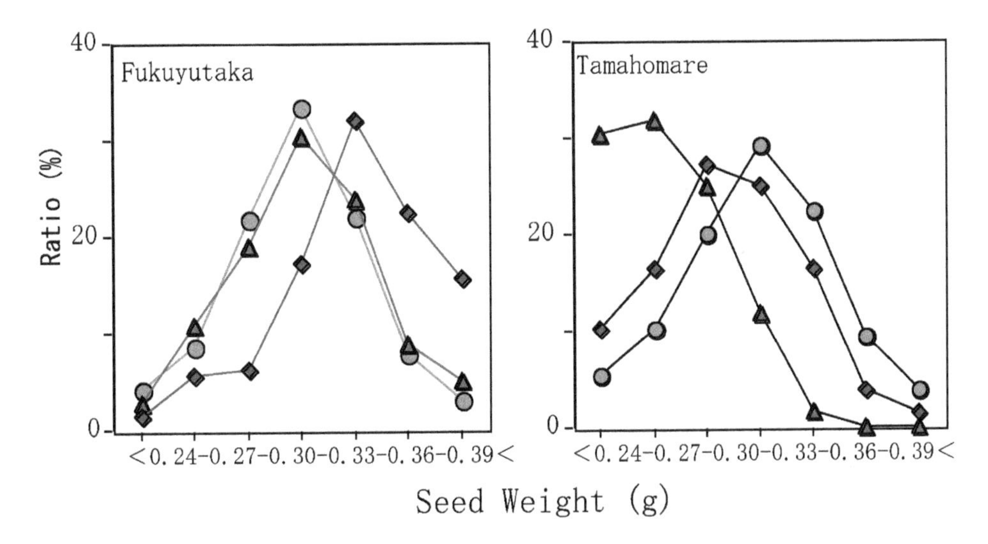

Figure13.
Differences in the seed weight distributions from plants grown on three types of UCSLFs between two cvs producing seeds of high and low protein contents. Left and right panels show ratios of seed counts per seed weight ranges as a cv Fukuyutaka producing seeds of high protein content and a cv Tamahomare producing those of low protein content, respectively. Symbols indicate the soil types as follows; rhombus, L-L; triangle, H-L; circle, H-H.

3. Carbon metabolism in maturing seeds is affected by the supply of N from source organs-role of phospho*enol*pyruvate carboxylase (PEPC) in the accumulation of protein

3.1 Background for the research on the role of PEPC in the synthesis of storage protein in soybean seeds

Seed storage compounds in soybean, proteins and oils, are synthesized from substances transported from vegetative tissues, leaves, roots, and nodules. The primary forms of the substances are sucrose, ureides, and Asn [3]. Amino acids constituted storage protein are synthesized by introducing amino acid residues into organic acids in cotyledon during seed maturation. This fact implied two factors: amounts of organic acids and amino residues formed from the imported substances affect the protein content of soybean seeds. Since organic acids are utilized as the substrate for the synthesis of both fatty and amino acids, the inverse correlation between contents of protein and oil in soybean seeds [1] could be caused by the competition of fatty and amino acid synthesis. Non-photosynthetic type Phospho*enol*pyruvate carboxylase (EC 4.1.1.31, PEPC) was thought to play the anaplerotic role in the supply of organic acids [16]. Immature crop seeds have a high activity of the enzyme, and its role in maturing seeds was assumed to refix CO_2 formed from respiration in seeds [17].

Multiple types of PEPC isogenes were encoded in plant genomes. These enzyme genes were categorized into plant-type and bacterial-type isogenes [18]. Plant-type isogenes were further subdivided into C_4 and other (C_3) type. Subsequent sections present our research on soybean seed PEPC over time and discuss the role of PEPC isozymes in the synthesis of storage proteins.

3.2 Relationships of PEPC activity and contents of storage compounds, protein and oil, in mature soybean seeds

High CO_2 fixing activity was observed in immature cotyledons of soybean seeds under an unilluminated condition in a study [K. Tanaka, personal communication], which evaluates the photosynthetic activity of immature soybean seeds [19]. In cotyledon, ribulose 1,5-bisphosphate carboxylase (RuBPCase) gradually decreased its activity during seed maturation, whereas PEPC kept its activity high during seed maturation. PEPC enzyme rapidly decreased its activity between 3 and 9 days after germination [20]. These results implied the engagement of PEPC in the accumula-tion of storage compounds. As the enzyme activity was kept in matured soybean seeds, the enzyme activity and contents of storage compounds, i.e., protein and oil, were compared among seeds from plants of 13 types of seeds in 11 CVs grown in 12 prefectures of Japan (**Figure 14**) [21]. The enzyme activity was positively and negatively proportional to the protein and oil contents of seeds, respectively. This observation suggested PEPC in immature soybean seeds plays a role in the supply of carbon skeleton to synthesze of amino acids.

PEPC activity in maturing rice seeds increased its activity by the addition of N fertilizer [22]. In soybean, N fertilizer application was thought to be insufficient to the enzyme activity in seeds because soybean plants had nodules and supply N compounds in nodules to maturing seeds [3]. When N contents in leaves and PEPC activity in maturing seeds simultaneously were compared, they were proportional with each other (**Figure 15**) [10]. Leaves gradually lose their greenness by exporting amino acids to developing seeds during seed maturation, which means leaves lose photosynthetic activity. This observation suggested that PEPC in soybean seeds changed its activity responding to N supply from vegetative tissues and that PEPC plays an essential role in the amino acid synthesis for storage protein. One of PEPC

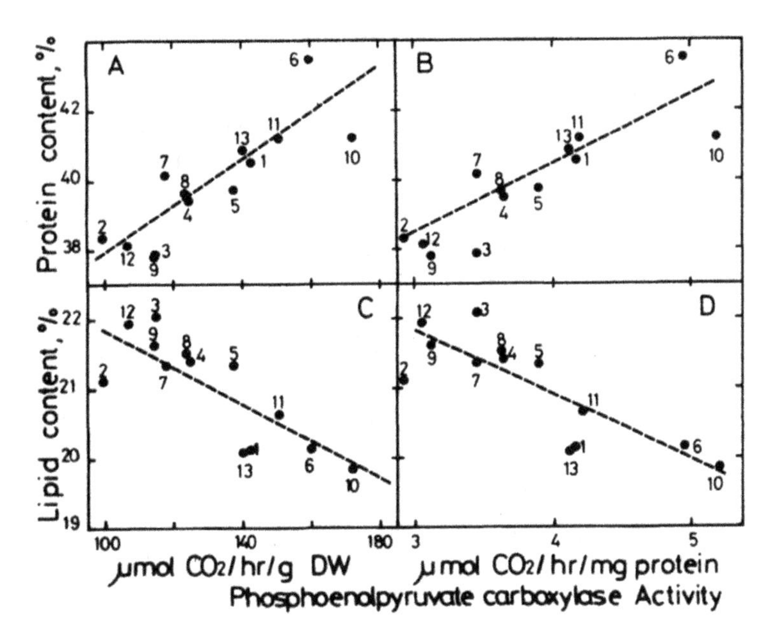

Figure 14.
Relationships between PEPC activity and contents of protein and lipid in soybean seed. Left-upper panel A: Relationship between PEPC activity per 1 g of dry seed (X) and protein content (Y) in soybean seed. Right-upper panel B: Relationship between PEPC activity per 1 mg of soluble protein (X) and protein content (Y) in soybean seed. Left-lower panel C: Relationship between PEPC activity per 1 g of dry seed (X) and lipid content (Y) in soybean seed. Right-lower panel D: Relationship between PEPC activity per 1 mg of soluble protein (X) and lipid content (Y) in soybean seed. Cultivar names and produced prefectures of soybean seeds harvested in 1986 are as follows: 1, Fukuyutaka (Fukuoka); 2, Tamahomare (Yamaguchi); 3, Akishirome (Yamaguchi); 4, Akishirome (Hiroshima); 5, Akiyoshi (Kagawa); 6, Enrei (Nagano); 7, Enrei (Niigata), 8, Tachisuzunari (Tochigi); 9, Suzuyutaka (Yamagata); 10, Miyagishirome (Miyagi); 11, Shirosennari (Akita); 12, Nanbushirome (Iwate); 13, Okushirome (Aomori). Equations and their correlation coefficients of the X-axis and Y-axis coefficient lines for A, B, C and D panels, respectively are as follows: $Y = 0.062X + 31.727$ $(r = 0.8395)$; $Y = 1.99X + 32.29$ $(r = 0.8660)$; $Y = -0.029X + 24.884$ $(r = -0.8411)$; *and* $Y = -0.91X + 24.35$ $(r = -0.8494)$.

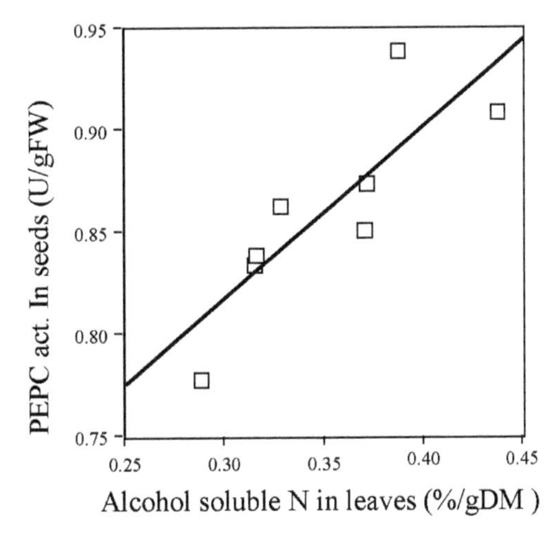

Figure 15.
Relationship between alcohol soluble nitrogen (amino acids) content in leaves and PEPC activity in immature seeds. The coefficient line and its coefficient of determination are as follows: $Y = 0.846X + 0.564$ $(r^2 = 0.703)$.

isogenes was expressed in immature seeds and other vegetative tissues of soybean plants [23]. Another PEPC isogene expressed in soybean maturing seeds was identified together with the isogene mentioned above [24]. In recent, it appeared that ten

PEPC isogenes were encoded in the soybean genome [25]. It might be likely that multiple PEPC supports soybean seed metabolism during seed development.

3.3 Roles of several PEPC isoforms in maturing soybean seeds starch is a significant carbon source of carbon in protein biosynthesis during the late maturation period in soybean seeds

3.3.1 *Comparison of PEPC activity, contents of protein and oil in seeds in high- and low-protein CVs using principal component analysis*

We applied principal component analysis (PCA) to evaluate the interrelationships among the four factors of maturing seeds, contents of protein and oil, PEPC activity, and seed weight (**Figure 16**) [26]. The enzyme activity was significantly

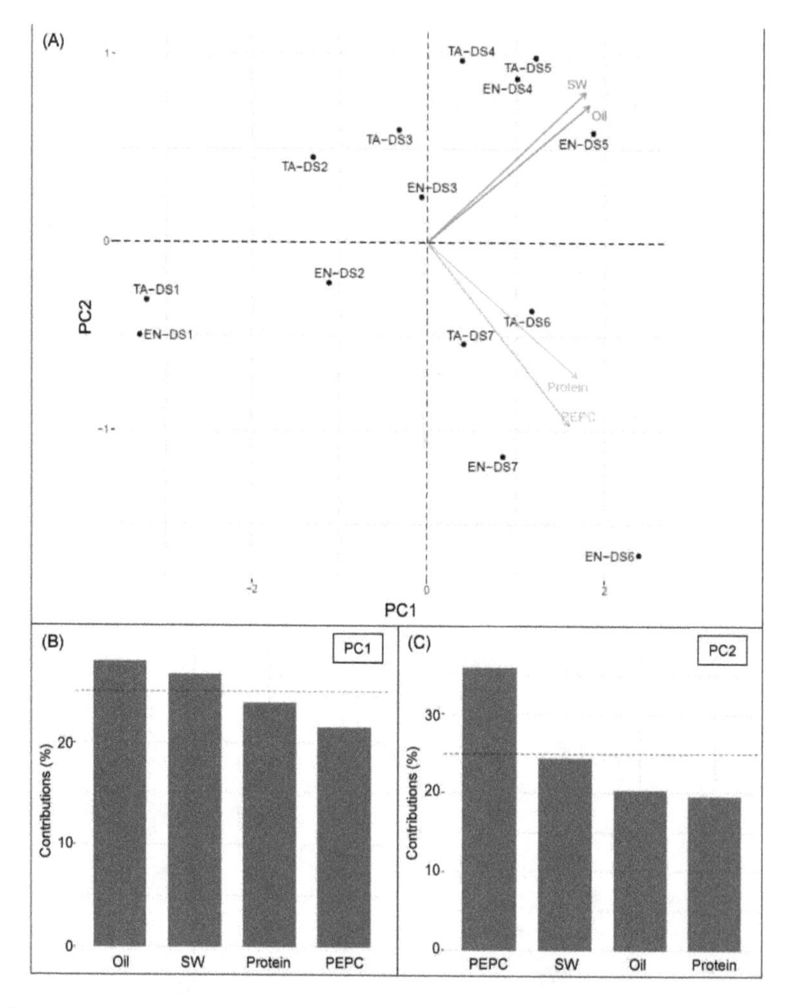

Figure 16.
The results of PCA in seed compositions, protein, oil, seed weight and PEPC activities during seed maturation. Characters and numerals show the name of cultivars and stage of samples, respectively. The characters EN and TA indicate Enrei and Tamahomare, respectively. The numerals with DS indicate the stage of samples. A: A biplot of the two major principal components, representing a distribution of the samples analyzed. Marks in black indicate samples, and arrows indicate the directions of protein content (protein), oil content (oil), seed fresh weight (SW), and PEPC activity per seed (PEPC). Horizontal- and vertical- axes represent principal component 1 (PC1) and principal component 2 (PC2), respectively. PC1 and PC2 explain 73.6 and 16.4% of the data variances, respectively. B and C: Contributions of the variables on PC1 and PC2, respectively. Dot lines indicate the averaged values.

associated with protein content but not with oil content. The oil content was associated with seed weight. Immunological assay on PEPC protein contents using antibodies for some PEPC isozymes showed plant-type ones expressed during all stages of seed maturation, and Gmppc2 expressed at the late maturation stage (**Figure 17**). The most critical period in the seed maturation period on the relation between protein content and PEPC activity was the late stage of seed maturation, DS6, in our discriminating seed growth stages.

Characteristic physiological changes of soybean plants are the loss of leaves and decrease in starch contents of seeds [27]. We observed that oil accumulation ceased at the late seed maturation period, and protein accumulation continued till seed maturation. Together, we proposed that PEPC plays a role in the supply of carbon skeleton of amino acids (organic acids) formed by the degradation of starch in seeds.

3.3.2 PEPC isogenes exhibits divergent expression patterns during seed maturation-Gmppc2 isogene is possibly a useful marker for improving seed protein content

The ten soybean PEPC isogenes showed different gene expression characteristics in developing seeds from each other [28]. Notably, one PEPC isogene *Gmppc2*

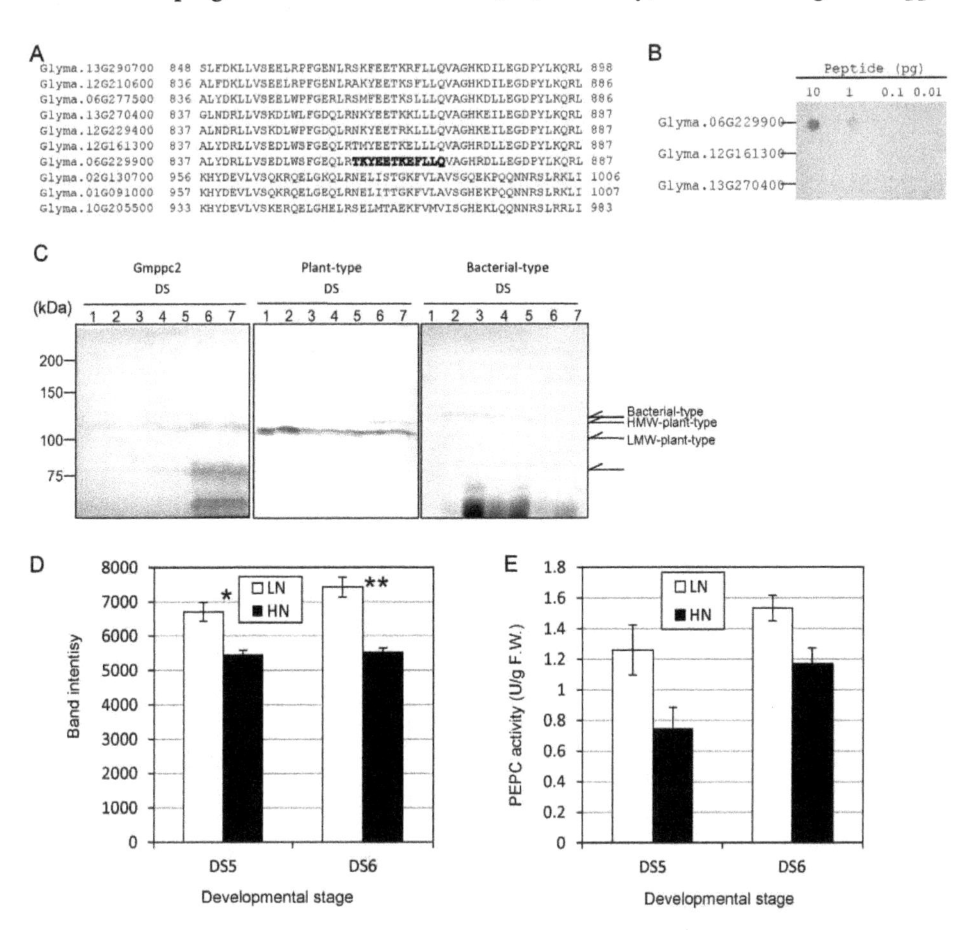

Figure 17.
*The expression of PEPC in immature soybean seeds. A: Partial sequence alignment of the ten PEPC isoforms of the C-termini. Bold letters: The region that was used to raise a Gmppc2-specific polyclonal antibody. B: Dot blot assay to determine the specificity of the Gmppc2 antibody. C: The protein expression patterns of Gmppc2, plant-type PEPC, and bacterial-type PEPC in developing whole seeds during seed maturation from DS1 to DS7. D: The effect of nitrogen application on the expression of Gmppc2 protein. The measurement was duplicated, and the average of the values is shown. Error bar: Standard error (SE). Significance at *10% and **5% by Student's t-test. E: The effect of nitrogen application on PEPC activity. Error bar: SE.*

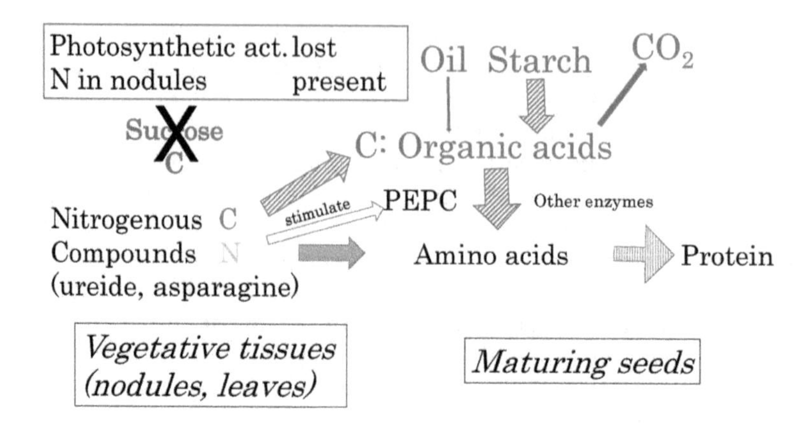

Figure18.
Schematic presentation on the role of PEPC in the synthesis of storage protein on the late stage of seed growth maturation. Events in vegetative tissues and seeds at the late maturing stage of a CV, Enrei were drawn. Photosynthetic activity of leaves dramatically decrease by the decreases of N and chlorophyll contents, and resultantly supply of photosynthate (sucrose) to maturing seeds cease. On the other hand, nitrogenous compounds are supplied from degrading nodules to maturing seeds. Nitrogenous compounds are once separated into organic acids and amino residues in maturing seeds. Amino residues are introduced to newly synthesized organic acids of which carbon source is starch in maturing seeds to form amino acids which are substrates for the synthesis of storage protein. Carbon metabolizing enzymes including PEPC work to synthesize those organic acids. Activity of PEPC respond to the N supply to maturing seeds, thus affecting the protein content of seeds. Plant-type isozymes of PEPC work in the synthesis of amino acids through maturation period of seeds, an isozyme of Gmppc 2 was expressedat the late maturing stage of seeds.

exhibited a distinct expression pattern during soybean seed maturation. Namely, soybean seeds kept the expression of *Gmppc2* at a low level until the soybean seed is going to be matured (until DS4). In contrast, soybean seeds let the expression levels of *Gmppc2* at DS4 up-regulated drastically until seed maturation. Especially the expression level at DS6 in cotyledon was high. As mentioned in the previous section [3.3.1], DS6 is the critical stage at which seed protein accumulation varied between the two representative CVs [26]. To verify whether *Gmppc2* attributes to the dif-ferent PEPC activity and protein content in the two CVs, expression of Gmppc2 protein was analyzed in soybean maturing seeds with the presence and absence of a nitrogen fertilizer, which suppressed nodulation and nitrogen fixation activity (**Figure 17**). The results indicated that soybean seeds highly expressed Gmppc2 protein at DS5 to DS7, and the expression was concordant with PEPC activity at DS5 and DS6 in response to the nitrogen fertilization. We illustrate the role of PEPC isozymes in the accumulation of protein (**Figure 18**). *Gmppc2* is the potential PEPC isogene, explaining the variation of seed protein content and PEPC activity among soybean CVs observed previously (**Figure 14**).

4. Conclusion: Our views of how nodules and their N_2 fixation activity affect protein and oil contents in soybean seeds

We carried out physiological experiments on the response of soybean plants to different types of N fertilizers. These experiments were designed from the view-point that there might be differences on the accumulation of storage protein and oil between soybean plants having different activities of N_2 fixation and N assimila-tion. We summarize our views on the characteristics of the accumulation of protein and oil in soybean seeds as follows.

1. Storage protein accumulation is independently regulated from storage oil ac-cumulation in seeds.

We showed the accumulation profiles of protein and oil during seed maturation were quite different from each other (**Figures 9** and **10**). In immature seeds, PEPC plays a role in the protein accumulation (**Figure 16**), and its activity responds to the supply of N (**Figure 17**).

2. N_2 fixation results in decreasing the amount of oil per seed and leads to the decrease in oil content of seeds.

Seeds of nodulated CVs with high N_2 fixation activities have less oil content than those with low N_2 fixation activities (**Figure 4; Table 4**).

3. A high pseudo negative correlation between protein and oil was observed in seeds from plants which were grown on soil applied no N fertilizer.

High negative correlations were observed between contents of protein and oil in seeds of both nodulated and non-nodulated CVs grown under low N supply at the late maturation period of plants (**Figure 6**). Amounts of oil per seed were almost constant, and those of protein per seed were variable among plants in the cases of nodulated plants grown on L-L and non-nodulated ones grown on H-L (**Figure 7**). These results suggested that the observed high negative correlations between protein and oil contents in seeds were caused by the differences in the amount of accumulated protein in seeds among respective plants, and not related to oil accumulation in seeds.

4. Plants utilize soil N during the late growth stage as the source for seed protein, which implies the importance of N fertilization during the reproductive phase of soybean plants.

Based on the estimated amounts of fertilizer-derived N in seed protein, N excreted at the late period of plant growth was highly incorporated to seed protein in plants of both the nodulated and non-nodulated CVs (**Tables 2** and **3**). The higher seed yield of plants of regular nodulated CV grown on L-H plot than that grown on L-L and H-H plots suggested N supplied from both nodules and fertilizer at the late plant growth period increased seed yields (**Figure 2**).

5. Plant-type PEPC gene family plays a role in the synthesis of amino acids for storage protein. Gmppc2 isogene may have a crucial role in accumulating protein on the late maturation period of seeds.

Pattern analysis suggested that PEPC promotes protein accumulation but not oil accumulations (**Figure 16**). As the increase of acetyl Co-A carboxylase activity in immature rapeseed by gene engineering increased seed oil content [29], this enzyme might play a key role in oil accumulation in soybean seeds. These enzymes might work to synthesize respective storage compounds, protein and oil, independently with each other. Plant-type PEPC isogenes were expressed in immature seeds through the seed maturation period (**Figure 17C**). *Gmppc2*, an isogene of plant-type PEPC, was expressed in seeds at the late maturation period, and its PEPC protein expression level was higher in seeds of low N condition than in those grown on high N one (**Figure 17D**). These observations coincides with the observations that seeds of low N soil (L-L plot) had higher protein content than those grown on high N soil (H-H plot) (**Figures 3** and **6**).

We observed that the ratio between changing amounts of protein and oil in seeds was almost the same among plants grown on soils fertilized with different types of CUSLNFs (**Figure 7**). This observation implied that the allocation ratio of C to protein and oil was controlled by some mechanism.

Genetic studies on the storage compounds of soybean seeds have made progress. A quantitative trait locus analysis mapped regions controlling the contents of storage compounds in soybean seeds on the contents of protein and oil on the soybean genome [30]. Concerts between the physiological and genetic approaches are useful to elucidate the mechanism of how contents of storage compounds are controlled in soybean seeds.

Acknowledgements

We thank Professor Takuji Ohyama of Tokyo University of Agriculture for giving us an opportunity to describe our work in this book. We also thank Dr. Yoshikiyo Oji (Emeritus Professor of Kobe Univ.), Kyoko Saio (Former Head of Protein Lab., National Institute for Food Sci. Japan), Kiyoshi Tanaka (Former Head of Plant Physiology Lab., National Institute for Environmental Studies, Japan), and Yukio Kawamura (Former Head of Protein Lab., National Institute for Food Sci. Japan) for their valuable advices and helpful encouragements on this study. One of authors (TS) thanks to members of Plant Nutrition Lab, Faculty of Agriculture, Kobe Univ. for their helpful discussions on this work.

Author details

Toshio Sugimoto[1,2,3]*, Naoki Yamamoto[3,4] and Takehiro Masumura[3,5]

1 Faculty of Agriculture, Graduate School of Agricultural Science, Kobe University, Kobe, Japan

2 Research Center for Food and Agriculture, Wakayama University, Wakayama, Japan

3 Graduate School of Life and Environmental Sciences, Kyoto Prefectural University, Kyoto, Japan

4 Key Laboratory of Southwest China Wildlife Resources Conservation (Ministry of Education), College of Life Science, China West Normal University, Nanchong, Sichuan, China

5 Biotechnology Research Department, Kyoto Prefectural Agriculture, Forestry, and Fisheries Technology Center, Kyoto, Japan

*Address all correspondence to: sugimoto@kobe-u.ac.jp

References

[1] Liu K 1997. 2 Chemistry and Nutritional Value of Soybean Components. In Soybeans, Chemistry, Technology, and Utilization, pp. 25-113, Chapman Hall, New York

[2] Yadav NF. 1996. Genetic modification of soybean oil quality. In: Verma DPS and Shoemaker RC(ed) Soybean: Genetics, Molecular Biology and Biotechnology. pp 165-188. CAB International, Wallingford, Oxon. UK

[3] Ohyama T, Minagawa R, Ishikawa S, Sueyoshi K, Sato T, Nakanishi Y, Asis Jr. CA, Ruamsungsri S, and Ando S. 2013. Soybean Seed Production and Nitrogen Nutrition. In: Board JE(ed) A Comprehensive Survey of International Soybean Research - Genetics, Physiology, Agronomy and Nitrogen Relationships, pp.115-157, InTechOpen. http://dx.doi.org/10.5772/56993

[4] Ledgard SF, Giller KE. 1995. Atmospheric N_2 fixation as an alternative N source. In Bacon PE (ed) Nitrogen Fertilization in the Environment. pp.443-486. Marcel Dekker, New York, Basel, Hong Kong.

[5] Hamaguchi H, Yamamoto N, Takeda A, Masumura T, Sugimoto T, Azuma T. Nitrogen fertilization affects yields and storage compound contens in seeds of field-grown soybeans cv Enrei (*Glycine max*. L) and its super-nodulatiing mutant En-b0-1 through changing N_2 fixation activity of plants. Soil Sci. Plant Nutri. 66, 299-307 2020. Doi: 10.1080/00380768.2019.1692636

[6] Hamaguchi H, Takeda A, Sugimoto T, Azuma T 2020: Pot experiment suggests that the low yields of a super-nodulating soybean (*glycine max* L. En-b0-1) was not caused by N_2 fixation activity of nodules itself. J. Crop Res., 65, 23-29. https://doi.org/10.18964/jcr.65.0_23

[7] Francisco PB, Akao S 1993: Autoregulation and nitrate inhibition of nodule formation in soybean cv. Enrei and its nodulation mutant. J Exp Bot. 44, 547-553. https://doi.org/10.1093/jxb/44.3.547

[8] Akao S, Kouchi H 1992: A supernodulating mutant isolated from soybean cultivar Enrei. Soil Sc Plant Nutri. 38, 183-187. https://doi.org/10.1080/00380768.1992.10416966

[9] Unkovich M, Herridge D, Peoples M, Cadisch G, Boddey B, Giller K Alves B, Chalk P. 2008 "^{15}N natural abundance method. In Measuring plant-associated nitrogen fixation in agricultural systems" ACIAR Monograph No.136, 131-162. Australian Center for International Agricultural Research, Austraria.

[10] Sugimoto T, Shiraishi N, Oji Y 2005: Daizu shusi hinshitsu no chissosehi niyoru hendo [Changes in contents of storage compounds in soybean seeds by application of nitrogen fertilizers]. In Daizu no seisan hinshitsu kojyo to eiyouseiri [Improvement of production and quality of soybean in relation to plant nutrition and physiology] edited by Ohyama T, 40-58. Hakuyusha Co. Ltd.: Tokyo.

[11] Streeter JG 1985: Nitrate inhibition of legume nodule growth and activity. I. Long term studies with a continuous supply of nitrate. Plant Physiol. 77, 321-324. https://doi: 10.1104/pp.77.2.321

[12] Moreira A, Morares LA, Schroth G, Becker FJ, Mandarino MG 2017: Soybean yield and nutritional status response to nitrogen sources and rates of foliar fertilization. Agron J., 109 1-7. https://doi.org/10.2134/agronj2016.04.0199

[13] Takahashi, Y., Chinushi, T., Nagumo, Y., Nakano, T. and Ohyama, T.1991:Effect of Deep Placement of

Controlled Release Nitrogen Fertilizer (Coated Urea) on Growth Yield, and N itrogen Fixation of Soybean Plants. Soil Sci Plant Nutri. 37, 223-231. https://doi.org/10.1080/00380768.1991.10415032

[14] Sugimoto T, Masuda R, Kito M, Shiraishi N, Oji Y. 2001. Nitrogen fixation and soil N level during maturation affect the contents of storage compounds of soybean seeds. Soil Sci. Plant Nutri., 47, 273-279. https://doi.org/10.1080/00380768.2001.10408391

[15] Sugimoto T, K Nomura, R Masuda, K Sueyoshi, Y Oji. J. 1998. Effect of Nitrogen Application at the Flowering Stage on the Quality of Soybean Seeds. J. Plant Nutri., 21, 2065-2075. https://doi. org/10.1080/01904169809365544

[16] O'Leary MH 1986. Phosphoenolpyruvate Carboxylase; An Enzymologist' View, Ann Rev Plant Physiol., 33, 297-315. https://doi.org/10.1146/annurev.pp.33.060182.001501

[17] Headley CL, Harvey DM, and RJ Kelly 1975. Role of PEP carboxylase during seed development in *Pisum sativum*. Nature, 258:352-354. https://doi.org/10.1038/258352a0

[18] O'Leary B, Park J, Plaxton WC. The remarkable diversity of plant PEPC (phospho*enol*pyruvate carboxylase): recent insights into the physiological functions and post-translational controls of non-photosynthetic PEPCs. Biochem J. 2011; 436:15-34. https://doi.org/10.1042/BJ20110078

[19] Sugimoto T, K Tanaka, M Monma, K Saio. 1987. Photosynthetic activity in the developing cotyledons of soybean seeds. Agric. Biol. Chem., 51, 1227-1230. DOI https://doi.org/10.1271/bbb1961.51.1227

[20] Sugimoto T, K Tanaka, M Monma, K Hashizume, K Saio. 1998. Physiological and enzymological properties of phospho*enol*pyruvate carboxylase in soybean (*Glycine max* L. cv Enrei) seeds. Mem. Grad. School Sci. Tech., Kobe Univ. 16-A. 77-87.

[21] Sugimoto T, K Tanaka, M Monma, K Hashizume, Y Kawamura, K Saio. 1989. Phospho*enol*pyruvate carboxylase level in soybean seed highly correlates to its contents of protein and lipid. Agric. Biol. Chem., 53, 885-887. https://doi.org/10.1080/00021369.1989.10869369

[22] Sugimoto T, Sueyoshi K, Oji Y 1997: Increase of PEPC activity in developing rice seeds with nitrogen application at flowering stage, Developments in plant and soil sciences, 78, 811-812.

[23] Sugimoto T, Kawasaki T, Kato T, Whittier RF, Shibata D, and Kawamura Y 1992: cDNA sequence and expression of a phosphoenolpyruvate carboxylase gene from soybean. Plant Mol. Biol. 20, 743-747. https://doi.org/10.1007/BF00046459

[24] Vazquez-Tello A, Whittier RF, Kawasaki T, Sugimoto T, Kawamura Y, Shibata D 1993: Sequence of a Soybean (*Glycine max* 1.) Phospho*enol*pyruvate Carboxylase cDNA. Plant Physiol. 103: 1025-1026. https://doi.org/10.1104/pp.103.3.1025

[25] Wang N, Zhong X, Cong Y, Wang T, Yang S, Li Y, Gai J 2016: Genome-wide Analysis of Phosphoenolpyruvate Carboxylase Gene Family and Their Response to Abiotic Stresses in Soybean. Sci. Rep. 6, 38448. DOI: 10.1038/srep38448

[26] Yamamoto N, Masumura T, Yano K, and Sugimoto T, 2019. Pattern analysis suggests that phospho*enol*pyruvate carboxylase in maturing soybean seeds promotes the accumulation of protein, Biosc. Biotech. Biochem., 83, 2238-2243 https://doi.org/10.1080/09168451.2019.1648205

[27] Liu K 1997. 3 Biological and Compositional Changes during Soybean

Maturation, Storage, and Germination. In Soybeans, Chemistry, Technology, and Utilization, pp. 114-136, Chapman Hall, New York

[28] Yamamoto N, Sugimoto T, Takano T, Sasou A, Morita S, and Yano K. 2020: The plant-type phospho*enol*pyruvate carboxylase Gmppc2 is developmentally induced in immature soy seeds at the late maturation stage: a potential protein biomarker for seed chemical composition. Biosci. Biotech. Biochem., 84, 552-562 https://doi.org/10.1080/09168451.2019.1696179

[29] Roesler K, Shintani D, Savage L, Boddupalli S, and Ohlrogge J 1997: Targeting of the Arabidopsis homomeric acetyl-Coenzyme A carboxylase to plastids of rapeseeds. Plant Physiol., 113, 75-81. https://doi.org/10.1104/pp.113.1.75

[30] Gunvant P, Tri D V, Sandip K, Babu V, Rupesh D, Chengsong Z, Xiaolei W, Yonghe B, Dennis Y, Fang Lu, Siva K, J Grover S, Rajeev K V, Henry T N. 2018: Dissecting genomic hotspots underlying seed protein, oil, and sucrose content in an interspecific mapping population of soybean using high-density linkage mapping. Plant Biotechnol J., 16, 1939- 1953. https://doi.org/10.1111/pbi.12929

14

Mitigation of Climate Change by Nitrogen Managements in Agriculture

Kazuyuki Inubushi and Miwa Yashima

Abstract

Soil is one of the important sources of nitrous oxide (N_2O), which is generally producing through soil microbial processes, such as nitrification and denitrification. Agricultural soils receive chemical and organic fertilizers to maintain or increase crop yield and soil fertility, but several factors are influencing N_2O emissions, such as types and conditions of soil and fertilizer, and rate, form, and timing of application. Mitigation of N_2O is a challenging topic for future earth by using inhibitors, controlled-release fertilizers, and other amendments, but the cost and side effects should be considered for feasibility.

Keywords: N_2O, nitrification, denitrification, mitigation, soil type

1. Introduction

Global warming is significant and the impact of human activities is no doubt, such as mining of fossil fuels and deforestation, over-grazing, and constant increase of nitrogen fertilizer, resulting in atmospheric concentrations of CO_2, methane (CH_4) and nitrous oxide (N_2O) keep increasing, respectively, as indicated by Intergovernmental Panel on Climate Change (**IPCC**), under the United National Framework Convention on Climate Change (**UNFCCC**) (**Figure 1**, [1]). CH_4 and N_2O are the main Short-Lived Climate Forcers (**SLCPs**) because these partici-pate in air pollution chemistry (ozone production, the oxidizing capacity of the atmosphere) and have very high Global Warming Potential (**GWP**) to compare with CO_2 = 1 as CH_4 GWP = ~28; N_2O GWP = ~298 (100 yr integration on per mole basis).

The Japanese government declared in 2020 that the year 2050 is the target of "Carbon Neutral Society", like other OECD countries. To achieve this target, we should reduce greenhouse gas emissions, not only CO_2 but also CH_4 and N_2O, both strongly related to food production and agriculture sectors.

Soil is one of the important sources of N_2O, which is generally producing through soil microbial processes, such as nitrification and denitrification. Agricultural soils receive chemical and organic fertilizers to maintain or increase crop yield and soil fertility. However, excess amount of chemical N fertilizer application may cause eutrophication and ground water pollution in the hydrosphere. Moreover, many factors are also influencing N_2O emission in the atmosphere, such as types and conditions of soil and fertilizer, and rate, form, and timing of application. Mitigation of N_2O emission to the atmosphere is a challenging topic in

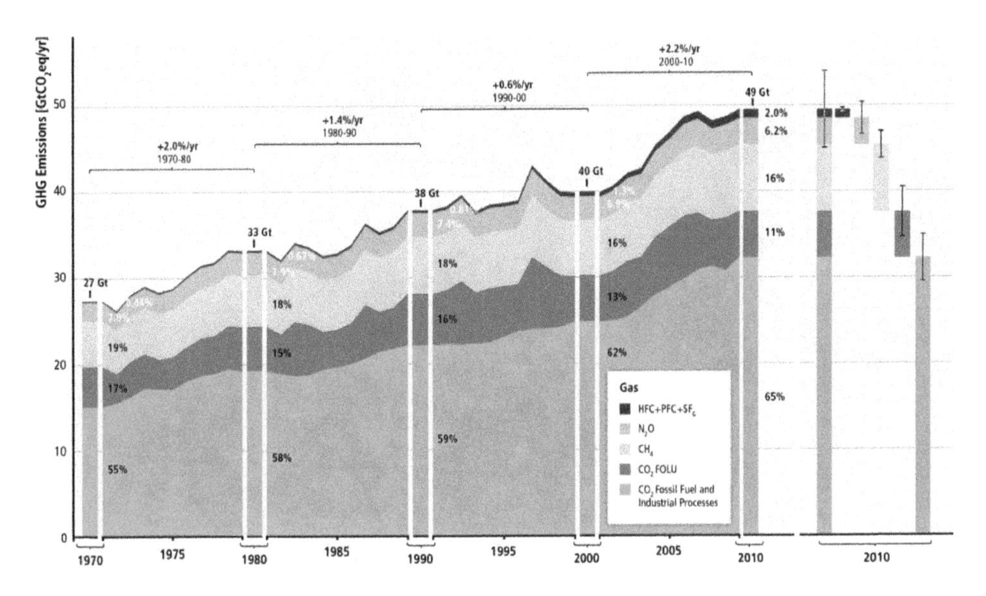

Figure 1.
Total annual anthropogenic GHG emissions by gases 1970–2010 (source: [1]).

sustainable agriculture, such as by using inhibitors, controlled-release fertilizers and other amendments, though the cost and side effects should be considered for feasibility. In this review, processes and influencing factors of N_2O production in the soil is reviewing and some trials for mitigation are introduced.

2. Global N_2O budget and production in soil

Global Carbon Project (GCP) published a comprehensive quantification of global nitrous oxide sources and sinks [2]. This reports details of the global N_2O budget in 21 natural and human sectors between 1980 and 2016 (**Figure 2**), indicating that

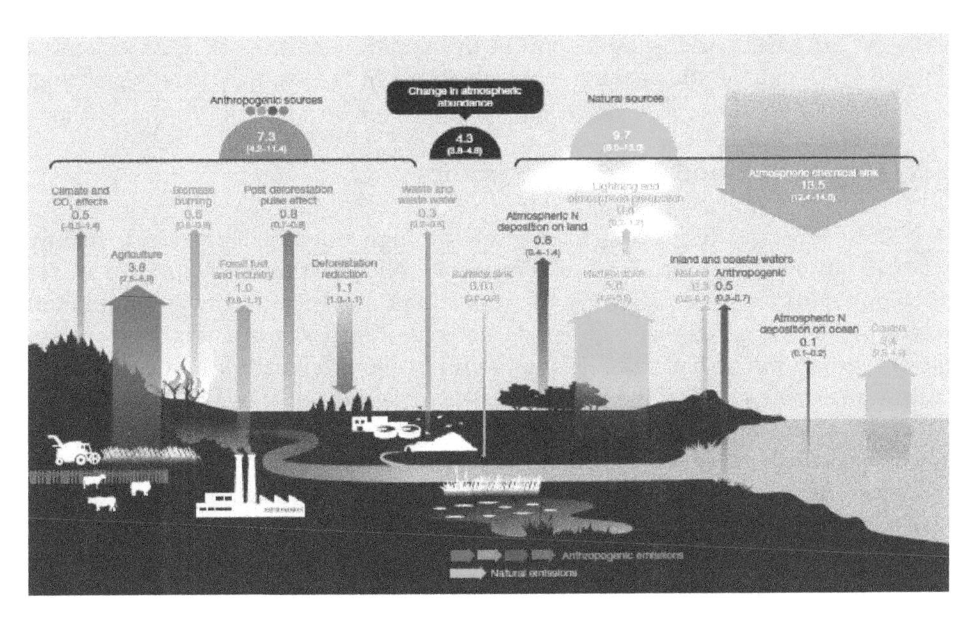

Figure 2.
Global N_2O budget [2].

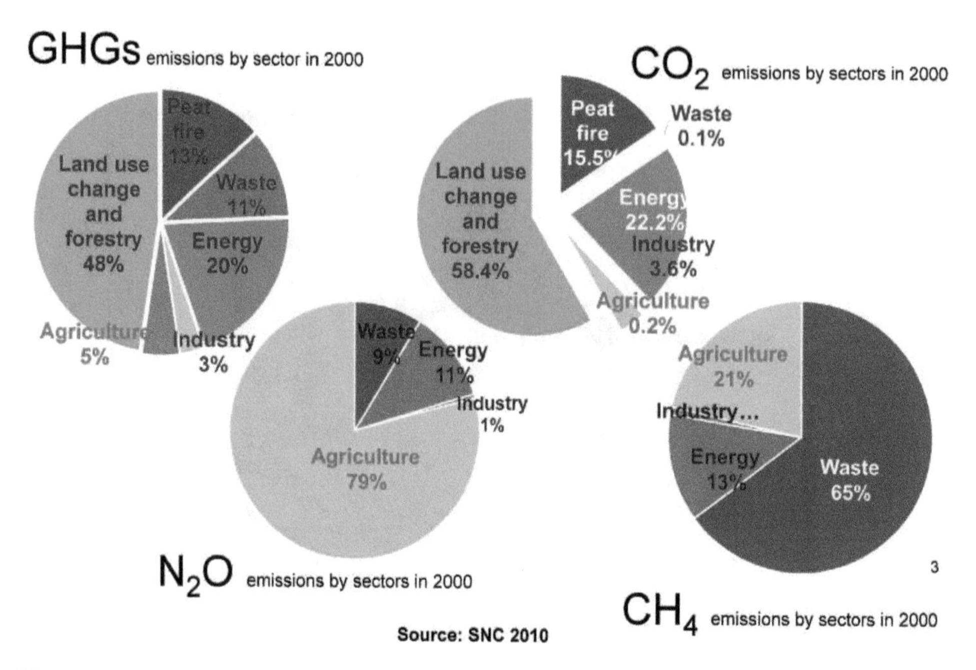

Figure 3.
National contribution of greenhouse gas emission by sector Indonesia source, [3].

natural and anthropogenic sources of N_2O were 57% and 43%, or 9.3 and 7.3 TgN yr^{-1} (1 Tg = 10^{12} g or 1 million ton), respectively and total as 17.0 (minimum 12.2 to maximum 23.5 Tg), while 13.5 Tg sink by atmospheric chemical reactions, resulting 3.5 Tg increase annually. By continental or regional estimates, Africa releases most (3 Tg yr^{-1}) due to large areas with tropical forests where high temperature and soil moisture, followed by Latin America and East Asia, where the agricultural contribution is largest. The annual increase of N_2O emission is more than 1%, and the agricultural sector is largest, especially in Asia, followed by Latin America, Africa, and particularly in East Asia, the input of chemical fertilizer and manure plus direct emission is increasing as more than double in past three decades. National inventory of greenhouse gas emission in developing countries such as Indonesia (**Figure 3**) [3] contributions of agricultural sectors in N_2O and CH_4 are bigger than other sectors.

3. N_2O production and its affecting factors

N_2O is generally producing in the soil through microbial processes, mainly via nitrification and denitrification (**Figure 4**). Nitrification is carried out under aerobic conditions by two groups of autotrophic nitrifiers, namely ammonium oxidizers and nitrite oxidizers, both do not require organic matters, not only in bacteria group but also archaea group. Autotrophic nitrification is the dominant process in aerobic soil (less than 60% water-holding capacity), while heterotrophic nitrification is negligible [4]. N_2O is producing as a byproduct during nitrite oxidation during nitrification. On the other hand, denitrification is carried out under wet and anaerobic conditions, such as in paddy soil and wetland soil, by heterotrophic denitrifiers, not only the bacteria but also fungi, both requires not only nitrate but also N-rich organic matter. N_2O is producing as an intermediate product dur-ing denitrification between nitrite and N_2. However, N_2O emission from flooded paddy soil is generally low, probably due to the high solubility of N_2O and complete denitrification to N_2. Chemical denitrification was also negligible [4]. Microbial

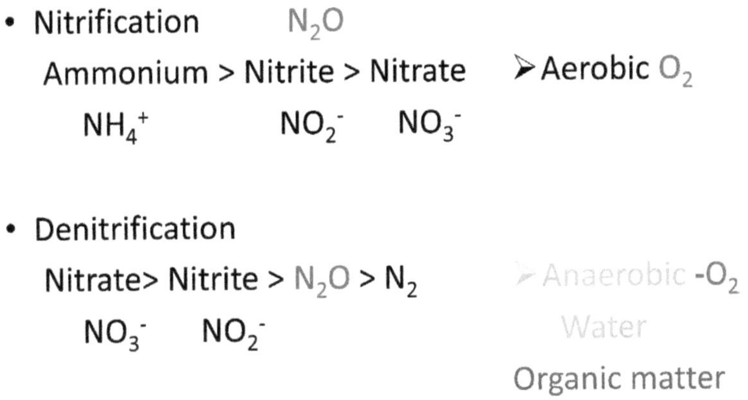

- Nitrification N_2O
  Ammonium > Nitrite > Nitrate ➢Aerobic O_2
  NH_4^+ NO_2^- NO_3^-

- Denitrification
  Nitrate> Nitrite > N_2O > N_2 ➢Anaerobic $-O_2$
  NO_3^- NO_2^- Water
  Organic matter

Figure 4.
Main processes of N_2O production in soil.

community structure was investigated also in tropical acid tea soil [5] and peat soil [6, 7]. Anaerobic ammonium oxidation (ANAMMOX) and dissimilatory nitrate reduction to ammonium (DNRA) are also focusing recently to see the possibility to relate N_2O production and contribution in soil N dynamics [8, 9].

Based on above knowledges about N_2O production processes, several mitigation technologies are proposed. To apply such mitigation technologies, it is important to understand factors affecting N_2O production, which are (1) Soil type and amendments such as manure, compost, and biochar, (2) Soil management and mitigation technologies such as controlled-release chemical fertilizers and nitrification inhibitors, and (3) Trade-off effects with other greenhouse gas mitigation such as water management in paddy field to reduce CH_4.

4. Effect of soil types and amendment on N_2O production and plant growth

Generally, soil with a large amount of soil organic matter (SOM) tended to produce more N_2O. However, in a case study using various soils in Japan and Hungary [10], Andosol, typical upland soil in Japan with higher SOM contents, produced less N_2O than Chernozem with lower SOM contents, typical upland soil in Europe, under the same incubation conditions, especially amended with chemical N fertilizer and biochar (**Figure 5**). However sandy soils with fewer SOM contents, N_2O production was small. Biochar is focused on soil C sequestration to build up C stock in soil, but also in Andosol, N_2O tended not to be increased with biochar and N fertilizer. Leafy vegetable (Komatsuna; *Brassica rapa*) growth and yield were also enhanced by amendments of chemical N fertilizer and biochar.

Effects of amendments on N_2O production were studied by many researchers, showing compost and N fertilizer generally increase N_2O production [11–16]. Under field conditions, N_2O emission to the atmosphere was much diversified in space and also soil depth [16, 17], and land-use [18]. Further study is needed for feasible soil and fertilizer management to meet sustainable developments.

5. Soil and fertilizer managements and mitigation technologies

To reduce N_2O emission, controlled-release chemical fertilizers and nitrification inhibitors have been examined [19, 20], and meta-analysis of 113 field experiment

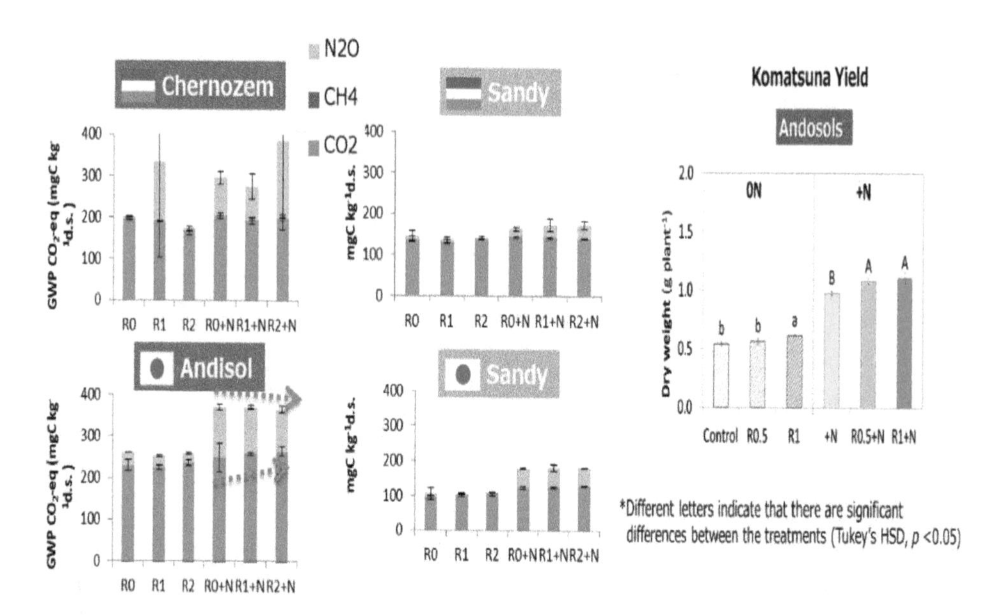

Figure 5.
Effect of soil types and amendment on N_2O production and plant growth (R: Rice husk biochar, 0, 1, 2: Application rate as 0, 1, 2%w/w, respectively, N: Urea).

datasets showed that polymer-coated fertilizers significantly reduced N_2O emissions (mean: −35%, 95% confidential interval: −58% to −14%) and nitrification inhibitors (−38%, −44% to −31%), respectively, depending on soil type and regions [21].

Controlled-release coated urea (CRCU) is a type of polymer-coated fertilizer. CRCU was examined to compare with conventional chemical fertilizer in tropical oil palm plantations over 340–580 days, where vast areas have been converted from rainforest and other plantations. Sakata et al. [22] reported the effect of CRCU compare with conventional fertilizer on N_2O emission and yield (**Figures 6** and 7). In Tungal sandy loam soil, controlled-release nitrogen fertilizer (CRNF; M) showed lower N_2O emission than conventional fertilizer (C), while in Simunjan sandy soil, N_2O was low in both M and C. On the other hand in Tatau peat soil, both M and C emitted a similar amount of N_2O, and much larger than other sites, and even from control (B; without fertilizer) (**Figure** 7). No significant effect on oil palm yield

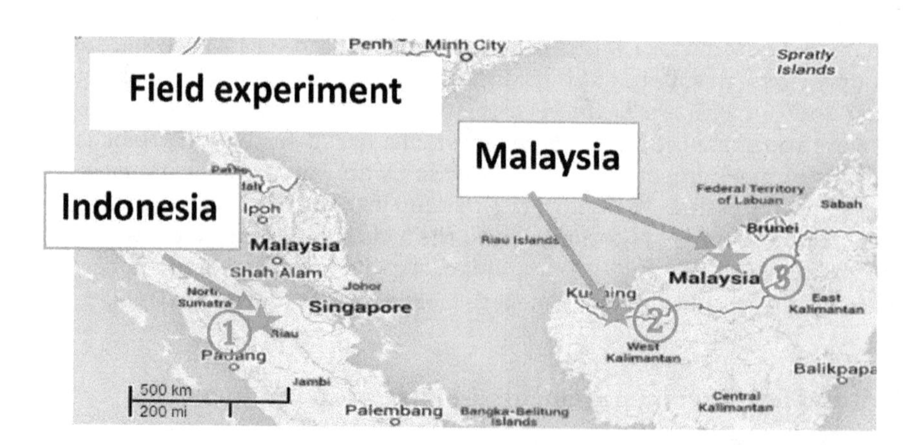

Figure 6.
Field experimental sites in Sumatra, Indonesia and Sarawak, Malaysia [22], ① Tunggal, ② Simunjan, ③ Tatau.

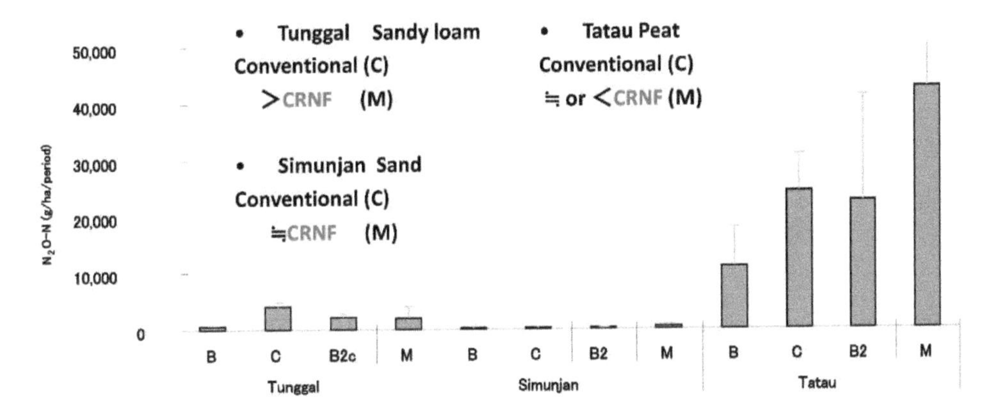

Figure 7.
N_2O emission from field experiments with different soil and fertilizer [22].

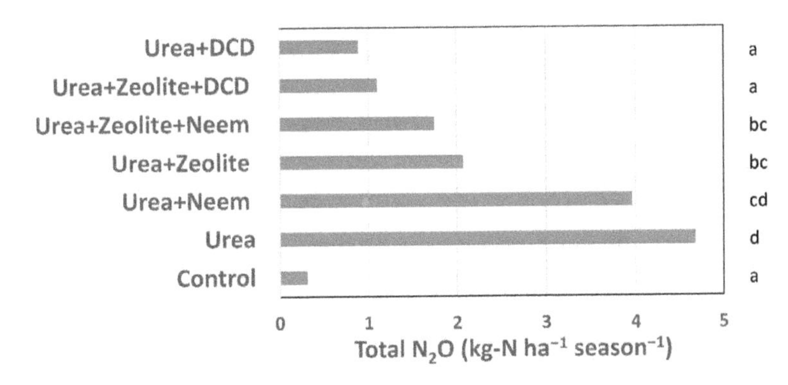

Figure 8.
Nitrification inhibitors on N_2O emission [29].

was observed even N application rate of M was half of C. Tropical peat soil has been pointed out as a significant N_2O emission source, even without fertilizer, strongly influenced by groundwater level [6, 7, 23–27]. Therefore CRCU has a significant impact even under tropical conditions to reduce N_2O in certain mineral soils, but not in organic peat soil. Long-term evaluation and cost–benefit analysis are impor-tant with yield evaluation in various soil types under diverse climate conditions.

Another mitigation option, nitrification inhibitors to stop ammonium oxida-tion have been also examined, typically DCD (Dicyandiamide; [20]) which is biologically and temperature-dependently decomposed [28]. However, it caused contamination in exposed milk powder in New Zealand, so NZ banned DCD from 2013. Nitrapyrin and 3,4-dimethyl pyrazole phosphate (DMPP) are other chemicals of nitrification inhibitors, but less effective. Neem cake is derived from natural compounds, so less expensive, but also less effective to compare with chemical inhibitors [21]. Combined effects of a nitrification inhibitor, including DCD, neem, and clay mineral (zeolite) on N_2O fluxes and corn growth were examined ([29]; **Figure 8**). Another biological nitrification inhibitor is also examined [30].

6. Trade-off with other mitigation

Mitigation options of other greenhouse gases, such as water management in the paddy field to reduce CH_4, have been examined in Japan [31]. The controlling irrigation water level was also examined in Indonesia [32]. Groundwater level

Varietyc	Treatment	CH_4 (kg ha^{-1})	N_2O (kg ha^{-1})	GWP (kg CO_2-eq/ha)	Reduction %
ADT 43	SRI	59.97 a	1.94 a	2077.4	**28.8%**
	MSRI	45.11 a	2.69 a	1929.4	**33.9%**
	CT	99.44 b	1.45 a	2918.1	
CO 51	SRI	51.73 a	2.09 a	1916.1	**27.1%**
	MSRI	50.76 a	1.53 a	1724.9	**34.3%**
	CT	88.86 b	1.36 a	2626.8	

SRI as a type of AWD with reduce seedling numbers, and MSRI as modified SRI with seedling age to compare with control CT [38].

Table 1.
Effects of water management and crop establishments on N_2O emission in field experimental site conducted at Tamil Nadu Rice Research Institute

control by alternate wet and drying (AWD) have established by IRRI and examined in Indonesia, the Philippines, Thailand, Vietnam [33–37] and India [38]. AWD has merit for saving labor and water. However, it may have a trade-off effect to increase N_2O, because of the removal of flooded water to expose anaerobic soil directly to the atmosphere. To examine this trade-off, they measured not only CH_4 but also N_2O emissions in the same field experiments and found that N_2O emission was mostly negligible without losing rice yield although CH_4 was significantly reduced (**Table 1**). Such trade-off should be examined not only for water management but also other soil managements including biochar for soil C sequestration.

7. Conclusions

Nitrogen is one of the most critical elements for food production, local and global environments. Nitrous oxide (N_2O) is an important greenhouse gas emitted from the soil via biological processes in N cycling. N_2O emission keeps increasing to induce global warming, climate change, and stratospheric ozone layer deple-tion. N_2O production in the soil is related to soil and fertilizer management and is influenced by many factors, such as soil conditions, chemical fertilizer, and organic manures. Mitigation is possible by appropriate soil and fertilizer management (controlled-release fertilizer and nitrification inhibitors), but exceptional soil such as peat soil should be careful. Feasibility is important to harmonize with yield and other factors (cost and economic merits, side effect, etc.). Under the COVID-19 pandemic, the balance of food production, human health, and environmental management become more and more crucial issues. Sustainable development goals become more important view-points than before [39, 40].

Author details

Kazuyuki Inubushi* and Miwa Yashima
Chiba University, Matsudo, Chiba, Japan

*Address all correspondence to: inubushi@faculty.chiba-u.jp

References

[1] IPCC, 2014. "AR5 Climate Change 2014: Mitigation of Climate Change." Working Group III to the Fifth Assessment Report of the Inter governmental Panel on Climate Change. Cambridge, UK: Cambridge University Press.

[2] Tian, H., Xu, R., Canadell, J.G., Yao, Y., *et al.* 2020. A comprehensive quantification of global nitrous oxide sources and sinks, Nature, 586, 248-256

[3] SNC 2010. Second National Communication of Indonesia to the UNFCCC (SNC), Jakarta.

[4] Inubushi, K., Naganuma, H. and Kitahara, S. 1996. Contribution of denitrification, autotrophic and heterotrophic nitrification in nitrous oxide production in andosols. Biology and Fertility of Soils, 23(3), 292-298.

[5] Jumadi, O., Hala, Y., Anas, I., Ali, A., Sakamoto, K., Saigusa, M., Yagi, K., and Inubushi, K., 2008. Community structure of ammonia oxidizing bacteria and their potential to produce nitrous oxide and carbon dioxide in acid tea soils, Geomicrobiology Journal, 25 (7-8), 381-389

[6] Arai, H., Hadi, A., Darung, U., Limin, S.H., Takahashi, H., Hatano, R., and Inubushi, K. 2014. Land use change affects microbial biomass and fluxes of carbon, dioxide and nitrous oxide in tropical peatlands, Soil Science and Plant Nutrition, 60: 423-434

[7] Hadi, A., Haridi, M., Inubushi, K., Purnomo, E., Razie, F., and H. Tsuruta, 2001. Effect of land-use change in tropical peat soil on the microbial population and emission of greenhouse gases, Microbes and Environments, 16 (2), 79-86.

[8] Martina, P., Schleusner P., Rütting, T., and Hallin, S. 2018. Relative abundance of denitrifying and DNRA bacteria and their activity determine nitrogen retention or loss in agricultural soil, Soil Biology and Biochemistry, 123, 97-104.

[9] Sato, Y., Ohta, H., Yamagishi, T., Guo, Y., Nichizawa, T., Rahman, M.H., Kuroda, H., Kato, T., Saito, M., Yoshinaga, I., Inubushi, K., and Suwa, Y. 2012. Detection of anammox activity and 16S rRNA genes in ravine paddy field soil, Microbes and Environments, 27(3), 316-319

[10] Saito, Y., Shiga, M., Sato, M., Bencsik, D., Kátai, J., Kovács, A.B., Tállai, M., Yashima, M.M., and Inubushi, K. 2021, Effect of biochar and soil moisture on nitrogen dynamics, greenhouse gas emissions, and Komatsuna (*Brassica rapa*) growth in Japanese and Hungarian soils with different fertilities, Special volume for Prof Kátai, University of Debrecen (ISBN 978-963-318-936-8).

[11] Inubushi, K., Goyal, S., Sakamoto, K., Wada, Y., Yamakawa, K. and Arai, T., 2000. Influence of application of sewage sludge compost on N_2O production in soils, Chemosphere-Global Change Sciences, 2, 329-334.

[12] Li, Xinhui, Inubushi, K., and Sakamoto, K., 2002a. N_2O concentration in the Andisol profile and emissions to the atmosphere as influenced by the application of nitrogen fertilizers and manure, Biology and Fertility of Soils, 35, 108-113.

[13] Li, Xinhui, T. Nishio, Y. Uemiya, and K. Inubushi, 2002b. Gaseous losses of applied nitrogen from a corn field determined by ^{15}N abundance of N_2 and N_2O, Communications in Soil Science and Plant Analysis, 33(15-18), 2715-2727.

[14] Singla, A., Dubey, S.K., Iwasa, H., and Inubushi, K. 2013. Nitrous oxide

flux from komatsuna (*Brassica rapa*) vegetated soil: a comparison between biogas digested liquid and chemical fertilizer, Biology and Fertility of Soils 49:971-976

[15] Zaman, M., Di, H.J., Sakamoto, K., Goto, S., Hayashi, H. and Inubushi, K. 2002. Effect of sewage sludge compost and chemical fertilizer applications on microbial biomass and N mineralization rates, Soil Science and Plant Nutrition, 48, 195-201.

[16] Zaman, M., Matsushima, M., Chang, S.X., Inubushi, K., Nguyen, L., Goto, S., Kaneko, F. and Yoneyama, T. 2004. Nitrogen mineralization, N_2O production and soil microbiological properties as affected by long-term applications of sewage sludge composts, Biology and Fertility of Soils, 40, 101-109.

[17] Yanai, J., Sawamoto, T., Oe, T., Kusa, K., Yamakawa, K., Sakamoto, K., Naganawa, T., Inubushi, K., Hatano, R. and Kosaki, T. 2003. Spatial variability of N_2O emissions and their soil-related determining factors in an agricultural field. Journal of Environmental Quality, 32(6), 1965-1977.

[18] Kong, Y.H., Nagano, H., Kátai, J, Vágó, I., Oláh, Á.Z., Yashima, M. and Inubushi, K.: 2013. CO_2, N_2O and CH_4 production/consumption potentials of soils under different land-use types in central Japan and eastern Hungary, Soil Science and Plant Nutrition, 59(3), 455-462

[19] Amkha, S., Inubushi, K., and Takagaki, M. 2007. Effects of controlled-release nitrogen fertilizer application on nitrogen uptake of a leafy vegetable (*Brassica campestris* L.), nitrate leaching and N_2O emission, *Japanese Journal of Tropical Agriculture*, 51(4), 152-159

[20] Hadi, A,, Jumadi, O., Inubushi, K., and Yagi, K., 2008. Mitigation options for N_2O emission from a corn field in Kalimantan, Indonesia: A case study, Soil Science and Plant Nutrition, 54 (4), 644-649

[21] Akiyama, H., Yan, X., and Yagi, K. 2010. Evaluation of effectiveness of enhanced-efficiency fertilizers as mitigation options for N_2O and NO emissions from agricultural soils: meta-analysis, Global Change Biology, https://doi.org/10.1111/j.1365-2486.2009.02031.x

[22] Sakata R., Shimada S., Arai H., Yoshioka N., Yoshioka R., Aoki H., Kimoto N., Sakamoto A., Melling L., and Inubushi K. 2015. Effect of soil types and nitrogen fertilizer on nitrous oxide and carbon dioxide emissions in oil palm plantations, Soil Science and Plant Nutrition, 61: 48-60

[23] Furukawa, Y., Inubushi, K., Ali, M., Itang, AM. and Tsuruta, H. 2005. Effect of changing groundwater levels caused by land-use changes on greenhouse gas emissions from tropical peatlands, Nutrient Cycling in Agroecosystems, 71, 81-91

[24] Hadi, A., K. Inubushi, E. Purnomo, F. Razie, K. Yamakawa and H. Tsuruta, 2000. Effect of land-use changes on nitrous oxide (N_2O) emission from tropical peatlands, Chemosphere-Global Change Sciences, 2, 347-358

[25] Hadi, A., Inubushi, K., Furukawa, Y., Purunomo, E., Rasmadi, M., and Tsuruta, H. 2005. Greenhouse gas emissions from tropical peatlands of Kalimantan, Indonesia, *Nutrient Cycling in Agroecosystems*, 71, 73-80.

[26] Inubushi, K., Furukawa, Y., Hadi, A., Purnomo, E., and Tsuruta, H. 2003. Seasonal changes of CO_2, CH_4 and N_2O fluxes in relation to land-use change in tropical peatlands located in coastal area of south Kalimantan, Chemosphere, 52(3), 603-608.

[27] Susilawati, H.L., Setyanto, P., Ariani, M., Hervani A., and Inubushi, K., 2016, Influence of water depth and

soil amelioration on greenhouse gas emissions from peat soil columns, Soil Science and Plant Nutrition, 62: 57-68

[28] Rajbanshi, S.S., Benckiser, G. and Ottow, J.C.G. 1992. Effects of concentration, incubation temperature, and repeated applications on degradation kinetics of dicyandiamide (DCD) in model experiments with a silt loam soil. Biology and Fertility of Soils **13,** 61-64. https://doi.org/10.1007/BF00337336

[29] Jumadi, O., Hala, Y., Iriany, R.N., Makkulawu, A.T., Baba, J., Hartono, Hiola St.F., and Inubushi, K. 2020. Combined effects of nitrification inhibitor and zeolite on greenhouse gas fluxes and corn growth. Environmental Science and Pollution Research, 27, 2087-2095.

[30] Subbarao, G.V., Nakahara, K., Hurtado, M.P., Ono, H., Moreta, D.E., Salcedo, A.F., Yoshihashi, A.T., Ishikawa, T., Ishitani, M., Ohnishi-Kameyama, M., Yoshida, M., Rondon, M., Rao, I.M., Lascano, C.E., Berry, W. L. O. Ito. 2009 Evidence for biological nitrification inhibition in *Brachiaria* pastures, Proceedings of the National Academy of Sciences 106 (41) 17302-17307; DOI: 10.1073/pnas.0903694106

[31] Yagi, K., H. Tsuruta, K. Kanda, and K. Minami. 1996. Effect of water management on methane emission from a Japanese rice paddy field: Automated methane monitoring. Global Biogeochemical Cycles 10 (2):255-267.

[32] Hadi, A., Inubushi, K., and Yagi, K., 2010. Effect of water management on greenhouse gas emissions and microbial properties of paddy soils in Japan and Indonesia, Paddy Water Environment, 8:319-324.

[33] Chidthaisong, A., N. Cha-un, B. Rossopa, C. Buddaboon, C. Kunuthai, P. Sriphirom, S. Towprayoon, T. Tokida, A. Tirol, and K. Minamikawa. 2018. Evaluating the effects of Alternate Wetting and Drying (AWD) on methane and nitrous oxide emissions from a paddy field in Thailand. Soil Science and Plant Nutrition, 64 (1): 31-38.

[34] Setyanto, P., A. Pramono, T. A. Adriany, H. L. Susilawati, T. Tokida, A. Tirol-Padre, and K. Minamikawa. 2018. "Alternate Wetting and Drying Reduces Methane Emission from a Rice Paddy in Central Java, Indonesia without Yield Loss." Soil Science and Plant Nutrition 64 (1): 23-30.

[35] Sibayan, E. B., K. Samoy-Pascual, F. S. Grospe, M. E. D. Casil, T. Tokida, A. Tirol-Padre, and K. Minamikawa. 2018. Effects of Alternate Wetting and Drying technique on greenhouse gas emissions from irrigated rice paddy in Central Luzon, Philippines. Soil Science and Plant Nutrition 64(1): 39-46.

[36] Tirol-Padre, A., K. Minamikawa, T. Tokida, R. Wassmann, and K. Yagi. 2018. Site-specific feasibility of Alternate Wetting and Drying as a greenhouse gas mitigation option in irrigated rice fields in Southeast Asia: A Synthesis. Soil Science and Plant Nutrition 64 (1):2-13.

[37] Tran, D. H., T. N. Hoang, T. Tokida, A. Tirol-Padre, and K. Minamikawa. 2018. Impacts of Alternate Wetting and Drying on greenhouse gas emission from paddy field in Central Vietnam. Soil Science and Plant Nutrition 64 (1): 14-22.

[38] Oo, A.Z., Sudo, S., Inubushi, K., Mano, M., Yamamoto, A., Ono, K., Osawa, T., Hayashida, S., Patra, P.K., Terao, Y., Elayakumar, P., Vanitha, K., Umamageswari, C., Jothimani, P., and Ravi, V., 2018. Methane and nitrous oxide emissions from conventional and modified rice cultivation systems in South India, Agriculture, Ecosystems and Environment 252, 148-158.

[39] Inubushi, K., 2021. Sustainable soil management in East, South and Southeast Asia, Soil Science and Plant Nutrition, 67:1, 1-9.

[40] Lal, R., 2020. Soil Science beyond COVID-19, Journal of Soil and Water Conservation, 1-3.

PERMISSIONS

All chapters in this book were first published by InTech Open; hereby published with permission under the Creative Commons Attribution License or equivalent. Every chapter published in this book has been scrutinized by our experts. Their significance has been extensively debated. The topics covered herein carry significant findings which will fuel the growth of the discipline. They may even be implemented as practical applications or may be referred to as a beginning point for another development.

The contributors of this book come from diverse backgrounds, making this book a truly international effort. This book will bring forth new frontiers with its revolutionizing research information and detailed analysis of the nascent developments around the world.

We would like to thank all the contributing authors for lending their expertise to make the book truly unique. They have played a crucial role in the development of this book. Without their invaluable contributions this book wouldn't have been possible. They have made vital efforts to compile up to date information on the varied aspects of this subject to make this book a valuable addition to the collection of many professionals and students.

This book was conceptualized with the vision of imparting up-to-date information and advanced data in this field. To ensure the same, a matchless editorial board was set up. Every individual on the board went through rigorous rounds of assessment to prove their worth. After which they invested a large part of their time researching and compiling the most relevant data for our readers.

The editorial board has been involved in producing this book since its inception. They have spent rigorous hours researching and exploring the diverse topics which have resulted in the successful publishing of this book. They have passed on their knowledge of decades through this book. To expedite this challenging task, the publisher supported the team at every step. A small team of assistant editors was also appointed to further simplify the editing procedure and attain best results for the readers.

Apart from the editorial board, the designing team has also invested a significant amount of their time in understanding the subject and creating the most relevant covers. They scrutinized every image to scout for the most suitable representation of the subject and create an appropriate cover for the book.

The publishing team has been an ardent support to the editorial, designing and production team. Their endless efforts to recruit the best for this project, has resulted in the accomplishment of this book. They are a veteran in the field of academics and their pool of knowledge is as vast as their experience in printing. Their expertise and guidance has proved useful at every step. Their uncompromising quality standards have made this book an exceptional effort. Their encouragement from time to time has been an inspiration for everyone.

The publisher and the editorial board hope that this book will prove to be a valuable piece of knowledge for researchers, students, practitioners and scholars across the globe.

LIST OF CONTRIBUTORS

Shweta Nandanwar
Princess Margaret Science Laboratories, Harper Adams University, Newport, Shropshire, United Kingdom

Yogesh Yele, Anil Dixit and Lalit Kharbikar
ICAR—National Institute of Biotic Stress Management, Raipur, India

Dennis Goss-Souza
Santa Catarina State University—UDESC, Florianópolis, Brazil

Ritesh Singh
Thakur Chhedilal Barrister College of Agriculture and Research Station, Bilaspur, India

Arti Shanware
Rajiv Gandhi Biotechnology Centre, RTM Nagpur University, Nagpur, India

Sérgio G. Quassi de Castro
Laboratório Nacional de Ciência e Tecnologia do Bioetanol (CTBE), Centro Nacional de Pesquisa em Energia e Materiais (CNPEM), Polo II de Alta Tecnologia, Campinas, SP, Brazil

Henrique C. Junqueira Franco
UNICAMP, Campinas, SP, Brazil
CROPMAN - Inovação Agrícola, Campinas, SP, Brazil

Murugaragavan Ramasamy, T. Geetha and M. Yuvaraj
Adhiparasakthi Agricultural College, Vellore, Tamil Nadu, India

Elizeu Monteiro Pereira Junior, Elaine Maria Silva Guedes Lobato, Beatriz Martineli Lima, Barbara Rodrigues Quadros, Allan Klynger da Silva Lobato, Izabelle Pereira Andrade and Letícia de Abreu Faria
Federal University of Rural Amazônia (UFRA), Paragominas, Brazil

Ivan dos Santos Pereira and Rogério Oliveira de Sousa
Federal University of Pelotas (UFPel), Pelotas, Brazil

Adilson Luis Bamberg, Carlos Augusto Posser Silveira and Luis Eduardo Corrêa Antunes
Brazilian Agricultural Research Corporation (EMBRAPA), Pelotas, Brazil

Mokhtar Rejili and Mohamed Mars
Laboratory of Biodiversity and Valorization of Arid Areas Bioresources (BVBAA), Faculty of Sciences of Gabes, Erriadh, Zrig, Tunisia

Mohamed Ali BenAbderrahim
Arid and Oases Cropping Laboratory, Arid Area Institute, Gabes, Tunisia

Eulene Francisco da Silva, Marlenildo Ferreira Melo, Kássio Ewerton Santos Sombra, Tatiane Severo Silva, Maria Eugênia da Costa, Eula Paula da Silva Santos, Larissa Fernandes da Silva and Paula Romyne de Morais Cavalcante Neitzke
Federal Rural University of the Semi-arid Region (UFERSA), Mossoró, Brazil

Diana Ferreira de Freitas
Federal Rural University Pernambuco (UFRPE/UAST), Serra Talhada, Brazil

Ademar Pereira Serra
Brazilian Agricultural Research Corporation (EMBRAPA), Campo Grande, Brazil

Muhammad Jamil Khan, Abida Saleem and Qudratullah Khan
Department of Soil Science, Institute of Soil and Environmental Sciences, Gomal University, Dera Ismail Khan, Khyber Pakhtunkhwa, Pakistan

Rafia Younas Mumtaz Khan and Rehan Ahmed
Department of Environmental Sciences, Institute of Soil and Environmental Sciences, Gomal University, Dera Ismail Khan, Khyber Pakhtunkhwa, Pakistan

Maria Luisa Tabing Mason
College of Agriculture, Central Luzon State University, Science City of Munoz, Nueva Ecija, Philippines

Yuichi Saeki
Faculty of Agriculture, University of Miyazaki, Miyazaki, Japan

Upendra M. Sainju
Northern Plains Agricultural Research Laboratory, US Department of Agriculture, Agricultural Research Service, Sidney, Montana, USA

Rajan Ghimire
Agricultural Science Center, New Mexico State University, Clovis, New Mexico, USA

Gautam P. Pradhan
Williston Research and Extension Center, North Dakota State University, Williston, North Dakota, USA

Gabriel Monteiro and Glauco Nogueira
Federal Rural University of Amazon (UFRA), Belém, Pará, Brazil

Cândido Neto and Joze Freitas
Institute of Agrarian Sciences, Federal Rural University of Amazon (UFRA), Belém, Pará, Brazil

Vitor Nascimento
Rede BIONORTE/UFPA, Belém, Pará, Brazil

Toshio Sugimoto
Faculty of Agriculture, Graduate School of Agricultural Science, Kobe University, Kobe, Japan
Research Center for Food and Agriculture, Wakayama University, Wakayama, Japan
Graduate School of Life and Environmental Sciences, Kyoto Prefectural University, Kyoto, Japan

Naoki Yamamoto
Graduate School of Life and Environmental Sciences, Kyoto Prefectural University, Kyoto, Japan
Key Laboratory of Southwest China Wildlife Resources Conservation (Ministry of Education), College of Life Science, China West Normal University, Nanchong, Sichuan, China

Takehiro Masumura
Graduate School of Life and Environmental Sciences, Kyoto Prefectural University, Kyoto, Japan
Biotechnology Research Department, Kyoto Prefectural Agriculture, Forestry, and Fisheries Technology Center, Kyoto, Japan

Kazuyuki Inubushi and Miwa Yashima
Chiba University, Matsudo, Chiba, Japan

Index

Printed in the USA
CPSIA information can be obtained
at www.ICGtesting.com
JSHW051258021023
49509JS00006B/54

9 781647 40350